Results and Problems in Cell Differentiation

Volume 77

Results and Problems in Cell Differentiation is an up-to-date book series that presents and explores selected questions of cell and developmental biology. Each volume focuses on a single, well-defined topic. Reviews address basic questions and phenomena, but also provide concise information on the most recent advances. Together, the volumes provide a valuable overview of this exciting and dynamically expanding field. This series is indexed in PubMed.

Malgorzata Kloc • Rafik Mark Ghobrial
Editors

Cellular and Molecular Mechanisms of Chronic Rejection in Organ Transplantation

Springer

Editors
Malgorzata Kloc
Houston Methodist Research Institute
and The Houston Methodist Hospital
Houston, TX, USA

Rafik Mark Ghobrial
Houston Methodist Research Institute
and The Houston Methodist Hospital
Houston, TX, USA

ISSN 0080-1844 ISSN 1861-0412 (electronic)
Results and Problems in Cell Differentiation
ISBN 978-3-032-07685-4 ISBN 978-3-032-07686-1 (eBook)
https://doi.org/10.1007/978-3-032-07686-1

This Springer imprint is published by the registered company Springer Nature Switzerland AG
The registered company address is: Gewerbestrasse 11, 6330 Cham, Switzerland

Preface

This volume comprises three intertwined and arbitrarily separated parts covering the molecular and cellular aspects of chronic rejection of solid and composite allografts. The first part provides a comprehensive overview and mechanistic understanding of the cellular and molecular pathways underlying chronic rejection, as well as their potential in identifying novel biomarkers and developing targeted anti-chronic rejection strategies. It discusses the role of various cytokines in regulating normal cell function versus acute and chronic allograft rejection, and how to induce transplant tolerance by boosting the function of regulatory T (Treg) cells while maintaining physiological cytokine levels. This part also explores the application of computational modeling, including artificial intelligence, machine learning, ordinary differential equations, partial differential equations, agent-based models, and gene network analysis for understanding chronic rejection of kidney, liver, heart, and lung allografts, and how such computational approaches can enhance diagnostic precision and identify novel therapeutic targets.

The second part of the volume focuses on immune cells, nonimmune cells, and extracellular molecules in the context of chronic rejection. It describes the regulatory role of allograft vascular endothelium, which is the first to encounter the recipient's blood, how repetitive injury by alloreactive and non-alloreactive stimuli damages endothelium and leads to chronic rejection, and potential therapies for preventing endothelial damage. It discusses how alloantibodies and complement induce endothelial cell surface expression and trans-presentation of IL-15, which recruits alloreactive T cells that trans-migrate into the intima, secrete interferon-γ, stimulate arterial wall cell proliferation, and result in allograft vasculopathy. It also describes the role of macrophages in chronic rejection and how interference with the RhoA pathway, which regulates the actin cytoskeleton, prevents chronic rejection. Additionally, it explores the usefulness of plasma concentration of donor-derived cell-free DNA (dd-cfDNA) for quantifying allograft injury.

The third part of the volume concentrates on shared and organ-specific features as well as therapeutic approaches in chronic rejection. It discusses transplant vasculopathy across all solid organ transplantations and the organ-specific features and challenges of heart, lung, kidney, liver, and facial vascularized composite allotransplantation. One chapter describes various aspects of transplant vasculopathy and therapeutic challenges for alleviating vascular remodeling. Another chapter explores myeloid-T cell interactions leading to fibrotic events in chronically rejecting lungs, and how the interactions between the donor and recipient monocytes/macrophages shape the early innate inflammatory environment of the lung allograft, laying the groundwork for subsequent alloimmune injury and chronic rejection. The next chapter describes cellular and molecular mechanisms of chronic rejection in facial transplants, highlighting diagnostic and therapeutic challenges of heterogeneous presentation and lack of standardized criteria. The following chapter focuses on antibody-mediated rejection as the primary cause of kidney allograft failure and how a better understanding of B-cell activation has led to the development of antibody- and plasma cell-directed therapeutics. The last chapter describes the typical and atypical histological features of liver chronic rejection and noninvasive biomarkers for monitoring liver allograft and guiding immunosuppression strategies to lengthen liver transplant health beyond 10 years post-transplantation.

We hope that this book, by covering various aspects of allograft chronic rejection, will not only provide an up-to-date overview of the mechanisms and challenges but also inspire the development of novel therapies for chronic rejection.

Houston, TX, USA

Malgorzata Kloc
Rafik Mark Ghobrial

Contents

Part I
Overview of Cellular and Molecular Effectors in Chronic Rejection and the Computational Approaches

Chapter 1
Cellular and Molecular Aspects of Chronic Rejection

Zhongcheng Mei and Alexander Sasha Krupnick

Abstract Chronic rejection is a principal cause of long-term graft failure in solid organ transplantation. Characterized by progressive fibrotic remodeling, chronic rejection results from complex interactions among alloimmune responses, persistent inflammation, and dysregulated tissue repair mechanisms. This chapter provides a comprehensive overview of the cellular and molecular mechanisms underlying chronic rejection, with a focus on the roles of innate immune cells, T and B lymphocytes, profibrotic cytokines, signaling pathways, and extracellular matrix remodeling. A detailed mechanistic understanding of these pathways is critical for the identification of novel biomarkers and the development of targeted therapeutic strategies to prevent or ameliorate chronic rejection, thereby improving long-term graft survival.

1.1 Introduction

Organ transplantation has revolutionized the management of end-stage organ failure, providing life-saving therapies for patients with irreversible dysfunction (Hamilton 2013). Over the past few decades, improved surgical techniques, organ preservation methods, and immunosuppressive regimens have significantly reduced early graft loss due to acute rejection (Watson and Dark 2012). As a result, short-term outcomes have dramatically improved across most solid organ transplants (Rana and Godfrey 2019).

Z. Mei (✉)
Department of Surgery, University of Maryland, Baltimore, MD, USA
e-mail: zmei@som.umaryland.edu

A. S. Krupnick (✉)
Department of Surgery, University of Maryland, Baltimore, MD, USA

Department of Microbiology and Immunology, University of Maryland, Baltimore, MD, USA
e-mail: akrupnick@som.umaryland.edu

M. Kloc, R. M. Ghobrial (eds.), *Cellular and Molecular Mechanisms of Chronic Rejection in Organ Transplantation*, Results and Problems in Cell Differentiation 77, https://doi.org/10.1007/978-3-032-07686-1_1

However, long-term graft survival remains limited by chronic rejection—a slow, insidious process that leads to progressive tissue remodeling, loss of graft function, and ultimately graft failure (Choudhary et al. 2017; Diebold et al. 2025). Unlike acute rejection, which is characterized by overt inflammation and is often reversible with intensified immunosuppression, chronic rejection is frequently subclinical in its early stages and resistant to therapy once fibrosis is established (Nankivell and Kuypers 2011; Moreso et al. 2012; Sellares et al. 2013; Lang et al. 2021). Chronic rejection represents a common pathological endpoint resulting from persistent low-grade alloimmune responses, dysregulated tissue repair, and immune-fibrotic cross-talk (Becker et al. 2016; Neil and Hübscher 2010). It manifests in organ-specific ways, but shares conserved cellular and molecular mechanisms across different graft types. Understanding these shared pathways is essential for developing novel diagnostics and therapeutic strategies aimed at extending graft longevity.

While chronic rejection has also been explored in vascularized composite allografts and small bowel transplants, this chapter focuses on the best-characterized forms of chronic rejection in lung, kidney, heart, and liver transplantation.

A graphical summary of the major immune and fibrotic mechanisms discussed in this chapter is provided in Fig. 1.1.

1.2 Clinical and Pathological Features of Chronic Rejection

Chronic rejection presents distinct clinical and histological manifestations depending on the transplanted organ; yet, all share the fundamental features of progressive fibrosis, vascular remodeling, and parenchymal injury. Clinical signs are often delayed, and histopathologic assessment remains the gold standard for diagnosis.

1.2.1 Lung

Chronic lung allograft dysfunction (CLAD), the umbrella term for chronic lung rejection, includes two phenotypes: bronchiolitis obliterans syndrome (BOS) and restrictive allograft syndrome (RAS). BOS is the most common form and is defined by progressive airflow obstruction with obliteration of small airways by fibrous tissue (bronchiolitis obliterans). RAS presents with a restrictive pattern on pulmonary function testing and is associated with pleural fibrosis and alveolar collapse (Verleden et al. 2015; Gauthier et al. 2018).

1.2.2 Kidney

Chronic rejection in kidney transplants manifests as transplant glomerulopathy (TG), interstitial fibrosis and tubular atrophy (IFTA), and transplant arteriopathy. TG is characterized by glomerular basement membrane duplication and mesangial

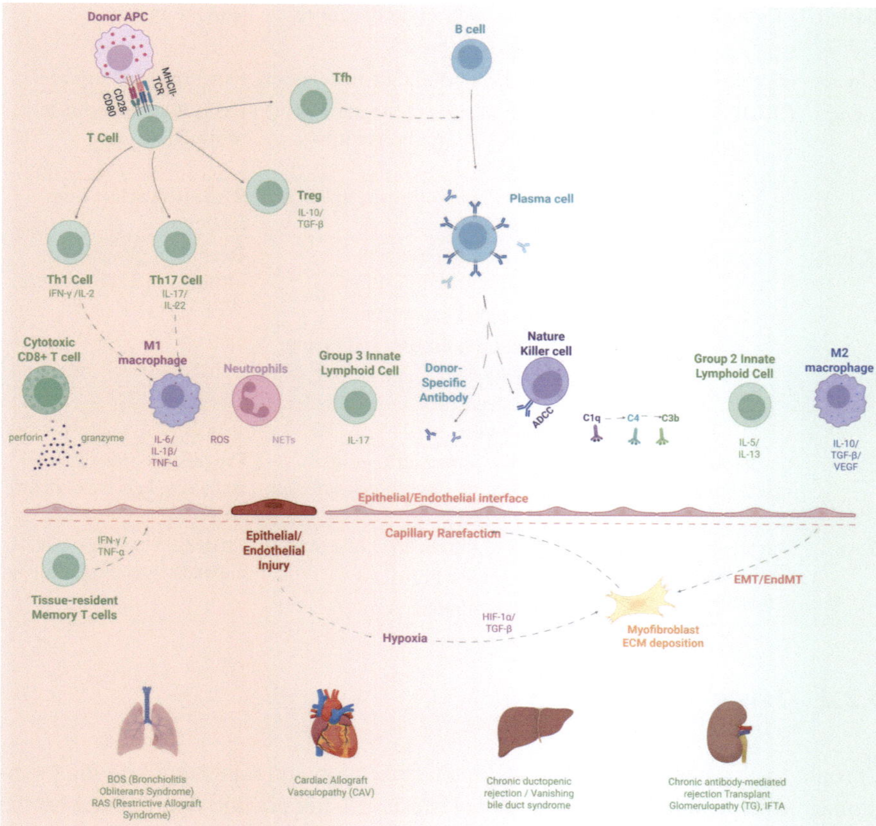

Fig. 1.1 Integrated schematic of chronic rejection mechanisms across organ transplants. The figure illustrates key immune cell subsets, cytokine signals, and tissue-level alterations that collectively contribute to chronic allograft injury, including epithelial/endothelial damage, vascular remodeling, and fibrosis. Representative clinical manifestations across organ systems are shown. (Created with BioRender.com. This figure is covered by a valid publication license)

expansion. Chronic antibody-mediated rejection (CABMR) is frequently associated with C4d deposition in peritubular capillaries (Hara, 2017; Diebold et al. 2025).

1.2.3 *Heart*

Cardiac allograft vasculopathy (CAV) is the hallmark of chronic rejection in heart transplants. It features diffuse, concentric intimal thickening of coronary arteries, leading to progressive ischemia and heart failure. CAV often remains clinically silent until late stages, making surveillance imaging and histopathology essential (Suzuki et al. 2010; Costello et al. 2013).

Table 1.1 Comparative histopathologic features of chronic rejection by organ system

Organ	Clinical entity/ terminology	Key histopathologic features	Functional consequences
Lung	CLAD: BOS and RAS	*BOS*: Fibrous obliteration of small airways (bronchiolitis obliterans)	BOS: Progressive airflow obstruction
		RAS: Pleural fibrosis, alveolar collapse	RAS: Restrictive lung function
Kidney	Chronic ABMR, TG, IFTA	Glomerular basement membrane duplication	Decline in glomerular filtration rate, proteinuria
		Mesangial expansion	
		Interstitial fibrosis and tubular atrophy	
		C4d deposition in peritubular capillaries	
Heart	Cardiac allograft vasculopathy (CAV)	Diffuse, concentric intimal thickening of coronary arteries	Progressive myocardial ischemia, late-onset heart failure
Liver	Ductopenic rejection/ Vanishing bile duct syndrome	Loss of intrahepatic bile ducts	Progressive cholestasis, elevated liver enzymes
		Perivenular fibrosis	
		Portal inflammation	

1.2.4 Liver

Chronic liver rejection includes ductopenic rejection and vanishing bile duct syndrome, which result in progressive cholestasis. Histologically, it is defined by the loss of intrahepatic bile ducts, perivenular fibrosis, and portal inflammation (Neil and Hübscher 2010; Choudhary et al. 2017).

Despite organ-specific differences in pathology, these phenotypes converge on shared mechanisms involving alloimmune activation, endothelial dysfunction, and maladaptive tissue repair (Table 1.1).

1.3 Cellular Mechanisms

1.3.1 Innate Immune Cells

Innate immune cells serve as the front line of defense against graft injury, acting not only as rapid responders to damage but also as crucial modulators of downstream adaptive immune responses. In the context of chronic rejection, these cells orchestrate a persistent inflammatory milieu, promote tissue remodeling, and facilitate fibrosis that ultimately leads to graft dysfunction and failure (Murphy et al. 2011; Ochando et al. 2019).

1.3.1.1 Macrophages

Macrophages are among the most abundant innate immune cells infiltrating chronically rejecting grafts. Over time, they undergo phenotypic polarization toward an M2-like state characterized by elevated expression of profibrotic and proangiogenic mediators. These include transforming growth factor-beta (TGF-β), interleukin-10 (IL-10), vascular endothelial growth factor (VEGF), and platelet-derived growth factor (PDGF). Through these mediators, M2 macrophages stimulate the proliferation and activation of fibroblasts, enhance angiogenesis, and drive the deposition of extracellular matrix (ECM) components such as collagen and fibronectin. Beyond their fibrogenic capacity, macrophages retain antigen-presenting capabilities, processing and presenting alloantigens to T cells, and they engage in bidirectional communication with adaptive immune cells, perpetuating chronic immune activation (Yu and Lu 2022; Salehi and Reed 2015).

1.3.1.2 Neutrophils

Traditionally viewed as short-lived effectors of acute inflammation, neutrophils have emerged as important contributors to chronic rejection, particularly in organ systems like the lung where epithelial barriers are repeatedly injured. Persistent or recurrent infiltration of neutrophils leads to continuous release of reactive oxygen species (ROS), proteolytic enzymes (including elastase and matrix metalloproteinases), and neutrophil extracellular traps (NETs). These mediators damage endothelial and epithelial structures, disrupt tissue integrity, and perpetuate cycles of inflammation. The sustained presence of neutrophils amplifies cytokine and chemokine networks, recruits additional immune cells, and exacerbates graft injury over the long term (Ley et al. 2018; Salehi and Reed 2015).

1.3.1.3 Natural Killer (NK) Cells

NK cells bridge innate and adaptive immunity by mediating antibody-dependent cellular cytotoxicity (ADCC). In the setting of chronic rejection, donor-specific antibodies (DSAs) bind to graft endothelial cells, flagging them for recognition by NK cell Fc receptors (CD16). Activated NK cells then release cytotoxic granules containing granzyme B and perforin and secrete interferon-gamma (IFN-γ), which induces endothelial apoptosis and upregulates adhesion molecules, further promoting leukocyte recruitment. This process not only contributes to direct graft injury but also sustains the inflammatory environment needed for the maintenance of chronic rejection (Benichou et al. 2011; Hamanda et al. 2021).

1.3.1.4 Dendritic Cells

Dendritic cells (DCs) are professional antigen-presenting cells that patrol peripheral tissues and migrate to secondary lymphoid organs upon activation. In chronic rejection, DCs sense damage-associated molecular patterns (DAMPs) released by stressed or dying graft cells. Upon activation, they upregulate costimulatory molecules and MHC expression, efficiently presenting alloantigens to T cells. They also secrete cytokines such as IL-12 and IL-23, which are critical for driving Th1 and Th17 polarization, respectively. This polarization promotes the expansion of effector T cells that infiltrate the graft and perpetuate chronic inflammation and tissue remodeling (Zhuang and Lakkis 2015).

1.3.1.5 Innate Lymphoid Cells (ILCs)

The role of ILCs in chronic rejection is an area of active investigation. Type 2 ILCs (ILC2s) produce cytokines such as IL-5 and IL-13, which have been implicated in epithelial repair, mucus production, and the limitation of excessive inflammation, especially in barrier organs like the lung. By contrast, dysregulated activity of type 3 ILCs (ILC3s), which secrete IL-17 and IL-22, may promote neutrophil recruitment, epithelial stress, and chronic inflammation. The balance between protective and pathogenic roles of ILC subsets likely determines their contribution to long-term graft health or deterioration (Charmetant et al. 2021; Mak et al. 2023).

Altogether, innate immune cells provide the initial inflammatory cues and lay the groundwork for sustained immune–fibrotic interactions that drive chronic rejection.

1.3.2 T Lymphocytes

T cells remain central players in chronic rejection, mediating both direct cytotoxicity and orchestrating broader immune responses (Short et al. 2023).

1.3.2.1 T Helper 1 (Th1) Cells

CD4+ Th1 cells are critical drivers of chronic rejection through secretion of interferon-gamma (IFN-γ). IFN-γ activates macrophages, enhances expression of MHC molecules on graft endothelial and parenchymal cells, and upregulates adhesion molecules, promoting infiltration of other effector cells. Sustained Th1 responses are linked with persistent low-grade inflammation, vascular injury, and promotion of alloantibody generation (Yamada et al. 2018).

1.3.2.2 T Helper 17 (Th17) Cells

Th17 cells produce IL-17A, IL-17F, and IL-22, which recruit neutrophils, induce epithelial stress responses, and activate fibroblasts. Th17-mediated pathways have been associated with increased fibrogenesis and airway remodeling in lung transplants, as well as interstitial fibrosis in other organs. Importantly, Th17 cells are relatively resistant to standard calcineurin-inhibitor-based regimens, posing challenges to long-term graft survival (Huang et al. 2023).

1.3.2.3 CD8+ Cytotoxic T Lymphocytes (CTLs)

CTLs recognize alloantigens presented by MHC class I molecules on graft cells and kill target cells via release of perforin and granzymes. Chronic CTL infiltration results in cumulative endothelial and epithelial damage, microvascular rarefaction, and tissue atrophy, creating hypoxic conditions that further drive fibrosis (Marino et al. 2016).

1.3.2.4 Tissue-Resident Memory T Cells (TRM)

TRM cells are increasingly recognized as critical mediators of chronic rejection. They persist in graft tissue for years, maintain expression of residency markers (e.g., CD69, CD103), and rapidly produce proinflammatory cytokines such as IFN-γ, TNF-α, and chemokines upon antigenic restimulation. These cells serve as a local reservoir sustaining inflammation, recruiting additional immune subsets, and are particularly resistant to systemic depletion by conventional immunosuppression. Recent studies also implicate TRM in crosstalk with local stromal and epithelial cells, linking adaptive immunity with tissue remodeling and fibrosis (Abou-Daya et al. 2021; Tian et al. 2022).

1.3.2.5 Regulatory T Cells (Treg)

Treg cells play a fundamental role in maintaining immune tolerance and suppressing alloimmune responses. They exert their effects through secretion of immunosuppressive cytokines (e.g., IL-10, TGF-β), cytolysis, metabolic disruption, and modulation of antigen-presenting cells. Reduced Treg numbers or functions have been linked with chronic rejection across multiple organ types. Conversely, strategies enhancing Treg activity have shown promise in prolonging graft survival. In organ transplantation, the balance between Th1/Th17 and Treg is particularly critical, and disruption of this axis contributes to chronic allograft dysfunction (Safinia et al. 2010; Huang et al. 2023).

Taken together, effector and regulatory T cell subsets dynamically shape the immune landscape of chronic rejection, balancing cytotoxic injury with attempts at immune regulation.

1.3.3 B Cells and Antibody-Mediated Injury

B cells and their secreted antibodies are pivotal in the evolution of chronic antibody-mediated rejection, providing a persistent humoral arm that complements cellular pathways. Unlike the rapid, transient responses seen in acute rejection, B cells in chronic rejection establish long-lived, high-affinity alloantibody responses that drive ongoing graft injury. Their role extends beyond mere antibody secretion, encompassing antigen presentation, cytokine production, and interaction with innate and adaptive immune cells to shape the fibrotic and inflammatory milieu characteristic of chronic rejection (Heeger et al. 2024).

1.3.3.1 Alloantibody Production

In chronic rejection, long-lived plasma cells continue to produce donor-specific antibodies (DSAs) that target HLA and non-HLA antigens on graft endothelial cells. These antibodies undergo affinity maturation and class switching, often persisting despite immunosuppression. Sustained DSA production results in chronic complement activation, endothelial cell injury, and recruitment of inflammatory cells that perpetuate graft damage (Pape et al. 2015; Diebold et al. 2025).

1.3.3.2 Germinal Center Reactions and Tertiary Lymphoid Structures

Germinal center reactions, typically occurring in secondary lymphoid organs, generate high-affinity, class-switched B cells with help from T follicular helper cells. In chronic rejection, similar processes occur within TLS that develop in graft tissue. These structures sustain local B cell maturation, plasma cell differentiation, and ongoing alloantibody production, creating a self-reinforcing loop of chronic injury (Kwun et al., 2017; Corsiero et al. 2019).

1.3.3.3 Complement Activation and Endothelial Injury

DSAs activate the classical complement pathway, leading to C4d deposition in peritubular capillaries—a diagnostic marker of CABMR. Complement activation induces endothelial expression of adhesion molecules, promotes procoagulant changes, and recruits monocytes and neutrophils, driving microvascular inflammation, intimal thickening, and eventual fibrotic vascular remodeling (Thurman et al. 2019).

1.3.3.4 Regulatory B Cells (Bregs)

A subset of B cells produces IL-10 and suppresses T cell activation, helping to dampen alloimmune responses. Decreased numbers or impaired function of Bregs have been correlated with worse graft outcomes, suggesting potential therapeutic targets for promoting graft tolerance (Cherukuri et al. 2021; Elias et al. 2024).

In summary, chronic antibody responses reinforce inflammation and vascular injury, integrating humoral immunity into the long-term process of graft deterioration.

1.4 Molecular Mechanisms

Chronic rejection arises not only from cellular immune responses but also from complex molecular networks that regulate inflammation, tissue remodeling, and fibrosis. These mechanisms involve persistent cytokine and chemokine signaling, activation of intracellular pathways, metabolic shifts, and extensive extracellular matrix (ECM) remodeling (Torres et al. 2014). Understanding the interplay among these molecular factors is critical for identifying biomarkers of disease progression, designing targeted therapies, and elucidating how immune injury transitions into irreversible fibrotic remodeling. This section examines key molecular mediators and pathways driving chronic rejection.

1.4.1 Cytokine Networks

1.4.1.1 TGF-β (Transforming Growth Factor Beta)

The central mediator of fibrosis, TGF-β promotes fibroblast activation, epithelial-to-mesenchymal transition (EMT), and myofibroblast differentiation. It also suppresses ECM degradation by downregulating MMPs. TGF-β signaling is often sustained by autocrine and paracrine loops involving immune and stromal cells. In addition to driving fibrosis, TGF-β exerts immunoregulatory functions that may paradoxically suppress effector T cell responses while promoting tissue remodeling (Borthwick et al. 2013).

1.4.1.2 IL-6 and IL-17

IL-6 is a pleiotropic cytokine that promotes B cell differentiation, plasma cell survival, and Th17 polarization. In the context of chronic rejection, IL-6 drives both humoral and cellular arms of immunity. IL-17, primarily secreted by Th17 and ILC3 cells, induces expression of chemokines like CXCL1 and CXCL8, leading to neutrophil recruitment. It also stimulates fibroblasts to produce IL-6 and type I collagen, reinforcing inflammatory and fibrotic circuits (Booth and Bishop 2010).

1.4.1.3 IL-5

Though classically associated with eosinophilic responses, IL-5 may have underappreciated roles in regulating granulocyte lifespan and trafficking. In chronic lung rejection, IL-5–producing ILC2s may help preserve epithelial integrity, though excessive or dysregulated IL-5 signaling could contribute to granulocyte-mediated epithelial damage (Guo et al. 2022).

1.4.1.4 Other Mediators

TNF-α, IL-1β, and GM-CSF are upregulated in chronic inflammation and further amplify immune cell recruitment and fibroblast activation. Their roles are particularly evident during episodes of subclinical rejection or infection-induced exacerbations (Li et al. 2019; Pandolfi et al. 2022).

These cytokine-driven circuits support persistent inflammation and fibrosis, driving the transition from immune activation to tissue remodeling.

1.4.2 Fibroblast Activation and EMT/EndMT

EMT and EndMT are critical transitions whereby epithelial and endothelial cells, respectively, lose polarity and adhesion markers (e.g., E-cadherin, VE-cadherin) and gain mesenchymal traits (N-cadherin, vimentin, α-SMA). These transitions expand the population of migratory, matrix-producing cells that fuel fibrosis. TGF-β, Notch, and Wnt signaling pathways cooperate to drive these phenotypic shifts. Concurrently, resident and bone marrow-derived fibroblasts differentiate into contractile myofibroblasts under the influence of PDGF, TGF-β, and mechanical stress, becoming the final effectors responsible for pathological ECM deposition and tissue remodeling (Hertig et al. 2010).

1.4.3 Intracellular Signaling Pathways

Multiple intracellular signaling pathways sustain the fibrotic and inflammatory loops in chronic rejection. mTOR integrates nutrient and cytokine signals to regulate fibroblast proliferation, protein synthesis, and T cell differentiation (Grimbert and Thaunat, 2017). Wnt/β-catenin signaling enhances fibroblast survival and ECM production, while Notch promotes fibroblast-to-myofibroblast transition and vascular remodeling (Majidinia et al. 2018). These pathways interact with TGF-β signaling in redundant circuits that reinforce fibrosis and immune activation.

1.4.4 ECM Remodeling and Metabolism

The extracellular matrix is both a structural scaffold and a dynamic signaling platform. ECM components such as fibronectin and tenascin-C are upregulated in fibrotic regions, facilitating fibroblast adhesion, migration, and survival. Matrix metalloproteinases (MMP-2, MMP-9) degrade ECM, but elevated tissue inhibitors of metalloproteinases (TIMPs) tip the balance toward net ECM accumulation (Barker et al. 2014). Simultaneously, chronic inflammation drives metabolic reprogramming: immune cells favor glycolysis to sustain cytokine output, while fibroblasts shift to glutaminolysis and lipid metabolism to fuel persistent collagen production (Degauque et al. 2018). These metabolic shifts further shape cytokine secretion profiles and epigenetic landscapes, perpetuating the fibrotic process.

1.5 Immune-Fibrotic Crosstalk: Integrative Pathways in Chronic Rejection

Chronic rejection is not simply a sum of isolated immune or fibrotic components—it is a progressive, self-perpetuating network of cellular and molecular interactions. In this section, we integrate the pathways described in previous chapters to illustrate how immune-mediated injury and fibrogenesis form a reinforcing pathological circuit.

1.5.1 Initiation: Danger Signals and Innate Immune Activation

Tissue injury from ischemia-reperfusion, infections, or early alloimmune events releases danger-associated molecular patterns (DAMPs). These activate innate immune cells—particularly dendritic cells and macrophages—leading to upregulation of costimulatory molecules and inflammatory cytokines (e.g., IL-1β, IL-6, TNF-α). Dendritic cells present alloantigens to T cells, priming both effector and memory responses. NK cells and neutrophils respond to antibody-coated endothelium, contributing to initial vascular damage (Barker et al. 2014; Braza et al. 2016; Li et al. 2019).

1.5.2 Propagation: Adaptive Immunity and Cytokine Amplification

Activated CD4+ and CD8+ T cells infiltrate the graft, producing IFN-γ, IL-17, and cytotoxic mediators that damage graft endothelium and parenchyma. Concurrently, B cells mature into plasma cells that produce donor-specific antibodies (DSAs), which fix complement and recruit innate effectors via Fc receptor engagement. These cytokines and antibodies further amplify macrophage and neutrophil activation, leading to a second wave of inflammation.

Importantly, IL-17 and IL-6 drive expansion of fibroblast and endothelial progenitor populations, blurring the boundary between immune injury and fibrotic transformation. Sustained IL-17 exposure also induces chemokines (CXCL1, CCL20) that perpetuate neutrophil and Th17 recruitment (Crescioli, 2016; Yamada et al. 2018; Huang et al. 2023; Short et al. 2023).

1.5.3 Transition to Fibrosis: Fibroblast Recruitment and Activation

TGF-β, secreted by M2 macrophages, T cells, and injured epithelial/endothelial cells, is the key molecular bridge between inflammation and fibrosis. It induces fibroblast proliferation, EMT/EndMT, and myofibroblast differentiation. IL-13, IL-17, and PDGF synergize with TGF-β to maintain a pro-fibrotic tissue environment. Fibroblasts activated in this milieu become highly secretory and resistant to apoptosis, favoring persistence and expansion (Hertig et al. 2010; Booth and Bishop 2010; Borthwick et al. 2013).

1.5.4 Pathologic Remodeling: ECM Accumulation and Vascular Rarefaction

Myofibroblasts deposit excessive ECM, creating a stiffened, hypoperfused matrix. This environment impairs nutrient and oxygen diffusion, exacerbates epithelial and endothelial stress, and triggers further inflammatory signaling. The loss of capillaries—particularly peritubular and peribronchial microvasculature—is a hallmark of chronic rejection, linking immune injury directly to oxygen deprivation and tissue attrition (Akhtar et al. 2014; Cardinal et al. 2018; Pasupneti and Nicolls 2019).

1.5.5 Feedback Loops and Chronicity

- Persistent alloantigen exposure sustains TRM and plasma cell niches in situ (Abou-Daya et al. 2021; Tian et al. 2022; Schäfer et al. 2023).
- DSA-mediated complement activation induces expression of fibrogenic genes in endothelial and mesenchymal cells (Stegall et al. 2012; Danobeitia et al. 2014).
- Hypoxia-inducible factors (e.g., HIF-1α) promote glycolytic metabolism and TGF-β expression (Akhtar et al. 2014; Pasupneti and Nicolls 2019).
- ECM remodeling during chronic rejection modifies immune responses by altering cell migration and transduction signaling (Crescioli, 2016).

Together, these factors reinforce chronic immune stimulation and fibrosis, forming the biological basis of persistent allograft rejection.

1.5.6 Organ-Level Integration: Example from Lung Transplantation

The lung provides a clear example of how chronic rejection results from the integration of immune-mediated injury and fibrotic remodeling. Chronic lung allograft dysfunction (CLAD) encompasses two main phenotypes—bronchiolitis obliterans syndrome (BOS) and restrictive allograft syndrome (RAS), each reflecting distinct patterns of pathology yet sharing core mechanisms (Verleden et al. 2015; Gauthier et al. 2018).

In BOS, early epithelial injury—often triggered or exacerbated by infections, aspiration, or gastroesophageal reflux—initiates the release of DAMPs, activating innate immune cells such as macrophages, neutrophils, and innate lymphoid cells (ILC2s and ILC3s) (Guo et al. 2022). These cells secrete mediators that promote local inflammation, recruit additional effector cells, and damage the epithelial barrier. Alloantigen presentation by dendritic cells activates T cells, particularly Th17 and Th2 subsets. Th17 cells produce IL-17, driving neutrophil recruitment and fibroblast activation, while Th2 cells and ILC2s secrete IL-5 and IL-13, supporting eosinophil survival and mucosal remodeling (Yamada et al. 2018).

Meanwhile, tissue-resident memory T cells persist in the graft, ready to rapidly produce proinflammatory cytokines upon antigen re-exposure. B cells generate donor-specific antibodies (DSAs) that bind to graft endothelium, activating complement and amplifying injury. In this inflamed microenvironment, fibroblasts respond to TGF-β and other profibrotic signals by differentiating into myofibroblasts, which deposit ECM that progressively narrows and obliterates small airways—the histological hallmark of BOS (Booth and Bishop 2010).

In contrast, RAS is characterized by parenchymal and pleural fibrosis extending beyond the bronchiolar tree. This may involve distinct molecular programs, including enhanced type I interferon signaling and a predominance of macrophage-driven fibrosis, although the precise mechanisms remain under investigation (Verleden et al. 2015).

Together, BOS and RAS illustrate how the convergence of innate and adaptive immune responses, coupled with dysregulated repair and fibrotic signaling, leads to organ-specific manifestations of chronic rejection.

1.6 Organ-Specific Molecular Signatures

While chronic rejection shares fundamental cellular and molecular mechanisms across organ types, accumulating transcriptomic and proteomic data have revealed unique, tissue-specific molecular signatures. These signatures reflect differences in immunologic microenvironments, tissue architecture, and injury response pathways that shape the trajectory of chronic rejection in each organ.

1.6.1 Role of Omics Technologies in Chronic Rejection

Advances in bulk and single-cell RNA sequencing, proteomics, spatial transcriptomics, and metabolomics have enabled high-resolution characterization of allograft tissue. These approaches have allowed for

- Identification of differentially expressed genes and proteins associated with inflammation, fibrosis, and rejection.
- Mapping of immune cell subsets and stromal populations within specific graft compartments.
- Discovery of candidate biomarkers for early, noninvasive diagnosis.
- Insights into tissue-specific therapeutic targets and pathways that could inform precision medicine (Kurian et al. 2014; Halloran 2015).

1.6.2 Molecular Signatures by Organ

Kidney

- Upregulated genes: COL1A1, FN1 (fibrosis), CXCL9/10 (T cell recruitment), HLA class II.
- Proteins: C4d, complement components, osteopontin.

- Chronic ABMR is associated with ongoing endothelial injury, microvascular inflammation, and tubular epithelial stress responses. Recent studies using spatial transcriptomics have revealed colocalization of immune cells and fibrotic lesions, underscoring the interplay between alloimmunity and fibrosis (Moreso et al. 2012; Sellares et al. 2013).

Lung

- Upregulated: IL1B, IL6, CXCL1, CXCL8, S100A8/A9 (neutrophilic inflammation).
- Downregulated: Epithelial junction and ciliary genes.
- CLAD subtypes exhibit distinct molecular patterns: BOS is characterized by airway remodeling gene signatures, while RAS shows enrichment of profibrotic and alveolar injury pathways. Emerging data suggests differential expansion of TRM and innate lymphoid cell subsets between BOS and RAS (Verleden et al. 2015; Gauthier et al. 2018).

Heart

- Upregulated: ACTA2 (smooth muscle), VCAM1 (endothelial activation), CD68 (macrophages).
- Gene signatures correlate with angiographic and IVUS measures of cardiac allograft vasculopathy (CAV) severity. Integrating gene expression with imaging modalities may enable earlier detection of CAV progression and stratify patients for targeted therapies (Suzuki et al. 2010; Costello et al. 2013).

Liver

- Downregulated: Bile acid transporters (e.g., ABCB11).
- Upregulated: TGF-β1, TIMP1, COL3A1 (fibrosis); MHC class II on endothelial and biliary cells.
- Molecular studies show that chronic ductopenia correlates with antigen presentation pathways and perivenular immune activation, highlighting potential intersections between alloimmune responses and bile duct injury (Neil and Hübscher 2010; Choudhary et al. 2017).

1.6.3 Translational Relevance

Organ-specific molecular signatures are increasingly informing precision medicine approaches in transplantation. High-throughput technologies have enabled the development of diagnostic tools that detect early molecular injury, stratify immunologic risk, and guide individualized therapy. For instance, noninvasive biomarkers such as donor-derived cell-free DNA (dd-cfDNA), urinary chemokines (e.g., CXCL9/10), and peripheral blood gene expression profiles are under active validation as sensitive indicators of subclinical rejection (Kurian et al. 2014; Halloran 2015; Bloom et al. 2017).

Table 1.2 Representative organ-specific molecular markers and technologies in chronic rejection

Organ	Upregulated markers	Downregulated markers	Key features/notes	Omics technology used
Kidney	COL1A1, FN1 (fibrosis); CXCL9/10 (T cell recruitment); HLA class II; C4d, osteopontin	–	Endothelial injury and tubular stress responses in chronic ABMR	Bulk RNA-seq, proteomics, spatial transcriptomics
Lung	IL1B, IL6, CXCL1, CXCL8, S100A8/A9	Epithelial junction and ciliary genes	BOS: airway remodeling; RAS: alveolar collapse and fibrosis	Single-cell RNA-seq, spatial transcriptomics
Heart	ACTA2, VCAM1, CD68	–	Expression correlates with CAV severity by imaging	Bulk RNA-seq, proteomics
Liver	TGF-β1, TIMP1, COL3A1; MHC II (endothelial, biliary)	Bile acid transporters (e.g., ABCB11)	Associated with ductopenia and perivenular immune activation	Bulk RNA-seq, proteomics, metabolomics

Integration of transcriptomic or proteomic data with clinical parameters may improve risk stratification by distinguishing stable allografts from those at heightened risk for immune-mediated damage (Sellares et al. 2013; Halloran 2015). Moreover, delineation of the dominant immune or fibrotic pathways in individual grafts may inform rational therapeutic selection—including the use of cytokine inhibitors, antifibrotic agents, and emerging cell-based tolerance strategies (Safinia et al. 2010; Heeger et al. 2024; Cherukuri et al. 2021). Representative markers and analytic technologies are summarized in Table 1.2.

1.7 Conclusion and Future Directions

Chronic rejection remains the leading cause of late graft failure in solid organ transplantation. Unlike acute rejection, which often presents with overt inflammation and responds to intensified immunosuppression, chronic rejection is driven by insidious and self-sustaining interactions between the immune system and tissue repair machinery. Central to this process are persistent alloantigen exposure, cytokine-driven inflammation, fibroblast activation, and extracellular matrix (ECM) accumulation—mechanisms that are increasingly well defined through high-resolution molecular and cellular studies.

The mechanistic insights outlined in this chapter have direct translational implications. Targeted disruption of specific immune pathways—such as interleukin-6 (IL-6), IL-17, or TGF-β signaling—is under active investigation. Tocilizumab, an IL-6 receptor antagonist, has shown clinical promise in reducing donor-specific antibodies and glomerular injury in chronic antibody-mediated rejection (Choi et al.

2017). Similarly, IL-17 inhibitors have demonstrated efficacy in animal models by attenuating neutrophil recruitment and fibrotic remodeling (Gupta et al. 2017).

Cell-based immunomodulatory therapies offer a mechanistically distinct and potentially synergistic approach. Regulatory T cells (Tregs), when expanded ex vivo and reinfused into transplant recipients, have demonstrated the ability to suppress effector immune responses and reduce subclinical inflammation. Beyond immunosuppression, Tregs also modulate antigen-presenting cell function and interfere with profibrotic signaling pathways, positioning them as a dual-action therapeutic candidate in chronic rejection (Romano et al. 2016). While clinical translation remains ongoing, several early-phase trials have underscored both the promise and the complexity of Treg therapy, highlighting the need for refined protocols tailored to specific organ types and immune contexts (Tang and Vincenti 2017).

Similarly, mesenchymal stromal cells (MSCs) have demonstrated potent immunoregulatory and antifibrotic properties across multiple models of solid organ transplantation. These multipotent cells can inhibit T cell proliferation, skew macrophage polarization toward anti-inflammatory phenotypes, and promote tissue repair through paracrine mechanisms. Preclinical and early-phase clinical studies support their ability to modulate alloimmune responses and reduce fibrotic progression, particularly in renal and lung transplantation settings (Podestà et al. 2021; Casiraghi et al. 2016). Given their dual potential to suppress alloimmunity and attenuate fibrosis, MSC-based therapies are now being investigated as adjunctive treatments for chronic rejection.

In parallel, gene-editing strategies—particularly CRISPR/Cas9—are being leveraged to engineer donor organs with reduced immunogenicity and enhanced compatibility (Kuscu et al. 2020). A notable example is xenotransplantation, which has brought these technologies into clinical application. In 2022, a landmark porcine-to-human heart transplant at the University of Maryland employed a donor pig with ten genetic modifications designed to minimize immune recognition and coagulation risk, marking a major milestone in genome-edited organ transplantation (Griffith et al. 2022). Collectively, these advances reflect a broader shift in the field—from generalized immunosuppression toward precision strategies informed by mechanistic understanding and molecular diagnostics. Achieving durable control of chronic rejection will ultimately depend on integrated approaches that restore immune-fibrotic balance in a tissue- and patient-specific manner.

Publishing Agreement

Acknowledgments The authors would like to thank Dr. Daniel Kreisel and Dr. Andrew Gelman (Washington University in St. Louis) and Dr. Elizabeth Jacobsen (Mayo Clinic) for their valuable discussions and insights related to chronic rejection and immunopathology. We also gratefully acknowledge the contributions of the University of Maryland, Baltimore Transplant Research Team. AI-based language tools were used to assist with grammar, rewording, and stylistic clarity.

References

Abou-Daya KI, Tieu R, Zhao D, Rammal R, Sacirbegovic F, Williams AL, Shlomchik WD, Oberbarnscheidt MH, Lakkis FG (2021) Resident memory T cells form during persistent antigen exposure leading to allograft rejection. Sci Immunol 6(57):eabc8122. https://doi.org/10.1126/sciimmunol.abc8122

Akhtar MZ, Sutherland AI, Huang H, Ploeg RJ, Pugh CW (2014) The role of hypoxia-inducible factors in organ donation and transplantation: the current perspective and future opportunities. Am J Transplant 14(7):1481–1487. https://doi.org/10.1111/ajt.12737

Barker CE, Ali S, O'Boyle G, Kirby JA (2014) Transplantation and inflammation: implications for the modification of chemokine function. Immunology 143(2):138–145. https://doi.org/10.1111/imm.12332

Becker LE, Morath C, Suesal C (2016) Immune mechanisms of acute and chronic rejection. Clin Biochem 49(4–5):320–323. https://doi.org/10.1016/j.clinbiochem.2016.02.001

Benichou G, Yamada Y, Aoyama A, Madsen JC (2011) Natural killer cells in rejection and tolerance of solid organ allografts. Curr Opin Organ Transplant 16(1):47–53. https://doi.org/10.1097/MOT.0b013e32834254cf

Bloom RD, Bromberg JS, Poggio ED, Bunnapradist S, Langone AJ, Sood P, Matas AJ, Mehta S, Mannon RB, Sharfuddin A, Fischbach B, Narayanan M, Jordan SC, Cohen D, Weir MR, Hiller D, Prasad P, Woodward RN, Grskovic M, Sninsky JJ, Yee JP, Brennan DC (2017) Cell-free DNA and active rejection in kidney allografts. J Am Soc Nephrol 28(7):2221–2232. https://doi.org/10.1681/ASN.2016091034

Booth AJ, Bishop DK (2010) TGF-β, IL-6, IL-17 and CTGF direct multiple pathologies of chronic cardiac allograft rejection. Immunotherapy 2(4):511–520. https://doi.org/10.2217/imt.10.33

Borthwick LA, Wynn TA, Fisher AJ (2013) Cytokine mediated tissue fibrosis. Biochim Biophys Acta 1832(7):1049–1060. https://doi.org/10.1016/j.bbadis.2012.09.014

Braza F, Brouard S, Chadban S, Goldstein DR (2016) Role of TLRs and DAMPs in allograft inflammation and transplant outcomes. Nat Rev Nephrol 12(5):281–290. https://www.nature.com/articles/nrneph.2016.41

Cardinal H, Dieudé M, Hébert MJ (2018) Endothelial dysfunction in kidney transplantation. Front Immunol 9:1130. https://doi.org/10.3389/fimmu.2018.01130

Casiraghi F, Perico N, Cortinovis M, Remuzzi G (2016) Mesenchymal stromal cells in renal transplantation: opportunities and challenges. Nat Rev Nephrol 12(4):241–253. https://doi.org/10.1038/nrneph.2016.7

Charmetant X, Bachelet T, Déchanet-Merville J, Walzer T, Thaunat O (2021) Innate (and innate-like) lymphoid cells: emerging immune subsets with multiple roles along transplant life. Transplantation 105(12):e322–e336. https://doi.org/10.1097/TP.0000000000003782

Cherukuri A, Mohib K, Rothstein DM (2021) Regulatory B cells: TIM-1, transplant tolerance, and rejection. Immunol Rev 299:31–44. https://doi.org/10.1111/imr.12933

Choi J, Aubert O, Vo A, Loupy A, Haas M, Puliyanda D, Kim I, Louie S, Kang A, Peng A, Kahwaji J, Reinsmoen N, Toyoda M, Jordan SC (2017) Assessment of tocilizumab (anti–interleukin-6 receptor monoclonal) as a potential treatment for chronic antibody-mediated rejection and transplant glomerulopathy in HLA-sensitized renal allograft recipients. Am J Transplant 17(9):2381–2389. https://doi.org/10.1111/ajt.14228

Choudhary NS, Saigal S, Bansal RK, Saraf N, Gautam D, Soin AS (2017) Acute and chronic rejection after liver transplantation: what a clinician needs to know. J Clin Exp Hepatol 7(4):358–366. https://doi.org/10.1016/j.jceh.2017.10.003

Corsiero E, Delvecchio FR, Bombardieri M, Pitzalis C (2019) B cells in the formation of tertiary lymphoid organs in autoimmunity, transplantation and tumorigenesis. Curr Opin Immunol 57:46–52. https://doi.org/10.1016/j.coi.2019.01.004

Costello JP, Mohanakumar T, Nath DS (2013) Mechanisms of chronic cardiac allograft rejection. Tex Heart Inst J 40(4):395–399

Crescioli C (2016) Chemokines and transplant outcome. Clin Biochem 49(4–5):355–362. https://doi.org/10.1016/j.clinbiochem.2015.07.026

Danobeitia JS, Djamali A, Fernandez LA (2014) The role of complement in the pathogenesis of renal ischemia–reperfusion injury and fibrosis. Fibrogenesis Tissue Repair 7:16. https://doi.org/10.1186/1755-1536-7-16

Degauque N, Brosseau C, Brouard S (2018) Regulation of the immune response by the inflammatory metabolic microenvironment in the context of allotransplantation. Front Immunol 9:1465. https://doi.org/10.3389/fimmu.2018.01465

Diebold M, Mayer KA, Hidalgo L, Kozakowski N, Budde K, Böhmig GA (2025) Chronic rejection after kidney transplantation. Transplantation 109(4):610–621. https://doi.org/10.1097/TP.0000000000005187

Elias C, Chen C, Cherukuri A (2024) Regulatory B cells in solid organ transplantation: from immune monitoring to immunotherapy. Transplantation 108(5):1080–1089. https://doi.org/10.1097/TP.0000000000004798

Gauthier JM, Ruiz-Pérez D, Li W, Hachem RR, Puri V, Gelman AE, Kreisel D (2018) Diagnosis, pathophysiology and experimental models of chronic lung allograft rejection. Transplantation 102(9):1459–1466. https://doi.org/10.1097/TP.0000000000002250

Griffith BP, Goerlich CE, Singh AK, Rothblatt M, Mohiuddin MM (2022) Genetically modified porcine-to-human cardiac xenotransplantation. N Engl J Med 387(1):35–44. https://doi.org/10.1056/NEJMoa2201422

Grimbert P, Thaunat O (2017) mTOR inhibitors and risk of chronic antibody-mediated rejection after kidney transplantation: where are we now? Transpl Int 30(7):647–657. https://doi.org/10.1111/tri.12975

Guo Y, Mei Z, Li D, Banerjee A, Khalil MA, Burke A, Krupnick AS (2022) Ischemia reperfusion injury facilitates lung allograft acceptance through IL-33–mediated activation of donor-derived IL-5 producing group 2 innate lymphoid cells. Am J Transplant 22(8):1963–1975. https://doi.org/10.1111/ajt.17084

Gupta PK, Wagner SR, Wu Q, Shilling RA (2017) IL-17A blockade attenuates obliterative bronchiolitis and IFN-γ cellular immune response in lung allografts. Am J Respir Cell Mol Biol 56(6):708–715. https://doi.org/10.1165/rcmb.2016-0154OC

Halloran PF (2015) Molecular assessment of disease states in kidney transplant biopsy samples. Nat Rev Nephrol 11(3):142–154. https://doi.org/10.1038/nrneph.2014.268

Hamanda S, Dubois V, Koenig A, Thaunat O (2021) Allograft recognition by recipient's natural killer cells: molecular mechanisms and role in transplant rejection. Transpl Immunol 59:101343. https://doi.org/10.1111/tan.14332

Hamilton D (2013) A history of organ transplantation: ancient legends to modern practice. University of Pittsburgh Press, Pittsburgh, PA

Hara S (2017) Current pathological perspectives on chronic rejection in renal allografts. Clin Exp Nephrol 21(6):943–951. https://doi.org/10.1007/s10157-016-1363-0

Heeger PS, Carrera Haro M, Jordan S (2024) Translating B cell immunology to the treatment of antibody-mediated allograft rejection. Nat Rev Nephrol 20:218–232. https://doi.org/10.1038/s41581-023-00796-w

Hertig A, Flier SN, Kalluri R (2010) Contribution of epithelial plasticity to renal transplantation-associated fibrosis. Transplant Proc 42(9 Suppl):S7–S12. https://doi.org/10.1016/j.transproceed.2010.07.040

Huang DL, He YR, Liu YJ, He HY, Gu ZY, Liu YM, Liu WJ, Luo Z, Ju MJ (2023) The immunomodulation role of Th17 and Treg in renal transplantation. Front Immunol 14:1113560. https://doi.org/10.3389/fimmu.2023.1113560

Kurian SM et al (2014) Molecular classifiers for acute kidney transplant rejection in peripheral blood by whole genome gene expression profiling. Am J Transplant 14(5):1164–1172. https://doi.org/10.1111/ajt.12671

Kuscu C, Kuscu C, Bajwa A, Eason JD, Maluf D, Mas VR (2020) Applications of CRISPR technologies in transplantation. Am J Transplant 20(12):3285–3293. https://doi.org/10.1111/ajt.16095

Kwun J, Manook M, Page E, Burghuber C, Hong J, Knechtle SJ (2017) Crosstalk between T and B cells in the germinal center after transplantation. Transplantation 101(4):704–712. https://doi.org/10.1097/TP.0000000000001588

Lang S, Riazy M, Blydt-Hansen TD (2021) Rejection challenges: diagnosis and management. In: Challenges in pediatric kidney transplantation. Springer Nature, pp 41–117

Ley K, Hoffman HM, Kubes P, Cassatella MA, Zychlinsky A, Hedrick CC, Catz SD (2018) Neutrophils: new insights and open questions. Sci Immunol 3(30):eaat4579. https://doi.org/10.1126/sciimmunol.aat4579

Li J, Li C, Zhuang Q, Peng B, Zhu Y, Ye Q, Ming Y (2019) The evolving roles of macrophages in organ transplantation. J Immunol Res 2019:5763430. https://doi.org/10.1155/2019/5763430

Majidinia M, Aghazadeh J, Jahanban-Esfahlani R, Yousefi B (2018) The roles of Wnt/β-catenin pathway in tissue development and regenerative medicine. J Cell Physiol 233(8):5598–5612. https://doi.org/10.1002/jcp.26265

Mak ML, Reid KT, Crome SQ (2023) Protective and pathogenic functions of innate lymphoid cells in transplantation. Clin Exp Immunol 213(1):23–39. https://doi.org/10.1093/cei/uxad050

Marino J, Paster J, Benichou G (2016) Allorecognition by T lymphocytes and allograft rejection. Front Immunol 7:582. https://doi.org/10.3389/fimmu.2016.00582

Moreso F, Carrera M, Goma M, Hueso M, Sellares J, Martorell J, Grinyó JM, Serón D (2012) Early subclinical rejection as a risk factor for late chronic humoral rejection. Transplantation 93(1):41–46. https://doi.org/10.1097/TP.0b013e31823bb647

Murphy SP, Porrett PM, Turka LA (2011) Innate immunity in transplant tolerance and rejection. Am J Transplant 11(5):856–860. https://doi.org/10.1111/j.1600-065X.2011.01009.x

Nankivell BJ, Kuypers DRJ (2011) Diagnosis and prevention of chronic kidney allograft loss. Lancet 378(9800):1428–1437. https://doi.org/10.1016/S0140-6736(11)60699-5

Neil DAH, Hübscher SG (2010) Current views on rejection pathology in liver transplantation. Transpl Int 23(5):441–450. https://doi.org/10.1111/j.1432-2277.2010.01143.x

Ochando J, Ordikhani F, Boros P, Jordan S (2019) The innate immune response to allotransplants: mechanisms and therapeutic potentials. Cell Mol Immunol 16(4):350–356. https://doi.org/10.1038/s41423-019-0216-2

Pandolfi L, Bozzini S, Morosini M, Sposito B, Broggi A, Vertui V, Lettieri S, Saracino L, De Amici M, Zanoni I, Meloni F (2022) Significant upregulation of BAL-f IL-1β in lung transplant recipients during stability and CLAD. J Heart Lung Transplant 41(4 Suppl):S306. https://doi.org/10.1016/j.healun.2022.01.758

Pape L, Becker JU, Immenschuh S, Ahlenstiel T (2015) Acute and chronic antibody-mediated rejection in pediatric kidney transplantation. Pediatr Nephrol 30:417–424. https://doi.org/10.1007/s00467-014-2944-2

Pasupneti S, Nicolls MR (2019) Airway hypoxia in lung transplantation. Curr Opin Physio 7:21–26. https://doi.org/10.1016/j.cophys.2018.12.002

Podestà MA, Remuzzi G, Casiraghi F (2021) Mesenchymal stromal cell therapy in solid organ transplantation. Front Immunol 11:618243. https://doi.org/10.3389/fimmu.2020.618243

Rana A, Godfrey EL (2019) Outcomes in solid-organ transplantation: success and stagnation. Tex Heart Inst J 46(1):75–76. https://doi.org/10.14503/THIJ-18-6749

Romano M, Tung SL, Smyth LA, Lombardi G (2016) Treg therapy in transplantation: a general overview. Transpl Int 30(8):745–753. https://doi.org/10.1111/tri.12909

Safinia N, Afzali B, Atalar K, Lombardi G, Lechler RI (2010) T-cell alloimmunity and chronic allograft dysfunction. Kidney Int 78:252–258. https://doi.org/10.1038/ki.2010.416

Salehi S, Reed EF (2015) The divergent roles of macrophages in solid organ transplantation. Curr Opin Organ Transplant 20(4):446–453. https://doi.org/10.1097/MOT.0000000000000209

Schäfer AL, Ruiz-Aparicio PF, Kraemer AN, Chevalier N (2023) Crosstalk in the diseased plasma cell niche – the force of inflammation. Front Immunol 14:1120398. https://doi.org/10.3389/fimmu.2023.1120398

Sellares J, Reeve J, Loupy A, Mengel M, Sis B, Skene A, de Freitas DG, Kreepala C, Hidalgo LG, Famulski KS, Halloran PF (2013) Molecular diagnosis of antibody-mediated rejection in human kidney transplants. Am J Transplant 13(4):971–983. https://doi.org/10.1111/ajt.12150

Short S, Lewik G, Issa F (2023) An immune atlas of T cells in transplant rejection: pathways and therapeutic opportunities. Transplantation 107(11):2341–2352. https://doi.org/10.1097/TP.0000000000004572

Stegall MD, Chedid MF, Cornell LD (2012) The role of complement in antibody-mediated rejection in kidney transplantation. Nat Rev Nephrol 8(10):670–678. https://doi.org/10.1038/nrneph.2012.213

Suzuki J, Isobe M, Morishita R, Nagai R (2010) Characteristics of chronic rejection in heart transplantation: important elements of pathogenesis and future treatments. Circ J 74(2):233–239. https://doi.org/10.1253/circj.CJ-09-0809

Tang Q, Vincenti F (2017) Transplant trials with Tregs: perils and promises. J Clin Invest 127(7):2505–2512. https://doi.org/10.1172/JCI90598

Thurman JM, Panzer SE, Le Quintrec M (2019) The role of complement in antibody-mediated transplant rejection. Mol Immunol 112:240–246. https://doi.org/10.1016/j.molimm.2019.06.002

Tian Q, Zhang Z, Tan L, Yang F, Xu Y, Guo Y, Wei D, Wu C, Cao P, Zhao Y (2022) Skin and heart allograft rejection solely by long-lived alloreactive TRM cells in skin of severe combined immunodeficient mice. Sci Adv 8(4):eabk0270. https://doi.org/10.1126/sciadv.abk0270

Torres IB, Moreso F, Sarró E, Meseguer A, Serón D (2014) The interplay between inflammation and fibrosis in kidney transplantation. Biomed Res Int 2014:750602. https://doi.org/10.1155/2014/750602

Verleden GM, Vos R, Van Raemdonck D, Dupont L, Yserbyt J, Vanaudenaerde B, Verleden S (2015) Current views on chronic rejection after lung transplantation. Transpl Int 28(10):1131–1139. https://doi.org/10.1111/tri.12579

Watson CJ, Dark JH (2012) Organ transplantation: historical perspective and current practice. Br J Anaesth 108(Suppl 1):i29–i42. https://doi.org/10.1093/bja/aer384

Yamada Y, Brüstle K, Jungraithmayr W (2018) T helper cell subsets in experimental lung allograft rejection. J Surg Res. https://doi.org/10.1016/j.jss.2018.07.073

Yu S, Lu J (2022) Macrophages in transplant rejection. Transpl Immunol 71:101536. https://doi.org/10.1016/j.trim.2022.101536

Zhuang Q, Lakkis FG (2015) Dendritic cells and innate immunity in kidney transplantation. Kidney Int 87(4):712–718. https://doi.org/10.1038/ki.2014.430

Chapter 2
How Cytokines Regulate Immune Response Toward Chronic Allograft Rejection?

Stanislaw Stepkowski, Jared Oenick, Dulat Bekbolsynov, Beata Mierzejewska, Michael Rees, and Obi Ekwenna

Abstract Intensive research for 50 years did not find a way to predict the faith of organ transplants after successful transplantation. Current immunosuppressive regiments reduce kidney transplant acute rejection episodes to 6–11% for the first year, but 49% of them develop chronic allograft nephropathy in 5 years. Similarly, heart transplants have 8% acute rejection episodes in the first year and 50% develop cardiac vasculopathy in 5 years. Multiple cytokines regulate acute cellular T-cell-mediated rejection (TCMR) and acute antibody-mediated rejection (AMR), and both mechanisms contribute to chronic allograft rejection. In addition to common γ chain (cγ) cytokines [interleukin-2 (IL-2), IL-4, IL-7, IL-9, IL-15, and IL-21], other cytokines [IL-1, IL-6, IL-10, interferon-α, (IFN) α, INFβ, INFγ, tumor necrosis factor α, TNFα; tumor growth factor β, TGFβ] are all extensively involved in the regulation of acute and chronic allograft rejections. Interestingly, the same cytokines also regulate functions of normal cells and tissues. To induce transplant tolerance, it is necessary to maintain physiological cytokine levels as well as boost the function of T regulatory T (Treg) cells producing IL-10, IL-35, and TGFβ. This chapter emphasizes the necessity of maintaining a balance by cytokines to divert them from the inflammation path producing chronic rejection to normal function support combined with Treg-dominated transplantation tolerance.

S. Stepkowski (✉) · J. Oenick · D. Bekbolsynov · B. Mierzejewska
Department of Medical Microbiology and Immunology, University of Toledo Medical Center, Toledo, OH, USA
e-mail: Stanislaw.Stepkowski@utoledo.edu

M. Rees · O. Ekwenna
Department of Urology, University of Toledo Medical Center, Toledo, OH, USA

M. Kloc, R. M. Ghobrial (eds.), *Cellular and Molecular Mechanisms of Chronic Rejection in Organ Transplantation*, Results and Problems in Cell Differentiation 77, https://doi.org/10.1007/978-3-032-07686-1_2

Abbreviations

AID	Activation-induced cytidine deaminase
AMR	Antibody-mediated rejection
BAF	B-cell activating factor
BATF	Basic leucine zipper ATF-like transcription factor
BMP7	Bone morphogenetic protein-7
caAMR	Chronic "active" AMR
cAMR	Chronic antibody-mediated rejection
caTCMR	Chronic "active" TCMR
CAV	Chronic allograft vasculopathy
cTCMR	Chronic T-cell-mediated rejection
dar Tregs	Donor antigen-reactive/specific regulatory T cells
dnDSA	De novo donor-specific antibodies
DSA	Donor-specific antibody
Ebi3	Epstein-Barr-virus-induced protein 3
EC	Endothelial cell
eNOS	Endothelial NO
Foxp3	Forkhead box P3
HO-1	Heme oxygenase 1
ICAM-1	Intercellular adhesion molecule-1
ICOS	Inducible T cell costimulator
IFN	Interferon
IFTA	Interstitial fibrosis, tubular atrophy
i-IFTA	Interstitial fibrosis, tubular atrophy with inflammation
IL-21R.Fc	IL-21 receptor fusion protein
IL-2R	IL-2 receptor
IL-6	Interleukin-6
iNOS	Induced nitric oxide
IRF8	Interferon regulatory factor 8
LFA-3	Lymphocyte function-associated antigen–3
MCP-1	Monocyte chemoattractant protein-1
MVI	Microvascular inflammation/injury
NO	Nitric oxide
PD-1	Programmed cell death protein 1
PDGF	Platelet-derived growth factor
PDGFR-β	Platelet-derived growth factor receptor-beta
pSmad	Phosphorylated Smad
PTC	Peritubular capillaries
rAAV1IL-10	Recombinant adeno-associated viral vector (serotype 1) encoding IL-10
ROS	Reactive oxygen species
Stat5	Signal transducer and activator of transcription 5
Tcm T	Central memory

TCMR	T-cell-mediated rejection
Tem	T effector memory
Tfh	T follicular helper cell
TGFβ	Transforming growth factor beta
Th1	T helper 1 cell
t-IFTA	Interstitial fibrosis, tubular atrophy with tubulitis
TLO	Tertiary lymphoid organ
TNF	Tumor necrosis factor
TNFR1	TNFa receptor
Tregs	Regulatory T cells
Trm	Tissue-resident memory T cells
VCAM-1	Vascular cell adhesion molecule
VSMC	Vascular smooth muscle cell
WT	Wild type
γc	Common γ-chain

2.1 Introduction

Despite intensive research for over 50 years, nobody can predict the faith of organ transplants after successful transplantation. Extensive progress in immunosuppressive therapies reduced the number of acute rejection episodes to 6–11% in the 2020s compared to 10–20% in the 2000s and 30–40% in the 1990s among deceased donors of kidney transplant recipients (Tamargo and Kant 2023). The rejection rate is slightly lower for live donors. Unfortunately, despite extensive basic and applied science efforts, up to 49% of kidney transplants are continuously diagnosed with chronic allograft nephropathy and that includes chronic rejection (Gaston et al. 2010). Some transplants experience acute cellular- or/and acute antibody-mediated rejections, often increasing the risk of subsequent chronic rejections (or chronic nephropathy), whereas others experience only chronic rejections (Clayton et al. 2019). Unfortunately, a relatively small fraction of all transplants not only does not experience acute/chronic rejections, but they survive with perfectly functioning transplants even being rewarded by partial or complete withdrawal of immunosuppression (Massart et al. 2016; Brouard et al. 2012; Newell et al. 2010; Becker et al. 2012; Soulillou et al. 2013). A very small number of transplants develop a state of permanent acceptance or even tolerance documented by immunological realignment with potent regulation (Soulillou et al. 2013). However, within a European cohort of 27 kidney transplants with operational tolerance, 30% presented slow deterioration of graft function most likely because of a smoldering immune response as evidenced by donor-specific antibody (DSA) in the absence of C4d marker (Brouard et al. 2012).

Extensive investigations shed light on each of these clinical phenomena finding multiple similarities and differences, but the initiating signals toward certain types of rejections or effortless long-term survivals remain unclear (Becker et al. 2012).

However, there is a mechanistic characteristic that may be involved in explaining why transplant faith tilts toward unique clinical outcomes, and that is the cytokine environment created by recipients inside of allografts and in recipient's lymphoid organs (Walsh et al. 2004; Corris and Kirby 2005; Tay et al. 2009). An extensive body of knowledge already describes multiple T/B cell subsets as well as innate immune cells, endothelial cells, and other cells producing a variety of cytokines regulating almost every step of ongoing immune systems. While T and B cells remain major players because of their superior functions in response to alloantigens, local environments are created and maintained by the activity of many cells. Our goal is to organize these multiple cytokines into functional families leading to clinical outcomes associated with acute cellular rejection or acute antibody rejection predisposing to chronic allograft rejection (Walsh et al. 2004). We also would like to distinguish cytokines leading to permanent acceptance of organ allografts with partial or complete tolerance (Souffrant et al. 2025). It is also important to remember that normal tissues are extensively controlled by physiological cytokine expression. During exercise, muscles increase their production of interleukin-6 (IL-6) with pleiotropic functions in tissues (Pedersen et al. 2001; Steensberg et al. 2000). Indeed, IL-6 regulates hypertrophic muscle growth and myogenesis by regulating the proliferative capacity of muscle stem cells. It also regulates energy metabolism, which is related to the activity of contracting muscle cells. The entire immune system is carefully regulated by physiological levels of multiple cytokines. Consequently, we need to remember that the best protection of organ allografts and allogeneic tissues would be to maintain balance through physiological regulatory cytokines.

2.2 Advanced Classification of Chronic Kidney Allograft Rejection

In kidney transplantation as in other types of organ transplants, chronic rejection may be defined as a continuous persistent harmful alloimmune response. The term chronic reflects an ongoing pathophysiological process with lesions of most likely irreversible tissue injury. It takes a long time (months or years) for the clinical manifestation of chronic rejection. Furthermore, chronic rejection may correlate with the appearance of de novo donor-specific antibodies (dnDSAs) and it is more frequent with previous incidences of acute cellular (TCMR) and/or antibody-mediated rejection (AMR). Independently of the AMR and/or TCMR type(s), kidney transplants undergo progressive nephron loss followed by subsequent decline in renal function (elevated creatinine and decreased glomerular filtration rate, GFR) leading to an inevitable graft failure. Most advanced science and recent classifications abandon the dichotomy of AMR and TCMR terminology as both can coexist with overlapping pathology (Haas et al. 2018; Naesens et al. 2024; Mengel and Adam 2024). These observations explain better simultaneous similar mechanistic events including correlations with overlapping cytokine profiles. Diagnostic measures are still relying on histomorphography of standardized semiquantitative criteria described as interstitial fibrosis, tubular atrophy (IFTA), and arteriolitis (Roufosse et al. 2018; Naesens et al. 2024; Shimizu et al. 2025). Even with the advent of the most sophisticated transcriptomics, proteomics, metabolomics, and other advanced technologies, histometric analyses remained a cornerstone of diagnostic measurements.

The 16th Banff meeting in 2022 modified the definition of AMR into more precise components: (1) microvascular inflammation/injury (MVI); (2) C4d deposition in the peritubular capillaries (PTC); and (3) diagnosis of DSA (Shimizu et al. 2025). Based on these three elements, cAMR may be defined as negative even when MVI is above the threshold but without the presence of C4d and DSA. However, when MVI is below the threshold and C4d is negative while dnDSA is positive, the cAMR diagnosis may be called "probable." Acute tubular injury (ATI) without other coexisting causes, like ischemia, is excluded from the cAMR diagnosis. Again, these precise distinctions further clarify diagnosis but complicate mechanistic interpretation including relevant cytokine profiles.

These constant updates in classification require changes in the nomenclature, such as chronic "active" AMR (caAMR) and chronic "active" TCMR (caTCMR) (Shimizu et al. 2025; Naesens et al. 2024). Chronic injury above specific defined thresholds changes the classification with some categories such as caAMR or just cAMR (introduced in 2022). For example, with significant MVI as caAMR and without significant MVI as cAMR. In other example, the detection of DSA and complement split product C4d along peritubular capillaries (PTC) is essential for diagnosing AMR, but with a new subcategory of "probable" AMR. By analogy, caTCMR (introduced in 2017) is based on finding inflammation in atrophic areas and it is called i-IFTA or in association with tubulitis as t-IFTA (Roufosse et al. 2018). Again, changes in precise TCMR diagnosis should be followed by more adequate mechanistic studies including cytokine profiles responsible for these i-IFTA and t-IFTA differences.

The first mechanistic observation in AMR pathogenesis is the development of dnDSA requiring expansion of B cell clones in germinal centers and that may happen only in the presence of IL-21-producing T follicular helper (Tfh) cells (Heeger et al. 2024; La Muraglia 2nd et al. 2020). Circulating Tfh cells ($CD4^+CXCR5^+$) expanded in blood following transplantation and reflected the equivalent of graft-draining lymph node-derived Tfh cells with the ability to produce IL-21 for B cell expansion (Heeger et al. 2024). Circulating Tfh cells expanded further into inducible T cell costimulator $(ICOS)^+PD\text{-}1^+$ Tfh cells correlated with germinal center alloreactivity proceeding with the production of dnDSA. Costimulation blockade after the detection of $ICOS^+PD\text{-}1^+$ Tfh cells prevented dnDSA. Such experiments connected dnDSA production with the generation of both alloantigen-specific B cell clones and IL-21-producing Tfh clones. Affinity maturation of dnDSA in germinal centers also requires IL-21 as well as B-cell activating factor (BAF) and IL-6 for the development of B/Tfh memory cells and long-living plasma cells (Heeger et al. 2024). In the effector phase of MVI, dnDSA bind to HLA alloantigens on graft's endothelial cells, initiating intracellular signaling for their proliferation and production of cytokines [IL-1, IL-6, IL-8, and tumor necrosis factor (TNF)α] as well as chemokines [monocyte chemoattractant protein-1 (MCP-1), CX3CL1 (Fractalkine) and CCL5] and cell adhesion molecules [E-selectin, P-selectin, intracellular adhesion molecule-1 and vascular cell adhesion molecule (VCAM-1)] (Pober and Sessa 2007). Activated endothelial cells and activated inflammatory macrophages produce several reactive oxygen species (ROS) responsible for tissue injury (observed as interstitial inflammation, tubulitis, IFTA) and arteritis (Laskin et al. 2011). The combination of these processes is regulated by cytokines, which are responsible for different histopathological outcomes reflected in the distinct i-IFTA and t-IFTA pictures (Shimizu et al. 2025; Naesens et al. 2024). The most recent understanding is that these two IFTAs overlap and only dominance in signaling and cytokines produce certain histological pictures.

An underlying MVI process may be reflected by slowly "burning" inflammation without symptoms maintained by locally produced proinflammatory cytokines tilting an immune response to i-IFTA or t-IFTA pathology (Shimizu et al. 2025). Proinflammatory cytokines (IL-1β and TNF-α) contribute to kidney fibrosis in both types of IFTAs as well as chemokines [MCP-1 and RANTES (CCL5)]. Inflammation in areas of fibrosis is part of active TCMR (i/t-IFTA ≥2) (Helgeson et al. 2021). There was no association between either AMR or TCMR and the development of i-IFTA; in fact, i-IFTA was connected to both cellular (inflammation, tubulitis, peritubular capillaritis) and humoral (C4d and DSA) markers (Helgeson et al. 2021). Biopsies with chronic active TCMR scored i-IFTA ≥1 should be a threshold for i-IFTA diagnosis as i-IFTA ≥2 may occur without previous TCMR. The complexity of these details may be further explained by measurements of transcriptomics.

Transcriptomics measure gene expression patterns; thus, they are capable of distinguishing trends of gene expression including cytokines (Fu et al. 2019). Profiling of mRNA of IFTA with subclinical inflammation revealed that CD2, IL-7 receptor (IL-7R), and CCL5 molecules characterized i-IFTA when compared to stable kidney transplants. The high degree of accuracy was correlated with the predictive power of i-IFTA [area under the curve (AUC) of 0.91, $p = 0.0002$]. Furthermore, interferon regulatory factor 8 (IRF8) in the type I IFN signaling pathway could distinguish different classifications of graft biopsies. Kaplan-Meier and long-rank methods confirm that IRF8 could predict renal graft failure in biopsies without rejection (AUC = 0.75) and recipients with high IRF8 expression in biopsies had a high risk of graft loss ($p = 0.0001$) (Fu et al. 2019). These data demonstrate utilities for analyzing patterns of gene expression by transcriptomics for distinguishing i-IFTA and t-IFTA as well as diagnostic and predictive tools, including cytokine profiles.

Kidney allograft transplants in mice developed tertiary lymphoid organs with B cells in their germinal centers (Tse et al. 2015). Intra-graft $B220^{+}$ B cells comprised a combination of $IgM^{high}CD23^{-}$ B cells, $IgM^{low}CD23^{+}$ B cells, and $IgM^{low}CD23^{-}$ B cells with increased expression of CD86 marker. Depletion of all these B cells with anti-CD20 monoclonal antibodies abolished chronic allograft damage, but only when therapy was implemented after kidney transplantation. Intra-allograft B cells synthesized multiple cytokines, such as the most abundant GRO-α (CXCL1), RANTES (CCL5), IL-6, and MCP-1 (CCL2). T-cell accumulation within the allograft with TCMR was associated with tubular loss and development of interstitial fibrosis of the t-IFTA category; type III collagen deposition was observed in areas of F4/80^{+} macrophages and Platelet-Derived Growth Factor Receptor Beta (PDGFR-β)$^{+}$/transgelin^{+} fibroblasts. Following B-cell depletion, both macrophages and fibroblasts were disappearing. These results showed that intra-allograft B/T cells are key mediators of chronic allograft damage as well as that both B and T cells contribute to intra-allograft secretion of cytokines and chemokines regulating intra-graft injury (Tse et al. 2015). Most advanced technologies will allow us to distinguish different types of chronic allograft rejections and correlate them with specific patterns of cytokine profiles. Despite all problems, the predicted average graft survival is 11.7 years and patient survival is 19.2 years, as calculated for the period 2014–2017 (Poggio et al. 2021), with many transplants and patients surviving much longer.

2.3 Advanced Classification of Heart Acute and Chronic Allograft Rejection

Heart allografts are affected by acute rejection and then even more by chronic allograft rejection and/or vasculopathy (Bertini et al. 2022). Most advanced categorization from 2023 by the International Society for Heart and Lung Transplantation (ISHLT) distinguishes acute cellular rejection (TCMR) with a mononuclear inflammatory response infiltrating myocardial tissue with predominant lymphocytic cells. Specific ISHLT grades differentiate the extent and severity of inflammatory cell infiltration and myocyte damage: Grade 0R confirms no evidence of rejection; Grade 1R shows mild rejection characterized by focal or diffuse infiltrates of inflammatory cells (lymphocytes, macrophages, and sometimes eosinophils) with minimal or no myocyte damage; Grade 2R indicates moderate rejection with more significant inflammatory cell infiltrates and evidence of myocyte damage (focal or multifocal); and Grade 3R reveals severe rejection with diffuse and severe inflammatory infiltrates, extensive myocyte damage, and potentially edema, hemorrhage, or vasculitis. Additional immunostaining histology assessment identifies CD4^{+} and CD8^{+} T cells, high-affinity IL-2Rs, ICAM-1, VICAM-1 and HLA class II expression (Velleca et al. 2023).

Antibody-mediated rejection (AMR) is signified by intravascular macrophage accumulation with interstitial edema, hemorrhage, and neutrophilic infiltration in and around capillaries. The predominant feature of cardiac allograft vasculopathy is a diffuse, progressive thickening of the arterial intima that develops in the transplanted heart's epicardial and intramyocardial arteries (Velleca et al. 2023). Immunopathologic findings for acute AMR include positive immunofluorescent staining for C4d, C3d, and anti-HLA-DR or immunoperoxidase staining for C4d and CD68 (or C3d). Acute AMR for heart has the following grading: Grade 0R indicates negative histologic and immunopathologic findings; Grade 1R is characterized by the presence of positive histologic and immunopathologic findings; Grade 2R shows the presence of both histologic and immunopathologic findings; and Grade 3 is marked by the presence of severe histologic plus immunopathologic findings (Velleca et al. 2023). For the safety of patients, testing for rejection is performed as a routine scheduled procedures.

Distinct and very characteristic changes are associated with chronic cardiac vasculopathy (CAV), diagnosed as the narrowing and blockage of coronary arteries, which is progressing over time in heart transplant with incidence increasing from 30% at 5 years to 50% at 10 years post-transplant (Agarwal et al. 2014). CAV remains a leading cause of death after cardiac transplantation. Identifying CAV just by imaging surveillance is difficult because of the diffuse nature of changes on the coronary epicardial arteries, branch vessels, and microvasculature. In 2010, the ISHLT proposed an invasive coronary angiography as the gold standard for CAV diagnosis, and its standardized grading of severity (Mehra et al. 2010): CAV 0 has no detectable angiographic lesion; CAV 1 (Mild) has angiographic left main (LM) <50%, or primary vessel with maximum lesion of <70%, or any branch stenosis <70% (including diffuse narrowing) without allograft dysfunction; CAV 2 (Moderate) has angiographic LM <50%, a single primary vessel ≥70%, or isolated branch stenosis ≥70% in branches of 2 systems but without allograft dysfunction; CAV 3 (Severe) has angiographic LM ≥50%, or two or more primary vessels ≥70% stenosis, or isolated branch stenosis ≥70% in all 3 systems. CAV 1/CAV 2 with allograft dysfunction (defined as left ventricular ejection fraction) LVEF ≤45% usually in the presence of regional wall motion abnormalities or evidence of significant restrictive physiology (which is common but not specific). These abbreviated description of acute TMCR and AMR as well as CAV must be carefully followed in patients with heart allografts to minimize damage by corrective immunosuppressive therapies. As for kidney transplants, cytokines play a fundamental role in both extent and damage to heart allografts.

Incidence of acute heart allograft rejection (TMCR and/or AMR) increases the inflammation burden at the graft site predisposing to CAV (Stoica et al. 2006). Several reports presented strong evidence about an association between acute rejection and CAV (Vassalli et al. 2003; Brunner-La Rocca et al. 1998; Jimenez et al. 2001; Yamani et al. 2004). Most of these studies examined the effect of acute rejection that occurred in the first year after transplantation followed by CAV (Brunner-La Rocca et al. 1998; Dandel et al. 2003). One of the study calculated a relative risk for the onset of CAV to be significantly elevated by donor age (1.26 per 10 years, 95% confidence interval [CI] 1.12 to 1.42), male recipient (1.72, 95% CI 1.01 to 2.94), pretransplant recipient ischemic disease (1.53, 95% CI 1.14 to 2.06), and cumulative numbers of moderate and severe rejection episodes (1.10 per episode, 95% CI 1.03 to 1.18) (Stoica et al. 2006). An estimation for an annual CAV rate was 11.3% for patients with no moderate/severe rejection, 13.6% for those with two moderate/severe rejection episodes, and 18.0% for those with five such events. These results confirm elevated risks of CAV by previous acute rejection episodes.

An evaluating report from 2015 of 5000 heart transplants performed annually worldwide showed an improved survival of patients (Lund et al. 2015). Overall, CAV remained a major impediment for heart transplants, as registry data reported that a high CAV incidence rate evaluated over the past two decades decreased only slightly from 32% to 29% at 5 years and from 46% to 40% at 8 years post-transplant (Lund et al. 2015). Over the same time, 5-year survival for CAV diagnosed patients within 3 years post-transplant improved from 71% to 76% with 82% for those without CAV (Lund et al. 2015). The 2019 ISHLT report described the prevalence of CAV in patients with cardiac transplants as 8% at one year, 29% at 5 years, and 47% at 10 years (Khush et al. 2019). All these data confirm that CAV remained the major problem for heart transplantation. Despite CAV and other challenges, the average survival of a heart transplant is 12–13 years (Wilhelm 2015). One-year heart survival rates are 90% in United States with an annual mortality rate of 4% after the first year. Some patients live for decades after the transplant.

2.4 The Role of Common γ Chain (γc) Cytokines in Both Acute and Chronic Allograft Rejection

Acute allograft rejection is often a T-cell-mediated immune response (TCMR) to allografts usually within the first three months after transplantation (Cooper 2020). IL-2 plays a crucial role in TCMR by promoting the proliferation and activation of T cells. As depicted in Table 2.1, acute allograft rejection is mediated by multiple cytokines, but IL-2 plays a major role in initiating TCMR. Drugs that target the production of IL-2 by T cells, namely, cyclosporine and tacrolimus, are fundamental in immunosuppressive therapy (Ferrini et al. 1986; Herold et al. 1986; Rauch et al. 2009; Thomson et al. 1995). Blockade of the IL-2Rα chain by daclizumab or basiliximab prevents acute rejection by inhibiting this T cell activation (Vincenti et al. 1998; Berard et al. 1999). Acute rejection is a major risk factor for the development of chronic renal allograft failure, which is a major reason for late graft loss (Meier-Kriesche et al. 2000). An acute rejection episode within the first 6 months after transplantation was the most important risk factor for subsequent chronic rejection with a relative risk (RR) probability of 2.4-fold [confidence interval (CI) 2.3–2.5]. Evaluation of the period from 1988 to 1997 showed that the impact of acute rejection on chronic allograft failure increased five-fold ($P < 0.001$). Overall, an episode of acute rejection (TCMR and/or AMR) is a high risk for the development of chronic rejection in kidney transplant recipients (Meier-Kriesche et al. 2000).

Acute immune response after transplantation is regulated not only by IL-2 but also by the cytokines of the IL-2 family (IL-2, IL-4, IL-7, IL-9, IL-15, and IL-21) as each of these cytokines exhibits redundancy but also unique functions (Table 2.1 and Fig. 2.1a). They all utilize a common gamma chain (γc) to regulate the development, proliferation, and survival of T cells and other immune cells (Leonard et al. 2019). While IL-2 itself modulates Th1, Th2, and Th9 differentiation, it inhibits

Table 2.1 Cytokines involved in regulating acute and chronic allograft rejection

Cytokine	Acute rejection	References	Chronic rejection	References
IL-1	IL-1β, produced by macrophages, plays a significant role in acute allograft rejection. IL-1β is a proinflammatory cytokine upregulated during rejection; its levels correlate with the severity of the rejection. IL-1β contributes to inflammation and tissue damage and promotes the expression of adhesion molecules on endothelial cells. IL-1β facilitates immune cell infiltration into the graft.	Hoffmann et al. (1993), Xu et al. (2013)	IL-1β, produced by macrophages, plays a significant role in chronic allograft rejection. IL-1β is a proinflammatory cytokine upregulated during rejection; its levels correlate with the rejection severity. IL-1β contributes to inflammation and tissue damage and promotes the expression of adhesion molecules on endothelial cells. IL-1β facilitates immune cell infiltration into the graft. IL-1β, along TNF-α, is involved in endothelial cell activation.	Jones et al. (2021)
IL-2	IL-2 is produced by Th1 and is acting as the main growth factor for T cells; delivers third signal during T cell activation.	Liao et al., (2013) Fehervari et al. (2006)	IL-2 is produced by Th1 and is acting as the main growth factor for T cells; delivers third signal during T cell activation.	Liao et al., (2013), Fehervari et al. (2006)
IL-4	IL-4 is produced by Th2, and it regulates humoral immunity and allergic responses. It induces CD4+ T cell differentiation into the Th2 subtype. Important signal in dendritic cell (DC) maturation.	Tay et al. (2009)	IL-4 plays a complex role in chronic allograft rejection; Th2-mediated response contributes to chronic rejection by increasing humoral response, and the development of obliterative arteriolopathy.	Le Moine et al. (1999)
IL-5	IL-5 plays a significant role in acute allograft rejection, particularly in the context of eosinophilic infiltration and MHC class II disparity.	Goldman et al. (2001)	IL-5 plays a significant role in chronic allograft rejection, particularly in models involving skin and liver allografts.	Moine et al. (1999), Le Moine et al. (1999)
IL-6	IL-6 is an inflammatory cytokine, it induces differentiation of immune cells, endothelial cells, and stromal cells into inflammatory phenotypes. It suppresses Treg cells.	Jordan et al. (2020)	IL-6 is a proinflammatory cytokine that plays a crucial role in both acute and chronic allograft rejection after organ transplantation.	Jordan et al. (2020) Nickerson (2022) Miller and Madsen (2022)

(continued)

Table 2.1 (continued)

Cytokine	Acute rejection	References	Chronic rejection	References
IL-7	IL-7 provides survival signal for developing thymocytes. Modulates IL-7 signaling to restore tolerance.	Zeng et al. (2025)	IL-7 is influencing the development and survival of memory T cell. It regulates maturation of T and B cells.	Wang et al. (2006)
IL-10	IL-10 is a crucial cytokine in the induction and maintenance of immune tolerance. Secreted by Tregs and other cells to decrease inflammation by preventing the production of proinflammatory cytokines (e.g., IL-1, TNF-α, IL-6).	Akdis and Akdis (2014) Hsueh et al. (2018)	IL-10 plays a complex role in chronic allograft rejection, potentially offering both protective and detrimental effects.	Chen et al. (2005)
IL-17	IL-17A/IL-17F plays a significant role in acute allograft rejection, particularly in the early stages of transplantation.	Haouami et al. (2018)	IL-17A and IL-17F produced by Stat3-dependent Th17 cells play a major role in chronic rejection.	Itoh et al. (2010), Shilling and Wilkes (2011)
IL-22	IL-22 plays a dual role in both promoting and suppressing inflammation and an important role in the maintenance and regeneration of thymic epithelial cells after injury.	Sajiir et al. (2025)	IL-22 plays a dual role in both promoting and suppressing inflammation and an important role in the maintenance and regeneration of thymic epithelial cells after injury.	Sajiir et al. (2025)
IL-23	IL-23 suppresses the differentiation of Treg cells and promotes differentiation, survival, and expansion of pathogenic Th17 cells. Provides proapoptotic signal that promotes negative selection in the thymus.	Krueger et al. (2024)	IL-23 suppresses the differentiation of Treg cells and promotes differentiation, survival, and expansion of pathogenic Th17 cells. Provides proapoptotic signal that promotes negative selection in the thymus	Krueger et al. (2024)
IL-35	Involved in the development of tolerance and the production of regulatory cells that express IL-35.	Sullivan et al. (2020) Sawant et al. (2015)	A proinflammatory cytokine, plays a significant role in both acute and chronic allograft rejection. It is involved in the immune response that leads to the destruction of the transplanted organ. TNF-alpha is produced by various immune cells, including macrophages, and its levels are often elevated during rejection episodes.	Sullivan et al. (2020) Sawant et al. (2015)

(continued)

Table 2.1 (continued)

Cytokine	Acute rejection	References	Chronic rejection	References
TGF-β	A crucial cytokine involved in both promoting and regulating immune tolerance. Regulatory cytokine secreted by Tregs regulates immune cell proliferation, differentiation, survival, and activity.	Loehe and Schauer, (2005) Wang et al. (2023)	A proinflammatory cytokine plays a significant role in the development and progression of chronic allograft nephropathy.	Tyler et al. (2006)
TNF-α	Inflammatory cytokine that stimulates immune cell proliferation, survival, and activity.	Huber et al. (2017) Park et al. (2016)	Proinflammatory, plays a role in autoimmune response and chronic allograft rejection. It is involved in the destruction of the transplanted organ. It is produced by various immune cells, including macrophages, and its levels are elevated during chronic rejection.	Ernandez and Mayadas, (2009) Melk et al. (2003)
IFN-γ	Plays a complex and multifaceted role in acute rejection. It amplifies immune response. It may have protective effects; may prevent microcirculation damage and necrosis; Th1/IFN-γ/ macrophage effector phase.	Krouchev et al. (2018) Halloran et al. (2001)	Plays a role in chronic allograft nephropathy (CAN) and kidney transplant failure; plays a complex and somewhat contradictory role in chronic allograft rejection.	Shiu et al. (2017) Sharma et al. (1996) Koh et al. (2004)
CCL5 (RANTES)	RANTES attracts and activates immune cells, damaging the transplant; it recruits monocytes, T cells, and other immune cells to allograft, leading to inflammation and tissue damage.	Belperio et al. (2000)	RANTES attracts and activates immune cells, damaging the transplant; recruits monocytes, T cells, and other immune cells to allograft, leading to inflammation and tissue damage.	Belperio et al. (2000)

Th17 and Tfh differentiation. The other critical function of IL-2 is its modulation of Tregs by competing with other T cell subsets and thus establishing a higher threshold for their activation (Leonard et al. 2019; De Groot-Kruseman et al. 2002). Since Tregs are not producing IL-2, Th1 cells must produce enough IL-2 to saturate "hungry" IL-2Rs on Tregs and then an IL-2 excess activates other T cell subsets. IL-4 produced by Th2 promotes B cell differentiation and drives immunoglobulin class switching (Howard et al. 1982; Isakson et al. 1982; Vitetta et al. 1985), and augments the production of immunoglobulins, especially IgG1 and IgE subclasses. It

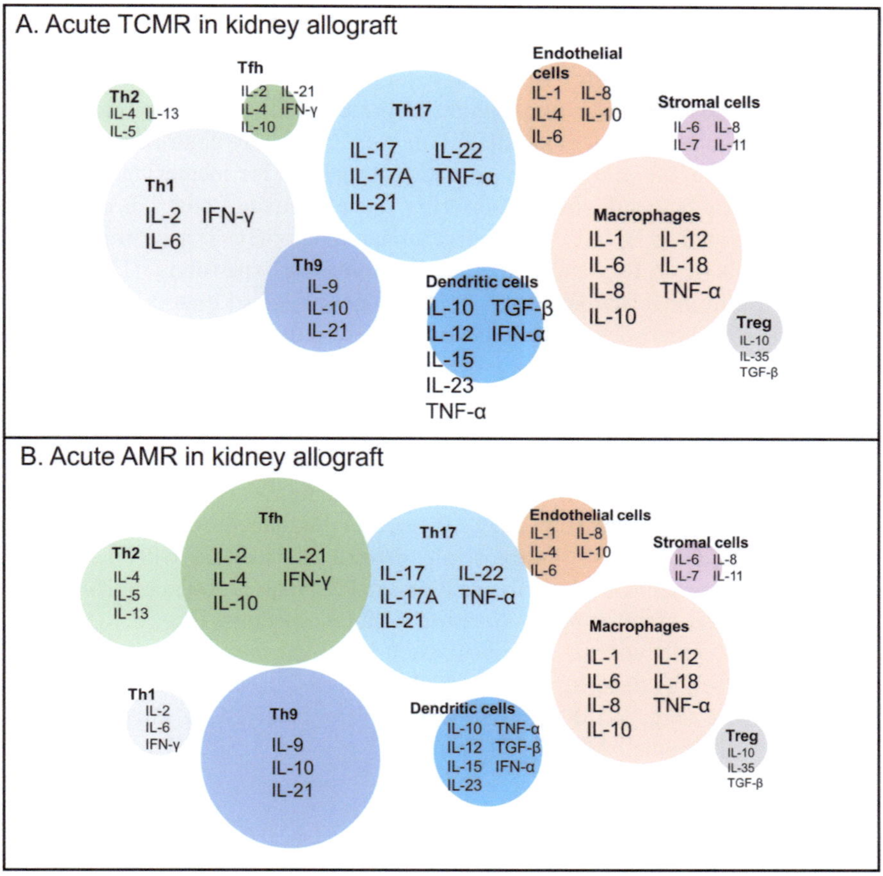

Fig. 2.1 Contribution of T cell subsets (Th1, Th2, Th9, Th17, and Treg) as well as innate immune cells (dendritic cells and macrophages) and other cells (endothelial cells) and their cytokines in acute cellular (T-cell-mediated rejection; TCMR) (Panel **a**) and/or acute antibody-mediated rejection; AMR) (Panel **b**). The size of subset circles reflects the relative number of each cell subset and the size of the circle for each cytokine produced amount

was shown that IL-4 drives Th2 differentiation (Le Gros et al. 1990; Swain et al. 1990) and promotes IL-9-producing Th9 differentiation (Van Dyken and Locksley 2013). IL-7, produced predominantly by stromal cells, creates a growing environment for B cells in the bone marrow and for T cells in the thymus (Goodwin et al. 1989; Namen et al. 1988). IL-7 is a critical factor for T/B cell development while, together with IL-15, mediates memory CD4/CD8 T cell homeostasis. IL-9 is a late T cell growth factor that also promotes memory T cells (Schmitt et al. 1989; Uyttenhove et al. 1988) as well as a growth factor for mast cells (Hultner et al. 1990). The IL-9-induced Th9 cells are further promoted by the environment reach in IL-2, IL-1, IL-4, IL-25, IL-33, and/or type 1 IFNs; in contrast, Th9 cells are inhibited by IL-23 and IL-27, and depending on the context, they are either promoted or inhibited by IL-6, IL-10, and/or IL-21 (Kaplan et al. 2015). IL-15 is a driving cytokine of memory T cells (Kennedy et al. 2000; Lodolce et al. 1998). Finally, IL-21 is produced by Tfh cells to regulate B cells in germinal centers, but it is also produced by Th17 cells, significantly contributing to acute and chronic allograft rejection (Leonard et al. 2019; Wei et al. 2007). By contributing to acute allograft rejection, all γc cytokines are then responsible for increasing the risk of chronic rejection by continuously "nurturing" development of new T/B cell clones (IL-7), by expanding B cell clones in germinal centers (IL-21), by rejuvenating memory T cells (IL-7, IL-9 and IL15), and by promoting expansion of T cell clones during acute/chronic rejections (IL-2). Indeed, Janus tyrosine kinase 3 (Jak3), which binds to the signal transducer and activator of transcription 5 (Stat5), was one of the most elegant targets to block all these cytokines in acute and chronic allograft rejection (Khattar et al. 2014; Guo et al. 2013; Stepkowski et al. 2002, 2005; Borie et al. 2003; Changelian et al. 2003). Overall, the γc-chain cytokines regulate multiple aspects of T-cell- and B-cell-mediated immune responses, contributing to both TCMR and AMR. The composition of the adaptive immune response, dominated by Th1, tilts response to acute TCMR (Fig. 2.1a), and dominated by Tfh tilts response to acute AMR (Fig. 2.1b). The explosive expansion of Th1 cells with IL-2, IFNγ, and IL6 cytokines activates dendritic cells and macrophages, producing IL-1, IL-6, and TNFα in the effector phase of acute cellular rejection (Fig. 2.1a). Strong response by Tfh cells with IL-21 drives potent B cell expansion and production of antibodies as well as dendritic cells and macrophages, producing IL-1, IL-6, and TNFα in AMR (Fig. 2.1b). Similar patterns are involved in chronic TCMR (Fig. 2.2A) and chronic AMR (Fig. 2.2b).

2.5 The Role of IL-17 in Acute and Chronic Allograft Rejection

As shown in Table 2.1, one of the most important cytokines involved in acute and chronic allograft rejection is IL-17 produced by Th17 cells (Haouami et al. 2018). This subset is unique among T cell subsets, because it depends on IL-17-driven Stat3 and that is very different from the IL-2-induced Stat5 in Th1 cells (Qin et al. 2024). However, the activation of Th17 is complex as IL-6/IL-23 cytokines (signal 3) are engaged only in a moderate level of Stat3 phosphorylation. In contrast, strong TCR stimulation (signal 1) induced Stat3 phosphorylation through Lck/Fyn, and then synergistically with IL-6/IL-23 (signal 3) induced robust Stat3 phosphorylation at Y705 amino acid (Qin et al. 2024). Active Th17 cells are involved in autoimmune diseases, tumor immunity as well as acute and chronic transplant rejection (Table 2.1). Intra-graft expression of IL-17 mRNA was evaluated together with a quantification of IL-17A, IL-17F, and IL-23 proteins in plasma (Qin et al. 2024). Recipients with active acute rejection displayed a significant increase in IL-17A mRNA expression levels at the kidney transplant site compared to controls

($P = 0.037$); high IL-17A plasma levels were associated with acute rejection at post-transplant day 1 ($P = 0.0002$) and day 7 ($P < 0.05$); IL-17F and IL-23 plasma levels were significantly higher in acute rejection only at day 7 ($P = 0.015$). Receiver-operating characteristic curves showed that IL-17A and IL-23 plasma levels at day 7 exhibited excellent sensitivities and specificities for predicting acute rejection (Qin et al. 2024). Based on these findings, IL-17 is a major regulatory cytokine of acute renal allograft rejection (Fig. 2.1a).

The role of IL-17 in accelerated rejection was evaluated in a model deficient of Th1 cells intended to evaluate CAV (Yuan et al. 2008). In such conditions, T-bet deficient CD4 Th17 cells were the primary mediators of rejection and neutralization of IL-17 prolonged heart allograft survivals. These data were confirmed in IL-17-deficient recipients with intact Th1/Th2 immunity (Itoh et al. 2010). While transforming growth factor-resident memory T cellsβ (TGFβ)-signaling dependent fibrosis was necessary for chronic rejection, IL-17 was also a required cytokine (Faust et al. 2009). A lung transplant model confirmed a requirement for IL-17A to mediate obliterative bronchiolitis, which was prevented by IL-10 (Fan et al. 2011). Patients with kidney transplant failure had elevated levels of Th17 cells in explanted kidney allografts, which correlated with the level of histologically evaluated chronic nephropathy/rejection. Every time, an elevated infiltration with Th17 cells was associated with increased IL-21 levels and the expression of activation-induced cytidine deaminase (AID) enzyme to regulate both antibody affinity maturation (somatic hypermutation) and immunoglobulin class switching in antigen-specific B cells (Deteix et al. 2010; Durandy 2003). These results suggest the role of IL-21- and IL-17A/IL-17F-producing Th17 cells in chronic kidney allograft rejection. In another investigation, an elevated number of IL-17^{+} total cell count in the renal transplant biopsies was associated with predicting late graft dysfunction, independent of C4d positivity. Some advocated for IL-17 immunostaining to be used as a predictive marker for chronic rejection (Yapici et al. 2011). Both Tfh cells with IL-21 and Th17 cells producing IL-17A/IL-17F and IL-21 are involved in chronic kidney allograft rejection (Fig. 2.2a).

2.6 The Role of Other Cytokines in Chronic Allograft Rejection

The role of TGF-β is well-established in chronic allograft nephropathy (Tyler et al. 2006). In fact, it has been argued that TGF-β can cause chronic allograft nephropathy by inducing the transition of epithelial cells to mesenchymal cells. Some studies suggested a reverse transition that was produced by bone morphogenetic protein-7 (BMP7). Normal tubular epithelial cells exposed to TGFβ and BMP7 expressed phosphorylated forms of both Smad2/3 and Smad1/5/8 within their nuclei. Phosphorylated (p)Smad2 and pSmad3 were expressed at increased levels by tubular epithelial cells during acute renal allograft rejection and chronic allograft

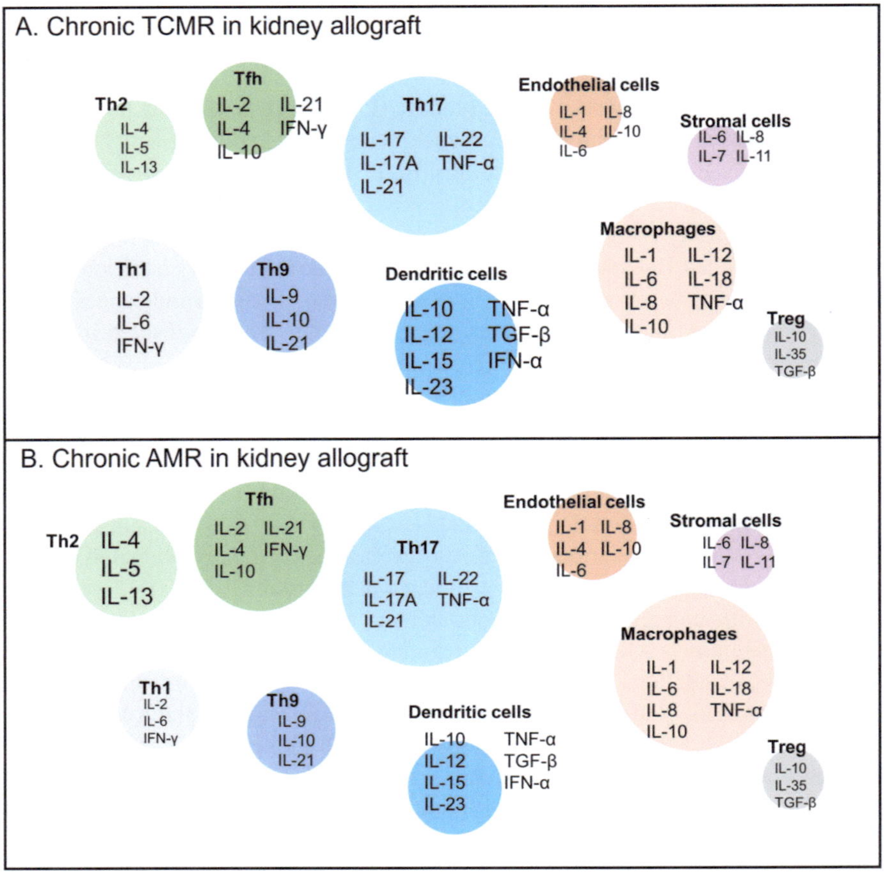

Fig. 2.2 Contribution of T cell subsets (Th1, Th2, Th9, Th17, and Treg) as well as innate immune cells (dendritic cells and macrophages) and other cells (endothelial cells) and their cytokines in chronic cellular (chronic T cell-mediated rejection; cTCMR) (Panel **a**) and/or chronic antibody-mediated rejection; cAMR) (Panel **b**). The size of subset circles reflects the relative number of each cell subset and the size of the circle for each cytokine produced amount

nephropathy, whereas pSmad1, pSmad5, and pSmad8 were expressed at very low levels. In vitro experiments documented the same pattern of Smad signaling was reproduced after TGFβ stimulation of tubular epithelial cells in the absence of BMP7. Addition of BMP7 to TGFβ-stimulated cells enhanced the expression of pSmad1, pSmad5, and pSmad8 (Tyler et al. 2006). These results explained the complexity of exogenous BMP7 regulation of the homeostatic balance of pSmad signaling and possibly preventing or reversing the development of chronic allograft nephropathy.

TNFα/TNFβ play a significant role in both acute and chronic allograft rejection (Wollin et al. 2009). Clinical observations and experimental data provide evidence that TNFα contributes to acute and chronic allograft rejection. Blockade of TNFα by infliximab, the chimeric anti-TNFα monoclonal antibody, significantly reduced luminal occlusion (arteriosclerosis) of C57BL/6 ($H2^b$) aortic grafts in a fully mismatched aortic allograft model. Under TNFα blockade, intercellular adhesion molecule-1 (ICAM-1) and platelet-derived growth factor (PDGF) mRNA expressions were strongly reduced within the graft, whereas induced nitric oxide (iNOS) mRNA expression was enhanced. These data showed that TNFα is involved in the development of transplant arteriosclerosis in allogeneic aortic grafts (Wollin et al. 2009). Destruction of transplanted kidneys through chronic allograft nephropathy (CAN) may be caused by viral infection with TNFα contribution (Krogerus et al. 2008). Among other viruses, CMV is a high-risk factor for chronic nephropathy. Rat kidney allogeneic transplants exposed to rat CMV showed a more intense TUNEL reaction in tubular epithelium as well as an earlier increase in the expression of TNFα in the vascular endothelium compared to noninfected transplant groups. This correlated with an increase in the tubular TNFα receptor (TNFR1) expression. These observations demonstrate that CMV-enhanced development of chronic nephropathy was associated with tubular apoptosis and concomitant increase of TNFα/TNFR1 signaling (Krogerus et al. 2008).

Th1-produced IFNγ is clearly involved in the development of allograft vascular dysfunction predisposing to arteriosclerosis, which eventually leads to graft loss. Transplanted human arteries were exposed to circulating allogeneic T cells in vivo using immunodeficient murine hosts (Koh et al. 2004). Within only 7 to 9 days, human arteries developed endothelial cell (EC) dysfunction, while in 14 days, grafts displayed impaired contractility and desensitization to nitric oxide (NO), both these symptoms confirmed vascular smooth muscle cell (VSMC) dysfunction. These responses of allogeneic human T cells correlated with loss of endothelial NO (eNOS) and expression of inducible NOS (iNOS). In reversing the effect experiments, neutralizing IFNγ function completely prevented both VSMC dysfunction and changes in iNOS/eNOS expression. Neutralization of TNFα reduced IFNγ production and inhibited VSMC dysfunction. In in vitro experiments, these effects required T cell activation by class II MHC alloantigens and costimulatory signal from lymphocyte function-associated antigen–3 (LFA-3) expressed by ECs, which then produced IFNγ and TNFα soluble mediators (Koh et al. 2004). These complex experiments demonstrated that IFNγ as well as TNFα play a central role in vascular dysfunction through dysregulation of NOS expression, thus linking dysfunction with arteriosclerosis.

Other complex experiments tried to study the link between cytokines and graft damage. It is a well-known fact that ischemia/reperfusion injury (IRI) is an important risk factor for accelerated cardiac allograft rejection and chronic graft dysfunction (Lutz et al. 2010). Syngeneic heart transplant model with IRI where IL-1 receptor (IL1R) antagonist (Anakinra) blocked caspase-1 and NF-kB signaling showed that an IL-1 blockade prevented IRI and mitigated CMV-accelerated chronic rejection(Jones et al. 2021). As exampled for IL-1, chronic rejection may involve IL-1 induced by alloantigen-dependent and alloantigenindependent mechanisms (Nadeau et al. 1995). Although isograft controls developed similar morphologic and histopathologic changes as allografts, they were quantitatively less intense and appeared later. During an initial T-cell-mediated allogeneic response, IL-2/IL-2 receptor (IL-2R), IL-4/IL4R, and IFN-γ/IFN-γR1/2 were upregulated, inducing the activation of macrophages, which produced IL-1α/IL-1β, TNFα, and IL-6 cytokines. To attract many other cells, chemokine CCL5 (RANTES; regulated upon activation, normal T-cell expressed and secreted) was found to peak at two weeks, while ICAM-1 was maximally expressed at 6 weeks. Multiple macrophage products such as monocyte chemoattractant protein (MCP-1) increased ten-fold at 16 weeks. In contrast, in isografts, ICAM-1 peaked at 24 weeks; MCP-1 was highly expressed at 52 weeks; and these changes coincided with a progressive increase in infiltrating macrophages. Since cytokine expression in the spleen of allograft and isograft recipients was insignificant, chronic rejection events happened predominantly locally with macrophage-driven immune response with up-regulation of Il-1, IL-6, TNFα, and MCP-1 as well as inducible nitric oxide synthase (Nadeau et al. 1995). These observations in syngeneic vs. allogeneic models stress the contribution of cytokines regulating pathophysiological changes in an alloantigen-independent fashion. Thus, similar changes may be caused by allogeneic chronic rejection but also independent of alloantigen stimulation causing dysfunction of cells and then organs. All cytokines presented in this section produced by T/B cells, macrophages, dendritic cells, endothelial cells, and stromal cells, contribute to chronic rejection through either TCMR (Fig. 2.2a) and/or AMR (Fig. 2.2b) dominated mechanisms.

2.7 The Role of IL-21 and Other Cytokines in Chronic Cardiac Vasculopathy (CAV)

Half of all heart transplants develop CAV within 10 years after transplantation (Mynbaev et al. 2023). The CAV process increases from 8% in the first year to 30% in 5 years, and 50% in 10 years, according to the ISHLT registry database. The CAV pathogenesis involves extensive regulation by cytokines with coronary remodeling, which proceeds through diffuse concentric hyperplasia to an ischemic graft failure (Salomon et al. 1991; Mitchell and Libby 2007; Hernandez et al. 2009). Several cytokines are involved in CAV development, including IL-1, IL-6, and TNFα, which induce smooth muscle cell proliferation that slowly leads to narrowing of the coronary arteries. Additionally, cytokines, like IL-2/IFNγ (Th1) inducing TCMR and IL-4/IL-5 (Th2) inducing AMR, are involved in the immune response that can lead to CAV (Kalil and Florescu 2022). Among cytokines listed as causing CAV, IL-6 is increasingly blamed for a critical proinflammatory involvement in allograft injury. Therapies that inhibit IL-6 signaling (anti-IL-6R monoclonal antibody, tocilizumab and anti-IL6 monoclonal antibody, clazakizumab) are undergoing clinical trials to see how far they are effective in preventing CAV (Miller and Madsen 2022) in heart transplant recipients (Kalil and Florescu 2022). Another major player in CAV is IL-17 produced by Th17 cells, which promotes neutrophil proliferation and migration, endothelial cell activation, and fibroblast activation/proliferation, all of which contribute to acute and chronic allograft rejection (Dong 2006). The interaction of IL-17 with IL-6 creates a positive feedback loop to stimulate monocytes/macrophages, endothelial cells, and other cells to produce proinflammatory cytokines (IL-1, IL6, and TNFα) (Ye et al. 2001). Graft-derived IL-6 enhanced T-cell responses contributing to CAV and allograft arteriosclerosis; IL-17 promoted the secretion of inflammatory cytokines by stromal cells and the maturation of macrophages and dendritic cells (Antonysamy et al. 1999). A reverse experiment showed that local expression of soluble IL-17R by gene therapy extended cardiac allograft survival preventing CAV lesions (Li et al. 2006).

While advances in immunosuppressive therapies greatly improved the survival of heart transplants, their long-term survival remains dismal and hindered by CAV (Mehra et al. 2010). Pathology of CAV with concentric intimal smooth muscle hyperplasia is worsened by local inflammatory infiltrates with macrophages and T/B cells (Mehra 2006). CAV is linked to anti-endothelial cell antibodies in AMR and combined AMR/TCMR as well as nonimmunologic risk factors such as hypertension (Jane-Wit et al. 2013; Vassalli et al. 2003; Pethig et al. 2000). Development of CAV is responsible for up to 32% mortality at 5–10 years post-transplant (Pober et al. 2021).

IL-21, produced by Tfh, regulates B cell proliferation and differentiation at germinal centers (Tian and Zajac 2016; Shi et al. 2016). Both Tfh and Th17 cells rely on Stat3 for their survival and that is different from Th1 and Th2 regulated by Stat5 (Ma et al. 2012; Yang et al. 2007). An IL-21-dependent chronic defense against viral infection was described in mice (Elsaesser et al. 2009) and in chronic autoimmune type-1 diabetes (Mcguire et al. 2011). In transplant recipients, IL-21 was associated with reduced allograft function and chronic rejection (Lindemann et al. 2018; Shi et al. 2016). Considering results indicating that IL-21 is a chronic phase cytokine (Elsaesser et al. 2009), our group examined the role of IL-21 in CAV using IL-21-deficient mice (IL-21−/−) (Nurieva et al. 2007). To examine whether IL-21 is needed for CAV development, we used a transplant model of chronic allograft rejection (Mckenzie et al. 1979; Fischbein et al. 2003). Heart allograft survived more than 100 days in IL-21-deficient mice in comparison to wild-type (WT) controls (Fig. 2.4a) and the difference was confirmed by heartbeat rates (Fig. 2.4b). While WT control mice showed characteristic changes of progressive CAV (Fig. 2.4c; left panel), IL-21-deficient (IL-21−/−) mice had heart transplants without CAV (Fig. 2.4c; right panel). The results of good heart graft function were confirmed by histological findings indicating that IL-21 regulates the development of CAV (Fig. 2.4d). At day 100 post-grafting, occlusion correlated with intensive infiltration by relatively organized clusters of macrophages, T-cells, and B-cells; both T and B-cells were abundantly present in areas around the occluded arteries in controls (Fig. 2.5a; left panel) but not in IL-21-deficient recipients (Fig. 2.5a; right panel). While T cells were proximal to the arteries, some T cells colocalized with nearby B-cells in controls (Fig. 2.5b), which also had more CD4, CD8, and B cells (Fig. 2.5c). We also examined the generation of IL-2- and IFNγ-producing Th1 cells during CAV; while both normal mice and IL-21-deficient mice had similar numbers of IL-2-producing Th1 cells, only normal mice showed higher numbers of IFNγ-producing Th1 cells (not shown). This suggests that IL-21 regulates the generation of IFNγ-producing Th1 cells, which enhances IFN-γ production contributing to CAV development in an IL-21-dependent fashion (Khattar et al. 2019).

The basic leucine zipper ATF-like transcription factor (BATF) regulates IL-21-dependent signaling (Xin et al. 2015). Since heart allografts survived more than 100 days in BATF-deficient recipients and maintained perfect heartbeat rates without CAV compared to allografts in normal recipients, IL-21 signaling through BATF participates in chronic rejection. To show therapeutic potential, an IL-21 receptor fusion protein (IL-21R.Fc) was used to prevent CAV (Young et al. 2007). Although an isotype control antibody-treated mice rejected heart allografts in 30–100 days, IL-21R.Fc therapy promoted survival for more than 100 days (Khattar et al. 2019). Treated mice with IL-21R.Fc maintained better heart functions than control mice. Functional changes in grafts of the control group correlated with severe CAV damage at days 35–45 and days 90–100; in contradistinction, allografts from the IL-21 R.Fc-treated recipients had reduced CAV. Our results documented the therapeutic potential of targeting IL-21 to prevent CAV.

Conceptually, our results about IL-21 in CAV (Khattar et al. 2019) are like those described in a chronic viral infection observed in mice (Elsaesser et al. 2009). Blocking IL-21 signaling using IL-21- or BATF-deficient mice or therapy with an IL-21R-Fc fusion protein prevented CAV (Khattar et al. 2019). Heart allografts that developed CAV lesions had tertiary lymphoid organ (TLO)-like structures mimicking ectopic germinal centers surrounded in small arteries with $CD23^+$ follicular dendritic cells, B/T cells, located near an occluded artery (Thaunat et al. 2005). It was previously described that IL-21-producing Tfh cells were associated with the formation of TLOs in human breast cancer patients in a CXCL13-dependent fashion (Gu-Trantien et al. 2017). Since TLO-like structures were absent in heart allografts of IL-21-, BATF-deficient mice as well as recipients treated with IL-21R.Fc, we proposed that CAV pathology develop TLOs in the presence of IL-21. A therapy with IL-21R.Fc fusion protein, preventing CAV and TLOs, may be promising for its clinical application. One study has shown that both IL-2 and IL-21 are extensively produced during acute allograft rejection (Baan et al. 2007), and that may influence the direction of TCMR and/or AMR for acute, accelerated, and/or chronic rejection. Our previous study showed that both CD4 and CD8 T-cell survival in the absence of IL-2 signaling was dependent on IL-21 production by CD4 T cells (Khattar et al. 2014). Our chronic allograft rejection model showed also IL-21-dependent T-cells producing IFNγ, which seems consistent with the finding that IL-21 sustains T-bet transcription factor in Th1 cells and cytotoxic T cells promoting IFNγ production (Fina et al. 2008; Sutherland et al. 2013). These results underline the main role of IL-21 signaling in mediating CAV.

As mentioned above, BATF plays a critical role in immune cell differentiation and function (Wang et al. 2025) and it is necessary for IL-21 production by Tfh and Th17 cells. In BATF-deficient mice, Tfh and Th17 cells did not mature, because BATF also controls the expression of Bcl-6 and c-Maf (Schraml et al. 2009; Ise et al. 2011). In our work, BATF- or IL-21-deficient mice had better heart graft survival without developing CAV (Khattar et al. 2019). These results suggest that CAV is mediated by IL-21-dependent Tfh and Th17 cells as BATF is necessary for the development of both subsets. In support of this conclusion, Th17 cells were found in the pathological remodeling of CAV in IL-17-deficient recipients, which had better heartbeat scores and reduced occlusion of vessels, similar to our findings (Itoh et al. 2010). We observed a striking decrease in B cell infiltration in allografts of IL-21-deficient mice, which correlated with the lack of TLOs. These observations showed that IL-21 is a fundamental player in the development of CAV with therapeutic potential for IL-21RFc fusion protein.

2.8 How Regulation of Acute Rejection by Cytokines Leads to Chronic Rejection

Declining rates of acute kidney allograft rejection and excellent 1-year graft survival in patients who had acute rejection episodes is misleading as these patients have an increased risk of chronic rejection (Jarrick et al. 2019). Patients who had acute rejection within the first 6 months after transplantation were analyzed for long-term outcomes between 1997 and 2017 using the database from the Australia and New Zealand Dialysis and Transplant (ANZDATA) Registry. Overall, recipients with early acute rejection were more likely to experience graft loss attributed to chronic allograft nephropathy (hazard ratio [HR], 1.39; 95% confidence interval [95% CI], 1.23 to 1.56) and recurrent acute rejection incidents (HR, 1.85; 95% CI, 1.39 to 2.46).

In a large study, 1699 patients with living donor transplants and 1795 patients with cadaver donor transplants were examined between January 1987 and November 1994 in the North American Pediatric Renal Transplant Cooperative Study Registry (Tejani et al. 1996). Graft failure from chronic rejection was observed in 4.5% (76) of pediatric recipients with living donor transplants and 8.3% (149) of pediatric recipients with cadaver transplants. Chronic rejection was the main cause of graft failure, namely 27% of recipients of living donors and 26.7% of recipients of cadaver donors. Univariate and multivariate analyses of risk factors revealed that the acute rejection episode (relative risk [RR] = 3.1, $P < 0.001$) and greater than two acute rejections (RR = 4.3, $P < 0.001$) were the most common correlates of chronic rejection. Additionally, late acute rejection, occurring at more than 365 days (RR = 2.3, $P < 0.001$), was a high-risk factor for cadaver donor source (RR = 1.6, $P = 0.001$) and African American race (RR = 1.6, $P < 0.003$). An analysis of cyclosporine dosing revealed that a dose of <5 mg/kg/day at day 30 was also a risk factor for chronic rejection (RR = 1.5, $P = 0.027$). The conclusion is that previous acute rejection is the single most critical element in the genesis of chronic rejection.

There is general belief that acute rejection episodes are associated with an increased risk for developing chronic allograft nephropathy, which is the major risk of long-term graft failure (Tejani et al. 1996; Jarrick et al. 2019). The most logical explanation for such increased risk is that after acute rejection patients have memory T and B cells (Nicosia et al. 2020), which are maintained by IL-7 and IL-15 cytokines (Tan et al. 2002; Kim et al. 2008; Chen et al. 2017). These cytokines produce a subclinical immune response directed against alloantigens by forging through macrophages the production of IL-1, IL-6, and TNFα cytokines activating endothelial cells and inflicting tissue injury by ROS activity. One group demonstrated that 44.4% of patients with initial biopsies showing acute subclinical rejection developed increased grades of chronic allograft nephropathy diagnosed in subsequent biopsies, compared to 29.6% in those patients with no subclinical rejection (Hart et al. 2017; Mclaren et al. 2000; Behrendt et al. 2000; Nankivell et al. 2003). Overall, the same active cytokines are involved in acute and chronic rejection but patients who had acute rejection episodes are equipped with alloantigen-specific T/B memory cells capable of readily producing effector cytokines (Figs. 2.1 and 2.2).

Cytokines IL-7, IL-9, and IL-15 play a crucial role in the development, maintenance, and function of memory T cell subsets. They influence their differentiation, survival, and effector functions, shaping the immune response towards Th1/Th17-type (TCMR) or Tfh/Th17-type (AMR) pathology in kidney allografts. Naïve and memory T cells have distinct surface markers and exhibit distinct kinetic behavior in vivo (Tough et al. 2000). Naïve T cells persist in a nondividing state, while memory T cells undergo cycling with a rapid rate of turnover. This different kinetic behavior of naïve and memory T cells reflects differential responsiveness to cytokines. It was described that memory but not naïve T cells were stimulated to proliferate by a variety of infection-induced cytokines (Tough et al. 2000). These results explain why cytokines contribute to the high background rate of turnover exhibited by memory T cells, which then produce subclinical chronic rejection and nephropathy.

Upon alloantigen stimulation (or any other cognate antigen), naive T cells undergo proliferation and differentiation into T effector cells (Th1, Th2, Th9, Th17, and Treg), followed by either their apoptosis or survival as precursors of long-lived memory (Tm) cells (Raeber et al. 2018). These early phases of a T-cell response and the ensuing maintenance of Tm cells are shaped by cytokines, such as IL-2, IL-7, and IL-15 that share the γc with specific α-chain receptors. Constant steady-state synthesis of IL-7 and IL-15 is required for repeated divisions in the homeostatic survival of $CD4^+$ and $CD8^+$ Tm cells. During a recall response, augmented levels of multiple cytokines (IL-2, IL-15, IL-21, IL-12, IL-18, and INFα/β) define the potential of effector $CD8^+$ T cells to become CD8+ Tm cells, whereas the elevated level of only two cytokines (IL-2 and IL-15) define the potential for aggressive response by $CD4^+$ and $CD8^+$ Tm cells. In environmental conditions with low levels of γc cytokines but high levels of IL-10 and IL-35 (produced by Treg cells), the generation of Tm cells may fail to become IL-7-/IL-15-dependent long-lived memory T cells (Raeber et al. 2018). In contrast, high levels of IL-7/IL-15 without Tregs may expand memory T cells, increasing chances for repeated acute rejection and persistent chronic rejection.

Humans and mice have two main memory T cell subsets (T central memory, Tcm; and T effector memory, Tem) with distinct capacities for their homing and effector functions (Lanzavecchia and Sallusto 2005; Sallusto et al. 1999; Masopust et al. 2001; Reinhardt et al. 2001). Human Tcm cells ($CD127^{+}CD62L^{+}CCR7^{+}CD45RA^{neg}CD45RO^{+}$) have homing receptors to migrate to lymph nodes and have limited effector function, but they can proliferate and then become T effector cells upon secondary stimulation. In contrast, Tem cells ($CD127^{neg}CD62L^{neg}CCR7^{neg}CD45RA^{neg}CD45RO^{+}$) migrate to peripheral tissues and upon contact with specific alloantigen (specific antigen) rapidly produce cytokines such as IFNγ mounting an aggressive immune response but have limited proliferative capacity (Klonowski et al. 2004). Human Tcm cells may differentiate spontaneously into Th1 or Th2 cells even in the absence of polarizing cytokines (defined as pre-Th1 and pre-Th2), which were identified according to the expression of CXCR3 and CCR4 chemokines (Rivino et al. 2004). In response to alloantigen (or other antigen) naïve $CD8^{+}$ T cells have very strong potential for proliferation to become $CD8^{+}$ T effector cells (days) which produce IL-2 and large amounts of IFNγ, IL-12, and IL-18; some of them become early (weeks/months) and then late (months/years) CD8 T memory cells which produce large amounts of IL-2/IFNγ/TNFα, IL-12, and IL-18 (Martin and Badovinac 2018). Additionally, specialized tissue-resident memory T cells (Trm; $CD69^{+}CD103^{+}$) were identified in humans and mice that undergo little or no circulation (Xu et al. 2023). Residing in the graft Trm produce IFNγ, TNFα, and IL17 as well as heightened levels of granzymes and perforin with potent cytolytic activities (Schenkel and Masopust 2014; Asada et al. 2022). In summary, through the generation of T memory cells, acute rejection incidence predisposes recipients to have both increased risk of repeated acute rejection, persistent subclinical rejection, and then chronic rejection. Following acute rejection, patients generate Tcm/Tem cells with unique abilities to respond rapidly and aggressively to alloantigens. While Tcm cells quickly proliferate and generate effector T cells, Tem do not need even to divide to produce IFNγ effector cytokine to mount an effector response. Cytokines IL-7 and IL-15 regulate periodic renewal of Tcm, Tem, and Trm cells, creating a constantly dangerous environment that dramatically increases the risk for chronic rejection.

2.9 Regulatory Cytokines Are Involved in Perfectly Functioning Allografts

In the last three decades, since the discovery of Treg cells (Stepkowski et al. 1987) and then $CD4^{+}Foxp3^{+}$ Treg cells (Issa and Wood 2010) in transplant recipients, immunologists are working hard to decode the nuances of the Treg-induced tolerance to allografts (Josefowicz et al. 2012). Harnessing their suppressive potential therapeutically to promote long-term allograft survival in transplant patients would solve many clinical problems. Approaches with different Treg therapies of ex vivo expanded Tregs produced many promising results but so far escaped practical applications (Ferreira et al. 2019). Paradoxically, IL-2 not only promotes an acute/chronic immune response but also is crucial in expanding Tregs, which continuously express large numbers of IL-2Rα-chains (IL-2Rα; CD25) out of the three combined IL-2Rα/β/γ chains (Chinen et al. 2016). While Treg cells abundantly express low-affinity IL-2Rα-chains, they require joining the common γ-chain (γc; CD132) and IL-2Rβ chain (CD122) for boosting of IL-2 affinity by ~1000 folds (Waldmann 1989). There is also a huge existential problem for Tregs, as they do not produce IL-2 and must rely on their survival by utilizing IL-2 produced by Th1 and other cells. Indeed, IL-2-induces the activation of a critical γc/Jak3/Stat5 signaling pathway (Rodriguez et al. 2023; Leonard et al. 2019), which is an indispensable step for inducing the expression of Forkhead box P3 (Foxp3) transcription factor for the differentiation and stability of Tregs (Furtado et al. 2002). IL-2Rβ and γc, shared between IL-2 and IL-15 receptors, can also contribute to the induction of Foxp3 expression (Guo et al. 2013). IL-2, in cooperation with TGFβ, is required for extrathymic Treg cell differentiation (Chinen et al. 2016; Chen et al. 2003). While it is possible to say that IL-2 controls the survival of $Foxp3^+$ Tregs, its exact mechanism is complex and requires more studies (Barron et al. 2010). Elimination of Bcl2 family member proapoptotic Bim protein in the IL-2/IL-2Rα-deficient mice restored numbers of Tregs but did not prevent their lethal autoimmune disease, thus showing IL-2-dependent survival of Tregs. Treatment with IL-2/anti-IL-2 antibody complexes increased the numbers and suppressor activity of IL-2-deprived Tregs. This observation confirmed that IL-2 therapy may be required at a threshold quantity to maintain Treg therapy.

One Study consortium found that Treg-based therapy was safe in living-donor kidney transplant recipients with fewer infection complications but similar rejection rates in the first year (Sawitzki et al. 2020). In a more advanced clinical study, donor antigen-reactive/specific (dar)Tregs (CD4 + CD25 + $CD127^{low}$) were administered to three recipients at 7 to 12 days after live donor renal transplants (Guinan et al. 2023). Recipients were kept on immunosuppression without induction therapy with tacrolimus, mycophenolate mofetil, and steroids. Steroids were withdrawn within 14 weeks. Patients had no rejection and all discontinued mycophenolate mofetil at 11 to 13 months post-transplant. One for-cause biopsy, 5 days after darTreg infusion, showed the absence of rejection but also the accumulation of Tregs in the kidney allograft. Furthermore, all three patients had Treg aggregates in biopsies scheduled by protocol at 8 months post-transplant. This successful therapy evidenced the potential in darTreg-based immunotherapy to prevent acute and chronic allograft rejection. These patients convincingly demonstrate that although it is very difficult, it is possible to utilize darTeg cells for clinical tolerance induction. We suggest that tolerant kidney allografts need to have darTreg/Treg present inside the graft to produce clinically sufficient cytokines such as TGFβ, IL10, and IL-35 to maintain tolerance.

IL-10 is a crucial anti-inflammatory cytokine important in regulating and terminating inflammatory responses (Iyer and Cheng 2012). It plays a central role in limiting immune response to pathogens to prevent damage inflicted by overzealous immune effects. Dysregulation of IL-10 is associated with autoimmune diseases (Iyer and Cheng 2012). A similarly important role was shown in allotransplantation (Hara et al. 2001). $CD4^+$ T cells contained within the $CD45RB^{high}$ fraction mediated skin allograft rejection when transferred to T cell-depleted mice. In contrast, the $CD4^+CD45RB^{low}$ and $CD4^+CD25^+$ Treg cells did not induce rejection, but rather inhibited rejection initiated by naive $CD4^+CD45RB^{high}$ T cells. This regulatory effect by $CD4^+CD45RB^{low}$ T cells in vivo was inhibited by neutralizing antibodies against IL-10, but not IL-4. $CD4^+CD45RB^{low}$ T cells from tolerant mice were immune suppressive in vitro, as they inhibited proliferation and Th1-type cytokine production. Thus, tolerance relied on IL-10 for functional activity (Hara et al. 2001). A single intramuscular injection of a recombinant adeno-associated viral vector (serotype 1) encoding IL-10 (rAAV1IL-10) reduced neointimal proliferation and graft infiltration with macrophages and T and B lymphocytes (Chen et al. 2005). This intervention underlined the protective effects of IL-10 in aortic allografts involved the heme oxygenase 1 (HO-1) because inhibition of HO activity reversed not only neointimal proliferation but also inflammatory cell infiltration. In conclusion, IL-10 attenuated neointimal proliferation and inflammatory infiltration by regulating HO-1 in chronic vascular rejection. Thus, IL-10 inhibits chronic allograft rejection.

Few cytokines are specializing in immunosuppression. The balance of inflammatory vs. anti-inflammatory events needs to be maintained through immunoregulatory cytokines (Ye et al. 2021). One of those is IL-35, an exclusively inhibitory cytokine that belongs to the family of IL-12 cytokines (IL-12, IL-23, IL-27, and IL-35); IL-35 suppresses T cell proliferation and induces IL-35-producing inducible Treg cells T cells (iTr35) (Collison et al. 2010). It is also secreted by the thymic and peripheral $CD4^+Foxp3^+$ Tregs (Pflanz et al. 2002). In fact, other cells also produce IL-35, such as Bregs, immature dendritic cells, vascular endothelial cells, epithelial cells, and smooth muscle cells in the human body (Shen et al. 2014; Dambuza et al. 2017; Haller et al. 2017; Mohd Jaya et al. 2019). Since its discovery in 2007 (Collison et al. 2007; Niedbala et al. 2007), IL-35 has been shown to regulate autoimmune diseases, immune-related disorders, infectious diseases, cancer response, and transplant tolerance (Ye et al. 2021; Guo et al. 2018). As a therapeutic cytokine, IL-35 showed promising results in preventing rheumatoid arthritis (Xin et al. 2021) and stabilized Tregs to protect heart allografts in mice (Huang et al. 2024). Because IL-35 has inhibitory activity it may be used for clinical immunotherapy (Ye et al. 2021) and potentially block chronic allograft rejection.

The IL-35 protein is composed of Epstein-Barr-virus-induced protein 3 (Ebi3) and IL-12α chain (p35) subunits (Sullivan et al. 2020). To test the function of IL-35-producing Treg cells, the TregEbi3 reporter mice were used to identify IL-35 producers expressing Ebi3TdTom gene reporter plus Ebi3 and p35 proteins. Both subunits of IL-35 were displayed on the surface of tolerogen-specific $Foxp3^+$ iTr35 cells. These IL-35 producers secreted Ebi3 and p35 on extracellular vehicles targeting a 25- to 100-fold higher number of T and B lymphocytes, causing them to acquire surface IL-35. This effect promoted exhaustion in T and B cells as a novel mechanism of infectious tolerance.

Mass cytometry analysis of intra-graft immune cells showed that IL-35 decreased $CD8^+$ T-cell infiltration and increased Foxp3 and IL-35 expression in Treg cells (Huang et al. 2024). In in vitro experiments, IL-35 promoted Treg phenotypic and functional stability with increased self-production of IL-35, thus generating a positive feedback loop. Tregs only with IL-35 exerted their suppressive effect on $CD8^+$ T cells. While IL-35 did not produce inhibitory effects in recipients depleted of Tregs, IL-35 maintained Treg stability via the gp130/signal transducer and activator of the transcription 1 signaling pathway. These findings shed light on the therapeutic potential of IL-35 treatment in delaying acute rejection and prevention of chronic rejection. Both Treg and iTr35 cells produce IL-35; thus, they may contribute to tolerance.

TGFβ plays an important role in promoting transplant tolerance by influencing Treg development and function. Specifically, TGF-β signaling is necessary to induce Foxp3 expression in Tregs, and stable Foxp3-expressing Tregs are critical for suppressing alloimmune responses against rejection in transplanted organs and mandatory for achieving tolerance (Waldmann et al. 2014). On the one hand, TGFβ itself mediates anti-inflammatory effects by suppressing immune cells and promotes Foxp3 expression in inducible $CD4^+$ Tregs cells. However, in the presence of inflammatory cytokines such as IL-6 or IL-1, TGF-β drives the differentiation of prorejection Th17 cells (Regateiro et al. 2011). Such a dual role may be important in a balancing act between acute/chronic rejection and tolerance to allografts (Table 2.2).

As presented in Fig. 2.3a, normal kidneys, like other tissues, produce low levels of cytokines, which are critical for physiological function (*Cell Res,* 16**,** 196–202). Tolerant kidney transplant requires constant presence of Tregs to induce and maintain tolerance to allograft. These cytokines, TGFβ, IL-10, and IL-35, are continuously needed to block all other T/B cells and innate immune cells (macrophages and dendritic cells) as well as endothelial cells from tilting their response toward proinflammatory posture (Fig. 2.3b).

2.10 Discussion

Organ and tissue transplantation became a standard medical procedure for patients with end-stage organ diseases (Grinyo 2013). Over the last 50 years, experimental work overcame several challenges by innovations in the technology of organ/tissue harvesting and preservation, advanced laboratory testing, and complex immunosuppression with induction and maintenance therapy. While results of graft survivals reached over 90–95% for kidney graft survival per year, 85% for heart transplants and 75% for liver transplants, chronic damage to these transplanted organs remains persistently high (Poggio et al. 2021; Wilhelm 2015; Kwong et al. 2024). While acute rejection episodes are now below 10% for kidney allografts, their long-term transplants show up to 50% gradual kidney injury at 5–10 years post-transplant in the form of chronic rejection, nephropathy, or graft dysfunction (Gaston et al. 2010). Vasculopathy rates for heart transplants are 8% in the first year, 29% in the 5 years,

Table 2.2 Cytokines involved in regulating transplantation tolerance

Cytokine	Function in tolerance	
IL-2	IL-2 plays a dual role in the immune system, acting as both a growth factor for T cells and a key regulator of immune tolerance. Promotes Treg cells.	Liao et al. (2013) Fehervari et al. (2006)
IL-4	IL-4 is important in humoral immunity and allergic responses. Induces CD4+ T cell differentiation into the Th2 subtype. Important signal in DC maturation.	Tay et al. (2009)
IL-6	IL-6 is an inflammatory cytokine, which induces differentiation of immune cells, endothelial cells, and stromal cells into inflammatory phenotypes. It suppresses Treg cells.	Jordan et al. (2022)
IL-7	IL-7 provides survival signal for developing thymocytes. Modulates IL-7 signaling to restore tolerance.	Zeng et al. (2025)
IL-10	IL-10 is a crucial cytokine in the induction and maintenance of immune tolerance. Secreted by Tregs and other cells to decrease inflammation by preventing the production of proinflammatory cytokines (e.g., IL-1, TNF-α, IL-6).	Akdis and Akdis, (2014) Hsueh et al. (2018)
IL-22	IL-22 plays a dual role in both promoting and suppressing inflammation. Important in maintenance and regeneration of thymic epithelial cells after injury.	Sajiir et al. (2025)
IL-23	IL-23 suppresses the differentiation of Treg cells and promotes differentiation, survival, and expansion of pathogenic Th17 cells. Provides proapoptotic signal that promotes negative selection in the thymus.	Krueger et al. (2024)
IL-35	IL-35 has shown promise in mitigating chronic allograft rejection; involved in the development of tolerance and the production of regulatory cells that express IL-35.	Baydoun and Camirand (2024) Sullivan et al. (2020) Sawant et al. (2015)
TGF-β	A crucial cytokine involved in both promoting and regulating immune tolerance. Regulatory cytokine secreted by Tregs regulates immune cell proliferation, differentiation, survival, and activity.	Loehe and Schauer (2005) Wang et al. (2023)
TNF-α	Repeated exposure to TNF-alpha reduces the sensitivity of cells to subsequent TNF-alpha stimulation. This tolerance is a protective mechanism to prevent excessive inflammation. Inflammatory cytokine that stimulates immune cell proliferation, survival, and activity.	Huber et al. (2017)
IFN-γ	Plays a complex role in immune tolerance, sometimes promoting it and sometimes hindering it, depending on the context. Inflammatory cytokine with broad effects on the immune system, notably macrophage activation. Secreted by Th1 cells and macrophages. Important in viral and mycobacterial immunity.	Rozman and Svajger (2018)

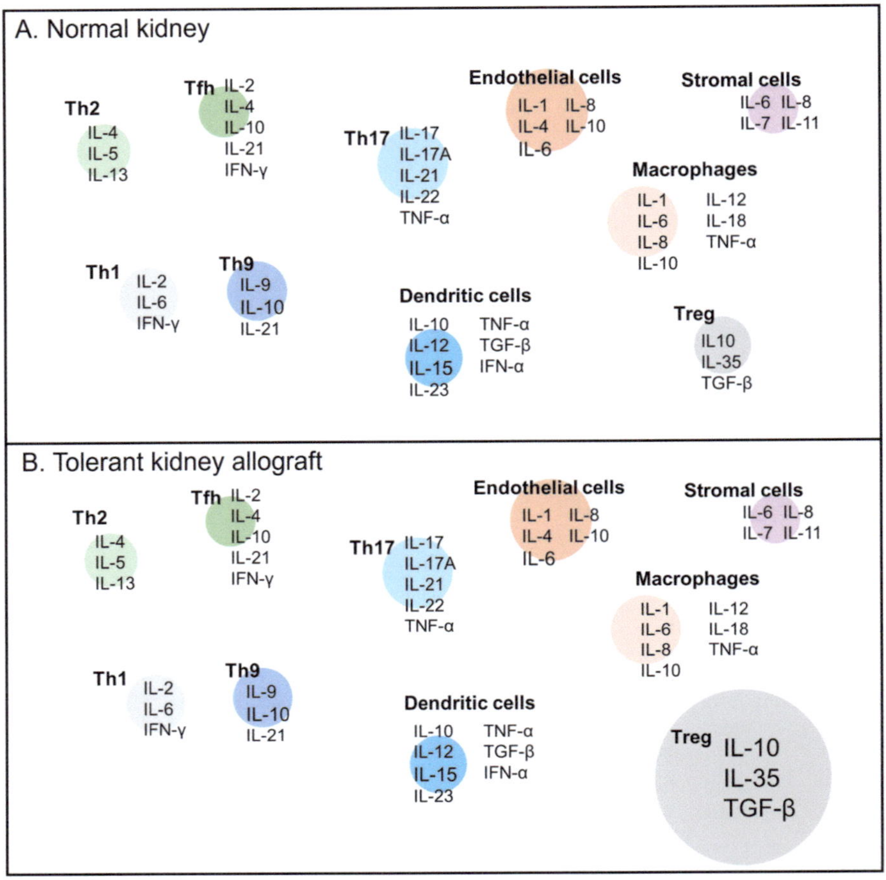

Fig. 2.3 Contribution of T cell subsets (Th1, Th2, Th9, Th17, and Treg) as well as innate immune cells (dendritic cells and macrophages) and other cells (endothelial cells) and their cytokines in normal kidneys (Panel **a**) and/or tolerant kidney allograft (Panel **b**). The size of subset circles reflects the relative number of each cell subset and the size of the circle for each cytokine produced amount

and 47% in the 10 years post-transplant (Khush et al. 2019). Despite progress in several aspects of organ transplantation, chronic rejection continues to be a major problem in organ transplantation.

Our understanding of chronic rejection/nephropathy based on the most modern histological classification may be correlated with patterns of cytokines controlling pathophysiological processes (Walsh et al. 2004; Tay et al. 2009; Corris and Kirby 2005). Two major injury mechanisms dominate, namely, TCMR and AMR; while their pathologies overlap with similarities, they may be recognized by very characteristic patterns of immune cells, chemokines, adhesion molecules, and cytokine profiles (Roufosse et al. 2018; Naesens et al. 2024). Independently of the mechanism for chronic rejection, kidney transplants progressively lose nephrons, and their

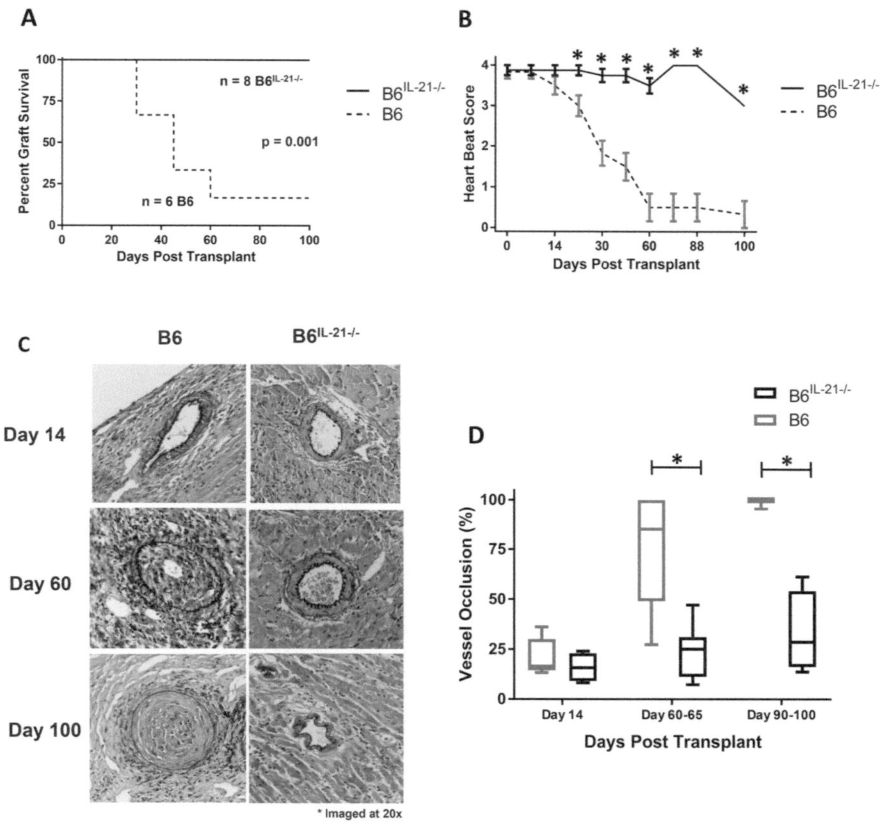

Fig. 2.4 Heart transplant graft evaluation in IL-21-deficient mice in a chronic rejection model (CAV). Graft survival (**a**), heartbeat scoring (**b**), and vessel occlusion (**c**, **d**) over time in normal and IL-21-deficient recipient mice of heart transplants from bm12 mice. Significance approximated by Mantel-Cox test (**a**), 2-way ANOVA with Bonferroni post-tests (**b**), and unpaired Student's t-test (**d**) indicates $p < 0.05$. (**c**) Grafts were recovered on the indicated day, cut, and stained for elastin. (**d**) Percent occlusion was determined by tracing the internal elastic lamina and lumen of each vessel and calculated by the formula [% occlusion = Area of internal lamina minus Area of lumen divided by Area of internal lamina times 100]. The number of mice and vessels were as follows: Day14 (2mice per group, 9–10 vessels); Day 35–45 (3–4 mice per group, 11–17 vessels); Day 60–65 (6–7 mice per group, 13–27 vessels); Day90–100 (2 mice per group, 6–8 vessels). (The figure is reproduced from the publication in Plos One 2019, Plos One 2019, 14:e0225624; https://doi.org/10.1371/journal.pone.0225624)

worsening function is displayed by elevated creatinine and decreased GFR values, predisposing to an inevitable kidney graft failure (Haas et al. 2018; Naesens et al. 2024; Mengel and Adam 2024). Chronic rejection diagnosis is measured by the standardized Banff semiquantitative criteria of interstitial fibrosis, tubular atrophy (IFTA), and arteriolitis (Roufosse et al. 2018; Naesens et al. 2024; Shimizu et al. 2025). The 16th Banff meeting in 2022 defined combining microvascular inflammation/injury (MVI) with C4d deposition in the peritubular capillaries (PTCs) and the

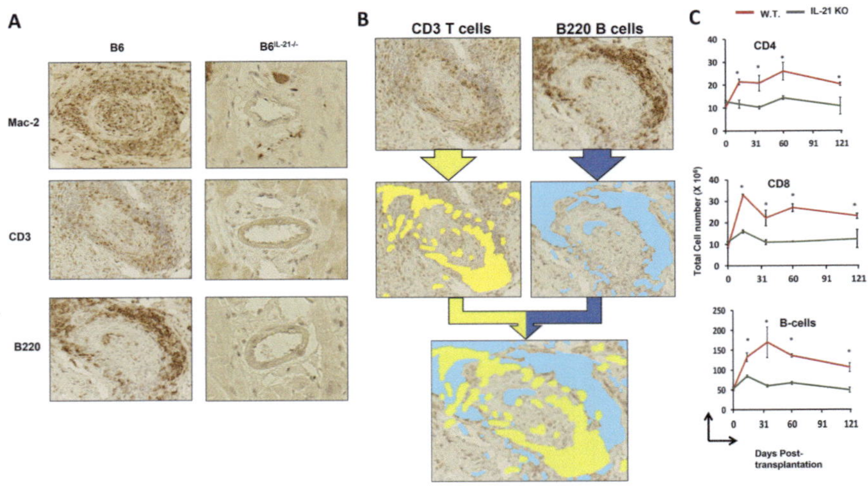

Fig. 2.5 Immune cell infiltration of grafts in IL-21-deficient mice in a CAV model. (**a**) Representative images of IHC stained vessels in grafts of heart transplants from bm12 mice in WT or IL-21-deficient recipient mice ($n = 3$ mice/group). The stains were as follows: MAC2, macrophages; CD3, T-cells; B220, B-cells. (**b**) Organization of CD3 stained T-cells and B220 stained B-cells near vessel with the possibility of a tertiary lymphoid organ-like development. (**c**) Accumulation of immune cells detected by flow cytometry of spleen homogenates. Significance approximated by unpaired Student's t-test indicates $p < 0.05$. (The figure is reproduced from the publication in Plos One 2019, 14:e0225624; https://doi.org/10.1371/journal.pone.0225624)

presence or absence of dnDSA (Shimizu et al. 2025). When MVI pathology is above the threshold, AMR is called negative without the C4d marker and dnDSA. In contrast, when MVI is below the threshold with negative C4d and positive dnDSA, the AMR is called "probable." These updates created chronic "active" AMR and TCMR pathology (Shimizu et al. 2025; Naesens et al. 2024). Chronic active TCMR was defined by inflammation in atrophic areas as i-IFTA or with tubulitis as t-IFTA (Roufosse et al. 2018). These qualitative categories are extensively regulated by unique patterns of cytokine, chemokine, and signaling profiles, which may be identifiable by transcriptomics, proteomics, and metabolomics (Heavey et al. 2022; Kupcova Skalnikova et al. 2017; Amir et al. 2022).

To correlate histology with the mechanism of AMR, investigators correlated dnDSA production with the expansion of B cell clones in germinal centers requiring IL-21 produced by Tfh cells (Heeger et al. 2024; La Muraglia 2nd et al. 2020). Expansion of blood circulating IL-21-producing Tfh cells ($CD4^+CXCR5^+$) followed kidney allograft transplantation in graft-draining lymph nodes. Additional markers identified $ICOS^+PD\text{-}1^+$ Tfh cells in germinal centers coincided with dnDSA synthesis. When costimulation was blocked for $ICOS^+PD\text{-}1^+$ Tfh cells, hosts did not produce dnDSA. During affinity maturation for dnDSA, B cells required IL-21, BAF, and IL-6 cytokines to change into functional plasma cells and Tfh memory cells (Heeger et al. 2024). In the active effector phase of MVI, dnDSA bound to allogeneic HLAs on endothelial cells, activating them. Proliferating endothelial cells produced IL-1, IL-6, IL-8, and TNFα cytokines, MCP-1, CXCL1, and CCL5 chemokines as well as E-selectin, P-selectin, ICAM-1, and VCAM-1 adhesion molecules (Pober and Sessa 2007). These inflammatory molecules activated macrophages, which then produced destructive power of reactive oxygen species (ROS; superoxide, O_{2-}, hydrogen peroxide, H_2O_2, hydroxyl radical, OH, and singlet oxygen, $1O_2$), which directly damaged tissue causing histological images of IFTA and arteritis (Naesens et al. 2024, Shimizu et al. 2025). All immune events are carefully regulated by cytokines with deliberate and identifiable steps resulting in unique changes of i-IFTA and t-IFTA. Combinations of many molecules may distinguish between two IFTAs, but many molecules are common. Modern methods measuring expression of these multiple molecules at RNA, protein, and metabolic levels may define specific histological images.

Although acute immune response engages potent response of Th1/Th2/Th9Th17 cells combined with activated macrophages and dendritic cells in TCMR and/or AMR mechanisms (Fig. 2.1a, b), slowly simmering MVI processes keep low levels of proinflammatory cytokines rearranging a chronic response into i-IFTA and/or t-IFTA injury (Fig. 2.2a, b) (Shimizu et al. 2025). Local proinflammatory IL-1β, IL-6, and TNFα signaling contribute to kidney fibrosis in both IFTAs; chemokines MCP-1 and RANTES attract even more immune cells and increase the release of ROS power. When chronically active AMR and/or /TCMR develop fibrosis, they also may become i-IFTA and/or t-IFTA, dependent on the localization of changes (Helgeson et al. 2021). Both these chronic forms may occur as cellular rejection with inflammation, tubulitis, and peritubular capillaritis as well as humoral chronic rejection with C4d and dnDSA (Helgeson et al. 2021). The sequence in the dynamics of these changes may require more sophisticated molecular markers, and research continues to unleash their predictive power to measure changes proceeding to the organ damage.

Transcriptomics, proteomics, and metabolomics may evaluate preemptive precise patterns of gene expression and transform them into mRNA and protein levels and their activation state to describe trends in volcano graphs (Fu et al. 2019). One such study found a profile mRNA expression for IFTA where subclinical inflammation was correlated with CD2, IL-7R, and CCL5 mRNAs associated with i-IFTA changes in comparison to stable renal transplants. Expression of IRF8 and IFNα correlated with specific findings in graft biopsies and predicted the risk of graft loss (Fu et al. 2019). We suggest that quantitative cytokines in combination with other gene families may be used to differentiate i-IFTA and t-IFTA prior to obvious clinical indicators, hopefully measured in urine and/or blood samples.

Experimental work in animals may help to define correlations between histological and molecular patterns, including the expression of cytokines (Tse et al. 2015). Kidney transplants in mice with TOS (tertiary lymphoid structures) had germinal centers loaded with B cells identified as $IgM^{high}CD23^-$ B cells, $IgM^{low}CD23^+$ B cells, and $IgM^{low}CD23^-$ B cell subsets. Elimination of these B cells removed chronic allograft damage. These B cell clones themselves produced regulatory cytokines, including CXCL1, RANTES, IL-6, and CCL2. Changes included TCMR resulting in IFTA image with type III collagen deposition, macrophages, and fibroblasts. Since experimental B-cell depletion removed macrophages and fibroblasts, both T and B cells were sufficient to produce active cytokines responsible for IFTA (Tse et al. 2015). As shown in Table 2.1, acute and chronic rejections rely on the same cytokines to produce damage and only internal regulatory patterns of cytokines induce acute TCMR and AMR (Fig. 2.1a, b) or chronic TCMR and AMR (Fig. 2.2a, b). We propose that cytokine patterns need to be combined with multiple other functional molecules to determine expected clinical outcomes.

In normal healthy kidney tissue, a chain of multiple physiological events (metabolism) do occur continuously that are in turn controlled by multiple cytokines (Fig. 2.3a). The levels of these cytokines are low as they regulate cell renewal, cell functions, and physiological cleaning processes (Shuai 2006). Otherwise, inflammatory in their nature, IL-1, IL-6, and TNFα regulate cell signals responsible for normal functions. In every healthy individual, multiple cytokines are crucial to maintain the normal immune system and physiological cellular signaling, as in perfectly normal conditions, cytokines are present in the serum and plasma at very low, sometimes almost undetectable quantities. Cytokines are always produced by a variety of cells, especially those involved in regulatory and immune events, but the list of cytokine producers includes endothelial cells, fibroblasts, and a variety of stromal cells. Indeed, cytokines serve as chemical messengers regulating multiple cell functions, including coordination of cell differentiation, cell proliferation, and renewal as well as multiple gene expression (Cui et al. 2024). Cytokines constantly bind to their specific cell surface receptors to activate signal transduction pathways regulating functional gene expression (Shuai 2006). Without cytokines, cells would be completely disorganized, and their life would often end, for example, by models of cytokine deficiencies (Cui et al. 2024). Cytokines activate four Janus tyrosine kinases (Jak1, Jak2, Jak3, and Tyk2) and seven Stats (Stat1, Stat2, Stat3, Stat4, Stat5a/b, and Stat6) as well as NF-kB as three important families of molecules activated by cytokines (Hu et al. 2021; Ghoreschi et al. 2009). Abnormal regulation of Jaks, Stats, and/or NF-κB activities has been identified and associated with severe human diseases. The protein inhibitor of the activated Stat (PIAS) protein family can interact with over 60 proteins, many of which are transcription factors involved in normal cell function and in the immune system (Shuai and Liu 2005). As Stats and NF-κB transcription factors regulate directly gene expression in cells, PIAS proteins regulate transcription activities through several mechanisms, including blocking the DNA-binding activity of other transcription factors, recruiting transcriptional corepressors, and promoting protein sumoylation (Shuai 2006; Maejima and Sadoshima 2014; Geiss-Friedlander and Melchior 2007)). This is only one example of multiple and complex functions regulated in cells by cytokines. An NF-κB family of dimeric transcription factors with the Rel family of DNA-binding proteins (Karin and Ben-Neriah 2000; May and Ghosh 1997). NF-κB itself may be activated by other signals delivered by physiological concentrations of IL-1, IL-6, and TNFα (Karin et al. 2004; Aggarwal 2003). Thus, normal kidneys and optimal physiology require threshold levels of cytokines for their normal functions, and these requirements are different for cytokines to induce and maintain transplantation tolerance to kidney allografts.

Transplantation tolerance to kidney allografts needs cytokines to maintain the physiology of kidney functions (like every normal tissue) as well as selective promotion of multiple Tregs (and other suppressors) to maintain the state of unresponsiveness (Sawitzki et al. 2020). As explained in the previous paragraph, normal tissues need cytokines for physiological operations (Shuai and Liu 2005). A tolerant state needs to be induced to avoid any initial damage to kidney allografts (to avoid triggers by damaged tissue), and to prevent the induction of memory T/B cells (which are unstoppable in their persistence to destroy allografts). On top of these conditions, the transplanted organ needs massive intra-graft donor-specific darTregs to continuously and actively regulate immune response (Guinan et al. 2023). Genomic DNA was extracted from blood samples of two patients who stopped their immunosuppressive therapy at a young age with HLA-DR matched and HLA-class I-mismatched donors (Burlingham et al. 2000). When they stopped immunosuppressants at 1–2 years, they retained kidney transplant function for the next 5 years. Both patients were a high responder genotype for TGFβ (promoting Tregs): one was a low responder genotype for TNFα and both were genotypically TNFα intermediate responders (inflammatory signals). These authors suggested the role of beneficial cytokine immunogenetics in the stability of their peripheral tolerance (Burlingham et al. 2000). The most successful deliberate tolerance induction was observed in three selected patients who received massive infusions of darTregs (Guinan et al. 2023). All three cases of achieved tolerance emphasize the enormous challenge of achieving a state where cytokines maintain normal levels of multiple cytokines for normal kidney function (Fig. 2.3a) and at the same time promote the constant presence of darTregs and perpetuated Tregs insight kidney allografts. Only these conditions meet requirements for stable tolerance, preventing both acute and chronic allograft rejection.

2.11 Conclusions

Cytokines regulate physiological cell functions and immune response to bacterial and viral infections as well as to self-antigens and foreign tissue alloantigens. The mechanism of acute, accelerated, and chronic allograft rejection requires multiple sequential regulatory events by the cytokine-mediated expansion of innate and adaptive immunity. When immunosuppressive agents block acute rejection, they usually rely on blocking cytokine production or their access to their receptors. This type of "firewall" is usually leaky and small amounts of cytokines initiate/mediate smoldering immune response inflicting allogeneic tissue injury, hence initiating/mediating chronic allograft rejection. Dominance by Th1- and/or Tfh-type immune response promotes acute/chronic TCMR and/or AMR, respectively. The γc-chain family of cytokines (IL-2, IL-4, IL-7, IL9, IL-15, and IL-21) regulates all T/B cell functions during TCMR/AMR responses. Blocking IL-21/Tfh/Th17 activity by

IL-21RFc fusion protein prevents cardiac vasculopathy in heart allografts. Multifunctional IL-6 cytokine (regulating T/B/macrophages/endothelial cells) emerged recently as the most appreciated target for chronic allograft rejection treatment /prophylaxis as it stimulates immune cell expansion, smooth muscle cell proliferation, and cell repair processes at the site of an allograft. Anti-IL-6R monoclonal antibody (tocilizumab) and anti-IL-6 antibody (clazakizumab) are in clinical trials for the prevention and treatment of chronic rejection. Transplantation tolerance relies on the function of darTreg cells producing IL-10, IL-35, and TGFβ cytokines. To induce tolerance to allografts, transplanted tissue must be protected from the beginning from injury by functionally potent donor-specific darTreg cells producing TGFβ, IL-10, and IL-35. Downregulation/inhibition of innate/adaptive immunity by these cytokines promotes anti-inflammatory while inhibiting proinflammatory functions. At the same time, normal levels of cytokines need to maintain an equilibrium in physiological metabolism between nonimmune and immune cells. Measurements using the most advanced technologies (transcriptomics, proteomics, and metabolomics) combined with cytokine expression patterns may serve as new tools to predict chronic allograft damage and graft loss in advance.

Contributions Material collection and selection of published papers: Stanislaw Stepkowski, Jared Oenick, Dulat Bekbolsynov, Beata Mierzejewska, Michael Rees, Obi Ekwenna.

Supervision: Stanislaw Stepkowski.

Visualization, figures and tables: Dulat Bekbolsynov, Stanislaw Stepkowski, Beata Mierzejewska.

Writing of original draft: Stanislaw Stepkowski, Jared Oenick, Dulat Bekbolsynov, Beata Mierzejewska.

Review and editing: Stanislaw Stepkowski, Jared Oenick, Dulat Bekbolsynov, Beata Mierzejewska, Michael Rees, Obi Ekwenna.

References

Agarwal S, Parashar A, Kapadia SR, Tuzcu EM, Modi D, Starling RC, Oliveira GH (2014) Long-term mortality after cardiac allograft vasculopathy: implications of percutaneous intervention. Jacc Heart Fail 2:281–288

Aggarwal BB (2003) Signalling pathways of the TNF superfamily: a double-edged sword. Nat Rev Immunol 3:745–756

Akdis CA, Akdis M (2014) Mechanisms of immune tolerance to allergens: role of IL-10 and Tregs. J Clin Invest 124(11):4678–4680. https://doi.org/10.1172/JCI78891. Epub 20141103. PubMed PMID: 25365074; PMCID: PMC4347251

Amir O, Goldberg Y, Mandel M, Bar-On YM, Bodenheimer O, Ash N, Alroy-Preis S, Huppert A, Milo R (2022) Protection following BNT162B2 booster in adolescents substantially exceeds that of a fresh 2-dose vaccine. Nat Commun 13:1971

Antonysamy MA, Fanslow WC, Fu F, Li W, Qian S, Troutt AB, Thomson AW (1999) Evidence for a role of IL-17 in organ allograft rejection: IL-17 promotes the functional differentiation of dendritic cell progenitors. J Immunol 162:577–584

Asada N, Ginsberg P, Gagliani N, Mittrucker HW, Panzer U (2022) Tissue-resident memory T cells in the kidney. Semin Immunopathol 44:801–811

Baan CC, Balk AH, Dijke IE, Korevaar SS, Peeters AM, De Kuiper RP, Klepper M, Zondervan PE, Maat LA, Weimar W (2007) Interleukin-21: an interleukin-2 dependent player in rejection processes. Transplantation 83:1485–1492

Barron L, Dooms H, Hoyer KK, Kuswanto W, Hofmann J, O'Gorman WE, Abbas AK (2010) Cutting edge: mechanisms of IL-2-dependent maintenance of functional regulatory T cells. J Immunol 185:6426–6430

Baydoun H, Camirand G (2024) IL-35 therapy boosts Treg in transplantation. Transplantation 108(1):14–15. https://doi.org/10.1097/TP.0000000000004708. Epub 20230719. PubMed PMID: 37464475; PMCID: PMC10796832

Becker LE, De Oliveira Biazotto F, Conrad H, Schaier M, Kihm LP, Gross-Weissmann ML, Waldherr R, Bierhaus A, Nawroth PP, Zeier M, Morath C (2012) Cellular infiltrates and NfkappaB subunit c-Rel signaling in kidney allografts of patients with clinical operational tolerance. Transplantation 94:729–737

Behrendt D, Ganz P, Fang JC (2000) Cardiac allograft vasculopathy. Curr Opin Cardiol 15:422–429

Belperio JA, Burdick MD, Keane MP, Xue YY, Lynch JP 3rd, Daugherty BL, Kunkel SL, Strieter RM (2000) The role of the CC chemokine, RANTES, in acute lung allograft rejection. J Immunol 165(1):461–472. https://doi.org/10.4049/jimmunol.165.1.461. PubMed PMID: 10861085

Berard JL, Velez RL, Freeman RB, Tsunoda SM (1999) A review of interleukin-2 receptor antagonists in solid organ transplantation. Pharmacotherapy 19:1127–1137

Bertini M, Mele D, Vitali F, Balla C, Malagu M (2022) Management of macro-reentrant right atrial tachycardia around multiple leads aided by high-density mapping. Rev Cardiovasc Med 23:1

Borie DC, Si MS, Morris RE, Reitz BA, Changelian PS (2003) Jak3 inhibition as a new concept for immune suppression. Curr Opin Investig Drugs 4:1297–1303

Brouard S, Pallier A, Renaudin K, Foucher Y, Danger R, Devys A, Cesbron A, Guillot-Guegen C, Ashton-Chess J, Le Roux S, Harb J, Roussey G, Subra JF, Villemain F, Legendre C, Bemelman FJ, Orlando G, Garnier A, Jambon H et al (2012) The natural history of clinical operational tolerance after kidney transplantation through twenty-seven cases. Am J Transplant 12:3296–3307

Brunner-La Rocca HP, Schneider J, Kunzli A, Turina M, Kiowski W (1998) Cardiac allograft rejection late after transplantation is a risk factor for graft coronary artery disease. Transplantation 65:538–543

Burlingham WJ, O'Connell PJ, Jacobson LM, Becker BN, Kirk AD, Pravica V, Hutchinson IV (2000) Tumor necrosis factor-alpha and tumor growth factor-beta1 genotype: partial association with intragraft gene expression in two cases of long-term peripheral tolerance to a kidney transplant. Transplantation 69:1527–1530

Changelian PS, Flanagan ME, Ball DJ, Kent CR, Magnuson KS, Martin WH, Rizzuti BJ, Sawyer PS, Perry BD, Brissette WH, Mccurdy SP, Kudlacz EM, Conklyn MJ, Elliott EA, Koslov ER, Fisher MB, Strelevitz TJ, Yoon K, Whipple DA et al (2003) Prevention of organ allograft rejection by a specific Janus kinase 3 inhibitor. Science 302:875–878

Chen W, Jin W, Hardegen N, Lei KJ, Li L, Marinos N, Mcgrady G, Wahl SM (2003) Conversion of peripheral CD4+CD25- naive T cells to CD4+CD25+ regulatory T cells by TGF-beta induction of transcription factor Foxp3. J Exp Med 198:1875–1886

Chen S, Kapturczak MH, Wasserfall C, Glushakova OY, Campbell-Thompson M, Deshane JS, Joseph R, Cruz PE, Hauswirth WW, Madsen KM, Croker BP, Berns KI, Atkinson MA, Flotte TR, Tisher CC, Agarwal A (2005) Interleukin 10 attenuates neointimal proliferation and inflammation in aortic allografts by a heme oxygenase-dependent pathway. Proc Natl Acad Sci USA 102:7251–7256

Chen Y, Chauhan SK, Tan X, Dana R (2017) Interleukin-7 and -15 maintain pathogenic memory Th17 cells in autoimmunity. J Autoimmun 77:96–103

Chinen T, Kannan AK, Levine AG, Fan X, Klein U, Zheng Y, Gasteiger G, Feng Y, Fontenot JD, Rudensky AY (2016) An essential role for the IL-2 receptor in T(reg) cell function. Nat Immunol 17:1322–1333

Clayton PA, Mcdonald SP, Russ GR, Chadban SJ (2019) Long-term outcomes after acute rejection in kidney transplant recipients: an ANZDATA analysis. J Am Soc Nephrol 30:1697–1707
Collison LW, Workman CJ, Kuo TT, Boyd K, Wang Y, Vignali KM, Cross R, Sehy D, Blumberg RS, Vignali DA (2007) The inhibitory cytokine IL-35 contributes to regulatory T-cell function. Nature 450:566–569
Collison LW, Chaturvedi V, Henderson AL, Giacomin PR, Guy C, Bankoti J, Finkelstein D, Forbes K, Workman CJ, Brown SA, Rehg JE, Jones ML, Ni HT, Artis D, Turk MJ, Vignali DA (2010) IL-35-mediated induction of a potent regulatory T cell population. Nat Immunol 11:1093–1101
Cooper JE (2020) Evaluation and treatment of acute rejection in kidney allografts. Clin J Am Soc Nephrol 15:430–438
Corris PA, Kirby JA (2005) A role for cytokine measurement in therapeutic monitoring of immunosuppressive drugs following lung transplantation. Clin Exp Immunol 139:176–178
Cui A, Huang T, Li S, Ma A, Perez JL, Sander C, Keskin DB, Wu CJ, Fraenkel E, Hacohen N (2024) Dictionary of immune responses to cytokines at single-cell resolution. Nature 625:377–384
Dambuza IM, He C, Choi JK, Yu CR, Wang R, Mattapallil MJ, Wingfield PT, Caspi RR, Egwuagu CE (2017) IL-12p35 induces expansion of IL-10 and IL-35-expressing regulatory B cells and ameliorates autoimmune disease. Nat Commun 8:719
Dandel M, Hummel M, Wellnhofer E, Kapell S, Lehmkuhl HB, Hetzer R (2003) Association between acute rejection and cardiac allograft vasculopathy. J Heart Lung Transplant 22:1064–1065
De Groot-Kruseman HA, Klepper M, Mol WM, Niesters HGM, Van Gelder T, Maat APWM, Balk A, Weimar W, Baan C (2002) Intragraft mRNA expression of the novel cytokine IL-21 during acute rejection after clinical heart transplantation. J Heart Lung Transplant 21:165–165
Deteix C, Attuil-Audenis V, Duthey A, Patey N, Mcgregor B, Dubois V, Caligiuri G, Graff-Dubois S, Morelon E, Thaunat O (2010) Intragraft Th17 infiltrate promotes lymphoid neogenesis and hastens clinical chronic rejection. J Immunol 184:5344–5351
Dong C (2006) Diversification of T-helper-cell lineages: finding the family root of IL-17-producing cells. Nat Rev Immunol 6:329–333
Durandy A (2003) Activation-induced cytidine deaminase: a dual role in class-switch recombination and somatic hypermutation. Eur J Immunol 33:2069–2073
Elsaesser H, Sauer K, Brooks DG (2009) IL-21 is required to control chronic viral infection. Science 324:1569–1572
Ernandez T, Mayadas TN. IMmunoregulatory role of TNFalpha in inflammatory kidney diseases. Kidney Int. 2009;76(3):262–276. doi: https://doi.org/10.1038/ki.2009.142.. Epub 20090513. PubMed PMID: 19436333
Fan L, Benson HL, Vittal R, Mickler EA, Presson R, Fisher AJ, Cummings OW, Heidler KM, Keller MR, Burlingham WJ, Wilkes DS (2011) Neutralizing IL-17 prevents obliterative bronchiolitis in murine orthotopic lung transplantation. Am J Transplant 11:911–922
Faust SM, Lu G, Marini BL, Zou W, Gordon D, Iwakura Y, Laouar Y, Bishop DK (2009) Role of T cell TGFbeta signaling and IL-17 in allograft acceptance and fibrosis associated with chronic rejection. J Immunol 183:7297–7306
Fehervari Z, Yamaguchi T, Sakaguchi S (2006) The dichotomous role of IL-2: tolerance versus immunity. Trends Immunol 27(3):109–111. https://doi.org/10.1016/j.it.2006.01.005. Epub 20060203. PubMed PMID: 16459146
Ferreira LMR, Muller YD, Bluestone JA, Tang Q (2019) Next-generation regulatory T cell therapy. Nat Rev Drug Discov 18:749–769
Ferrini S, Moretta A, Biassoni R, Nicolin A, Moretta L (1986) Cyclosporin-A inhibits IL-2 production by all human T-cell clones having this function, independent of the T4/T8 phenotype or the coexpression of cytolytic activity. Clin Immunol Immunopathol 38:79–84
Fina D, Sarra M, Caruso R, Del Vecchio Blanco G, Pallone F, Macdonald TT, Monteleone G (2008) Interleukin 21 contributes to the mucosal T helper cell type 1 response in coeliac disease. Gut 57:887–892

Fischbein MP, Yun J, Laks H, Irie Y, Oslund-Pinderski L, Fishbein MC, Bonavida B, Ardehali A (2003) Regulated interleukin-10 expression prevents chronic rejection of transplanted hearts. J Thorac Cardiovasc Surg 126:216–223

Fu Q, Liao M, Feng C, Tang J, Liao R, Wei L, Yang H, Markmann JF, Chen K, Deng S (2019) Profiling of mRNA of interstitial fibrosis and tubular atrophy with subclinical inflammation in recipients after kidney transplantation. Aging (Albany NY) 11:5215–5231

Furtado GC, Curotto De Lafaille MA, Kutchukhidze N, Lafaille JJ (2002) Interleukin 2 signaling is required for CD4(+) regulatory T cell function. J Exp Med 196:851–857

Gaston RS, Cecka JM, Kasiske BL, Fieberg AM, Leduc R, Cosio FC, Gourishankar S, Grande J, Halloran P, Hunsicker L, Mannon R, Rush D, Matas AJ (2010) Evidence for antibody-mediated injury as a major determinant of late kidney allograft failure. Transplantation 90:68–74

Geiss-Friedlander R, Melchior F (2007) Concepts in sumoylation: a decade on. Nat Rev Mol Cell Biol 8:947–956

Ghoreschi K, Laurence A, O'Shea JJ (2009) Janus kinases in immune cell signaling. Immunol Rev 228:273–287

Goldman M, Le Moine A, Braun M, Flamand V, Abramowicz D (2001) A role for eosinophils in transplant rejection. Trends Immunol 22(5):247–251. https://doi.org/10.1016/s1471-4906(01)01893-2. PubMed PMID: 11323281

Goodwin RG, Lupton S, Schmierer A, Hjerrild KJ, Jerzy R, Clevenger W, Gillis S, Cosman D, Namen AE (1989) Human interleukin 7: molecular cloning and growth factor activity on human and murine B-lineage cells. Proc Natl Acad Sci USA 86:302–306

Grinyo JM (2013) Why is organ transplantation clinically important? Cold Spring Harb Perspect Med 3:a014985

Guinan EC, Contreras-Ruiz L, Crisalli K, Rickert C, Rosales I, Makar R, Colvin R, Geissler EK, Sawitzki B, Harden P, Tang Q, Blancho G, Turka LA, Markmann JF (2023) Donor antigen-specific regulatory T cell administration to recipients of live donor kidneys: A ONE study consortium pilot trial. Am J Transplant 23:1872–1881

Guo ZY, Khattar M, Schroder PM, Miyahara Y, Wang GH, He XS, Chen WH, Stepkowski SM (2013) A dynamic dual role of IL-2 signaling in the two-step differentiation process of adaptive regulatory T cells. J Immunol 190:3153–3162

Guo H, Li B, Wang W, Zhao N, Gao H (2018) Mesenchymal stem cells overexpressing IL-35: a novel immunosuppressive strategy and therapeutic target for inducing transplant tolerance. Stem Cell Res Ther 9:254

Gu-Trantien C, Migliori E, Buisseret L, De Wind A, Brohee S, Garaud S, Noel G, Dang Chi VL, Lodewyckx JN, Naveaux C, Duvillier H, Goriely S, Larsimont D, Willard-Gallo K (2017) CXCL13-producing TFH cells link immune suppression and adaptive memory in human breast cancer. JCI Insight 2(11):e91487

Haas M, Loupy A, Lefaucheur C, Roufosse C, Glotz D, Seron D, Nankivell BJ, Halloran PF, Colvin RB, Akalin E, Alachkar N, Bagnasco S, Bouatou Y, Becker JU, Cornell LD, Duong Van Huyen JP, Gibson IW, Kraus ES, Mannon RB et al (2018) The Banff 2017 Kidney Meeting Report: revised diagnostic criteria for chronic active T cell-mediated rejection, antibody-mediated rejection, and prospects for integrative endpoints for next-generation clinical trials. Am J Transplant 18:293–307

Haller S, Duval A, Migliorini R, Stevanin M, Mack V, Acha-Orbea H (2017) Interleukin-35-producing CD8ALPHA(+) dendritic cells acquire a Tolerogenic state and regulate T cell function. Front Immunol 8:98

Halloran PF, Miller LW, Urmson J, Ramassar V, Zhu LF, Kneteman NM, Solez K, Afrouzian M (2001) IFN-gamma alters the pathology of graft rejection: protection from early necrosis. J Immunol 166(12):7072–7081. https://doi.org/10.4049/jimmunol.166.12.7072. PubMed PMID: 11390451

Haouami Y, Dhaouadi T, Sfar I, Bacha M, Gargah T, Bardi R, Abderrahim E, Goucha R, Ben Abdallah T, Gorgi Y (2018) The role of IL-23/IL-17 axis in human kidney allograft rejection. J Leukoc Biol 104:1229–1239

Hara M, Kingsley CI, Niimi M, Read S, Turvey SE, Bushell AR, Morris PJ, Powrie F, Wood KJ (2001) IL-10 is required for regulatory T cells to mediate tolerance to alloantigens in vivo. J Immunol 166:3789–3796

Hart A, Smith JM, Skeans MA, Gustafson SK, Stewart DE, Cherikh WS, Wainright JL, Kucheryavaya A, Woodbury M, Snyder JJ, Kasiske BL, Israni AK (2017) OPTN/SRTR 2015 annual data report: kidney. Am J Transplant 17(Suppl 1):21–116

Heavey MK, Durmusoglu D, Crook N, Anselmo AC (2022) Discovery and delivery strategies for engineered live biotherapeutic products. Trends Biotechnol 40:354–369

Heeger PS, Haro MC, Jordan S (2024) Translating B cell immunology to the treatment of antibody-mediated allograft rejection. Nat Rev Nephrol 20:218–232

Helgeson ES, Mannon R, Grande J, Gaston RS, Cecka MJ, Kasiske BL, Rush D, Gourishankar S, Cosio F, Hunsicker L, Connett J, Matas AJ (2021) i-IFTA and chronic active T cell-mediated rejection: A tale of 2 (DeKAF) cohorts. Am J Transplant 21:1866–1877

Hernandez JM, De Prada JA, Burgos V, Sainz Laso F, Valls MF, Vilchez FG, Llano M, Ruano J, Zueco J, Colman T, Duran RM (2009) Virtual histology intravascular ultrasound assessment of cardiac allograft vasculopathy from 1 to 20 years after heart transplantation. J Heart Lung Transplant 28:156–162

Herold KC, Lancki DW, Moldwin RL, Fitch FW (1986) Immunosuppressive effects of cyclosporin A on cloned T cells. J Immunol 136:1315–1321

Hoffmann MW, Wonigeit K, Steinhoff G, Herzbeck H, Flad HD, Pichlmayr R (1993) Production of cytokines (TNF-alpha, IL-1-beta) and endothelial cell activation in human liver allograft rejection. Transplantation 55(2):329–335. https://doi.org/10.1097/00007890-199302000-00019. PubMed PMID: 8094579

Howard M, Farrar J, Hilfiker M, Johnson B, Takatsu K, Hamaoka T, Paul WE (1982) Identification of a T cell-derived b cell growth factor distinct from interleukin 2. J Exp Med 155:914–923

Hsueh YH, Chen HW, Syu BJ, Lin CI, Leung PSC, Gershwin ME, Chuang YH (2018) Endogenous IL-10 maintains immune tolerance but IL-10 gene transfer exacerbates autoimmune cholangitis. J Autoimmun 95:159–170. https://doi.org/10.1016/j.jaut.2018.09.009. Epub 20180928. PubMed PMID: 30274824; PMCID: PMC6289859

Hu X, Li J, Fu M, Zhao X, Wang W (2021) The JAK/STAT signaling pathway: from bench to clinic. Signal Transduct Target Ther 6:402

Huang A, Liu K, Yin Z, Liu J, Wei H, Xing S, Qu Y, Huang L, Li L, Li C, Zhang L, Li X, Zheng C, Liu Q, Jiang K (2024) IL-35 stabilizes Treg phenotype to protect cardiac allografts in mice. Transplantation 108:161–174

Huber R, Bikker R, Welz B, Christmann M, Brand K (2017) TNF tolerance in monocytes and macrophages:characteristics and molecular mechanisms. J Immunol Res 2017:9570129. https://doi.org/10.1155/2017/9570129. Epub 20171108. PubMed PMID: 29250561; PMCID: PMC5698820

Hultner L, Druez C, Moeller J, Uyttenhove C, Schmitt E, Rude E, Dormer P, Van Snick J (1990) Mast cell growth-enhancing activity (MEA) is structurally related and functionally identical to the novel mouse T cell growth factor P40/TCGFIII (interleukin 9). Eur J Immunol 20:1413–1416

Isakson PC, Pure E, Vitetta ES, Krammer PH (1982) T cell-derived B cell differentiation factor(s). Effect on the isotype switch of murine B cells. J Exp Med 155:734–748

Ise W, Kohyama M, Schraml BU, Zhang T, Schwer B, Basu U, Alt FW, Tang J, Oltz EM, Murphy TL, Murphy KM (2011) The transcription factor BATF controls the global regulators of class-switch recombination in both B cells and T cells. Nat Immunol 12:536–543

Issa F, Wood KJ (2010) CD4+ regulatory T cells in solid organ transplantation. Curr Opin Organ Transplant 15:757–764

Itoh S, Nakae S, Axtell RC, Velotta JB, Kimura N, Kajiwara N, Iwakura Y, Saito H, Adachi H, Steinman L, Robbins RC, Fischbein MP (2010) IL-17 contributes to the development of chronic rejection in a murine heart transplant model. J Clin Immunol 30:235–240

Iyer SS, Cheng G (2012) Role of interleukin 10 transcriptional regulation in inflammation and autoimmune disease. Crit Rev Immunol 32:23–63

Jane-Wit D, Manes TD, Yi T, Qin L, Clark P, Kirkiles-Smith NC, Abrahimi P, Devalliere J, Moeckel G, Kulkarni S, Tellides G, Pober JS (2013) Alloantibody and complement promote T cell-mediated cardiac allograft vasculopathy through noncanonical nuclear factor-kappaB signaling in endothelial cells. Circulation 128:2504–2516

Jarrick S, Lundberg S, Welander A, Carrero JJ, Hoijer J, Bottai M, Ludvigsson JF (2019) Mortality in IgA nephropathy: a nationwide population-based cohort study. J Am Soc Nephrol 30:866–876

Jimenez J, Kapadia SR, Yamani MH, Platt L, Hobbs RE, Rincon G, Botts-Silverman C, Starling RC, Young JB, Nissen SE, Tuzcu M, Ratliff NB (2001) Cellular rejection and rate of progression of transplant vasculopathy: a 3-year serial intravascular ultrasound study. J Heart Lung Transplant 20:393–398

Jones IKA, Orloff S, Burg JM, Haese NN, Andoh TF, Chambers A, Fei SS, Gao L, Kreklywich CN, Streblow ZJ, Enesthvedt K, Wanderer A, Baker J, Streblow DN (2021) Blocking the IL-1 receptor reduces cardiac transplant ischemia and reperfusion injury and mitigates CMV-accelerated chronic rejection. Am J Transplant 21:44–59

Jordan SC, Ammerman N, Choi J, Kumar S, Huang E, Toyoda M, Kim I, Wu G, Vo A (2020) Interleukin-6: an important mediator of allograft injury. Transplantation 104(12):2497–2506. https://doi.org/10.1097/TP.0000000000003249. PubMed PMID: 32235253

Jordan SC, Ammerman N, Huang E, Vo A (2022) Importance of IL-6 inhibition in prevention and treatment of antibodymediated rejection in kidney allografts. Am J Transplant 22(Suppl 4):28–37. https://doi.org/10.1111/ajt.17207. PubMed PMID: 36453709

Josefowicz SZ, Lu LF, Rudensky AY (2012) Regulatory T cells: mechanisms of differentiation and function. Annu Rev Immunol 30:531–564

Kalil AC, Florescu DF (2022) Mortality in solid organ transplant recipients hospitalized for COVID-19. Am J Transplant 22:12–13

Kaplan MH, Hufford MM, Olson MR (2015) The development and in vivo function of T helper 9 cells. Nat Rev Immunol 15:295–307

Karin M, Ben-Neriah Y (2000) Phosphorylation meets ubiquitination: the control of NF-[kappa]B activity. Annu Rev Immunol 18:621–663

Karin M, Yamamoto Y, Wang QM (2004) The IKK NF-kappa B system: a treasure trove for drug development. Nat Rev Drug Discov 3:17–26

Kennedy MK, Glaccum M, Brown SN, Butz EA, Viney JL, Embers M, Matsuki N, Charrier K, Sedger L, Willis CR, Brasel K, Morrissey PJ, Stocking K, Schuh JC, Joyce S, Peschon JJ (2000) Reversible defects in natural killer and memory CD8 T cell lineages in interleukin 15-deficient mice. J Exp Med 191:771–780

Khattar M, Miyahara Y, Schroder PM, Xie AN, Chen WH, Stepkowski SM (2014) Interleukin-21 is a critical regulator of CD4 and CD8 T cell survival during priming under Interleukin-2 deprivation conditions. PLoS One 9:e85882

Khattar M, Baum CE, Schroder P, Breidenbach JD, Haller ST, Chen WH, Stepkowski S (2019) Interleukin 21 (IL-21) regulates chronic allograft vasculopathy (CAV) in murine heart allograft rejection. PLoS One 14:e0225624

Khush KK, Cherikh WS, Chambers DC, Harhay MO, Hayes D Jr, Hsich E, Meiser B, Potena L, Robinson A, Rossano JW, Sadavarte A, Singh TP, Zuckermann A, Stehlik J, International Society For, H, Lung T (2019) The international thoracic organ transplant registry of the International Society for Heart and Lung Transplantation: thirty-sixth adult heart transplantation report – 2019; focus theme: donor and recipient size match. J Heart Lung Transplant 38:1056–1066

Kim HR, Hwang KA, Park SH, Kang I (2008) IL-7 and IL-15: biology and roles in T-cell immunity in health and disease. Crit Rev Immunol 28:325–339

Klonowski KD, Williams KJ, Marzo AL, Blair DA, Lingenheld EG, Lefrancois L (2004) Dynamics of blood-borne CD8 memory T cell migration in vivo. Immunity 20:551–562

Koh KP, Wang Y, Yi T, Shiao SL, Lorber MI, Sessa WC, Tellides G, Pober JS (2004) T cell-mediated vascular dysfunction of human allografts results from IFN-gamma dysregulation of NO synthase. J Clin Invest 114:846–856
Krogerus L, Soots A, Loginov R, Bruggeman C, Lautenschlager I (2008) CMV increases tubular apoptosis through the TNF-alpha-TNF-R1 pathway in a rat model of chronic renal allograft rejection. Transpl Immunol 18:232–236
Krouchev R, Tamana S, Nguyen N, Yorke E, Harris D, Sampath S (2018) Total intracorporeal anastomosis for right hemicolectomy: experience from a Canadian center. Am J Surg 215(5):905–908. https://doi.org/10.1016/j.amjsurg.2018.02.013. Epub 20180227. PubMed PMID: 29587966
Krueger JG, Eyerich K, Kuchroo VK, Ritchlin CT, Abreu MT, Elloso MM, Fourie A, Fakharzadeh S, Sherlock JP, Yang YW, Cua DJ, McInnes IB (2024) IL-23 past, present, and future: a roadmap to advancing IL-23 science and therapy. Front Immunol 15:1331217. https://doi.org/10.3389/fimmu.2024.1331217. Epub 20240415. PubMed PMID: 38686385; PMCID: PMC11056518
Kupcova Skalnikova H, Cizkova J, Cervenka J, Vodicka P (2017) Advances in proteomic techniques for cytokine analysis: focus on melanoma research. Int J Mol Sci 18(12):2697
Kwong AJ, Kim WR, Lake JR, Schladt DP, Schnellinger EM, Gauntt K, Mcdermott M, Weiss S, Handarova DK, Snyder JJ, Israni AK (2024) OPTN/SRTR 2022 annual data report: liver. Am J Transplant 24:S176–S265
La Muraglia GM 2nd, Wagener ME, Ford ML, Badell IR (2020) Circulating T follicular helper cells are a biomarker of humoral alloreactivity and predict donor-specific antibody formation after transplantation. Am J Transplant 20:75–87
Lanzavecchia A, Sallusto F (2005) Understanding the generation and function of memory T cell subsets. Curr Opin Immunol 17:326–332
Laskin DL, Sunil VR, Gardner CR, Laskin JD (2011) Macrophages and tissue injury: agents of defense or destruction? Annu Rev Pharmacol Toxicol 51:267–288
Le Gros G, Ben-Sasson SZ, Seder R, Finkelman FD, Paul WE (1990) Generation of interleukin 4 (IL-4)-producing cells in vivo and in vitro: IL-2 and IL-4 are required for in vitro generation of IL-4-producing cells. J Exp Med 172:921–929
Le Moine A, Flamand V, Demoor FX, Noel JC, Surquin M, Kiss R, Nahori MA, Pretolani M, Goldman M, Abramowicz D (1999) Critical roles for IL-4, IL-5, and eosinophils in chronic skin allograft rejection. J Clin Invest 103(12):1659–1667. https://doi.org/10.1172/JCI5504. PubMed PMID: 10377172; PMCID: PMC408380
Leonard WJ, Lin JX, O'Shea JJ (2019) The gamma(c) family of cytokines: basic biology to therapeutic ramifications. Immunity 50:832–850
Li J, Simeoni E, Fleury S, Dudler J, Fiorini E, Kappenberger L, Von Segesser LK, Vassalli G (2006) Gene transfer of soluble interleukin-17 receptor prolongs cardiac allograft survival in a rat model. Eur J Cardiothorac Surg 29:779–783
Liao W, Lin JX, Leonard WJ (2013) Interleukin-2 at the crossroads of effector responses, tolerance, and immunotherapy. Immunity 38(1):13–25. https://doi.org/10.1016/j.immuni.2013.01.004. PubMed PMID: 23352221; PMCID: PMC3610532
Lindemann M, Korth J, Sun M, Xu S, Struve C, Werner K, Dornieden T, Horn PA, Witzke O, Wilde B (2018) The cytomegalovirus-specific IL-21 ELISpot correlates with allograft function of kidney transplant recipients. Int J Mol Sci 19(12):3945
Lodolce JP, Boone DL, Chai S, Swain RE, Dassopoulos T, Trettin S, Ma A (1998) IL-15 receptor maintains lymphoid homeostasis by supporting lymphocyte homing and proliferation. Immunity 9:669–676
Loehe F, Schauer RJ (2005) Surgical treatment of primary biliary cirrhosis and primary sclerosing cholangitis. Clin Rev Allergy Immunol 28(2):167–174. https://doi.org/10.1385/CRIAI:28:2:167. PubMed PMID: 15879622
Lund LH, Edwards LB, Kucheryavaya AY, Benden C, Dipchand AI, Goldfarb S, Levvey BJ, Meiser B, Rossano JW, Yusen RD, Stehlik J (2015) The Registry of the International Society for Heart

and Lung Transplantation: thirty-second official adult heart transplantation report–2015; focus theme: early graft failure. J Heart Lung Transplant 34:1244–1254
Lutz J, Thurmel K, Heemann U (2010) Anti-inflammatory treatment strategies for ischemia/reperfusion injury in transplantation. J Inflamm (Lond) 7:27
Ma CS, Avery DT, Chan A, Batten M, Bustamante J, Boisson-Dupuis S, Arkwright PD, Kreins AY, Averbuch D, Engelhard D, Magdorf K, Kilic SS, Minegishi Y, Nonoyama S, French MA, Choo S, Smart JM, Peake J, Wong M et al (2012) Functional STAT3 deficiency compromises the generation of human T follicular helper cells. Blood 119:3997–4008
Maejima Y, Sadoshima J (2014) SUMOylation: a novel protein quality control modifier in the heart. Circ Res 115:686–689
Martin MD, Badovinac VP (2018) Defining memory CD8 T cell. Front Immunol 9:2692
Masopust D, Vezys V, Marzo AL, Lefrancois L (2001) Preferential localization of effector memory cells in nonlymphoid tissue. Science 291:2413–2417
Massart A, Pallier A, Pascual J, Viklicky O, Budde K, Spasovski G, Klinger M, Sever MS, Sorensen SS, Hadaya K, Oberbauer R, Dudley C, De Fijter JW, Yussim A, Hazzan M, Wekerle T, Berglund D, De Biase C, Perez-Saez MJ et al (2016) The DESCARTES-Nantes survey of kidney transplant recipients displaying clinical operational tolerance identifies 35 new tolerant patients and 34 almost tolerant patients. Nephrol Dial Transplant 31:1002–1013
May MJ, Ghosh S (1997) Rel/NF-kappa B and I kappa B proteins: an overview. Semin Cancer Biol 8:63–73
Mcguire HM, Walters S, Vogelzang A, Lee CM, Webster KE, Sprent J, Christ D, Grey S, King C (2011) Interleukin-21 is critically required in autoimmune and allogeneic responses to islet tissue in murine models. Diabetes 60:867–875
Mckenzie IF, Morgan GM, Sandrin MS, Michaelides MM, Melvold RW, Kohn HI (1979) B6.C--H-2bm12. A new H-2 mutation in the I region in the mouse. J Exp Med 150:1323–1338
Mclaren AJ, Fuggle SV, Welsh KI, Gray DW, Morris PJ (2000) Chronic allograft failure in human renal transplantation: a multivariate risk factor analysis. Ann Surg 232:98–103
Mehra MR (2006) Contemporary concepts in prevention and treatment of cardiac allograft vasculopathy. Am J Transplant 6:1248–1256
Mehra MR, Crespo-Leiro MG, Dipchand A, Ensminger SM, Hiemann NE, Kobashigawa JA, Madsen J, Parameshwar J, Starling RC, Uber PA (2010) International Society for Heart and Lung Transplantation working formulation of a standardized nomenclature for cardiac allograft vasculopathy-2010. J Heart Lung Transplant 29:717–727
Meier-Kriesche HU, Ojo AO, Hanson JA, Cibrik DM, Punch JD, Leichtman AB, Kaplan B (2000) Increased impact of acute rejection on chronic allograft failure in recent era. Transplantation 70:1098–1100
Melk A, Henne T, Kollmar T, Strehlau J, Latta K, Offner G, Jhangri GS, Ehrich JH, Von Schnakenburg C (2003) Cytokine single nucleotide polymorphisms and intrarenal gene expression in chronic allograft nephropathy in children. Kidney Int 64(1):314–320. https://doi.org/10.1046/j.1523-1755.2003.00050.x. PubMed PMID: 12787424
Mengel M, Adam BA (2024) Emerging phenotypes in kidney transplant rejection. Curr Opin Organ Transplant 29:97–103
Miller CL, Madsen JC (2022) Targeting IL-6 to prevent cardiac allograft rejection. Am J Transplant 22(Suppl 4):12–17
Mitchell RN, Libby P (2007) Vascular remodeling in transplant vasculopathy. Circ Res 100:967–978
Mohd Jaya FN, Garcia SG, Borras FE, Chan GCF, Franquesa M (2019) Paradoxical role of Breg-inducing cytokines in autoimmune diseases. J Transl Autoimmun 2:100011
Mynbaev OA, Zhanabayev NS, Idrissov KS, Botabayeva RY, Baimagambetov AK (2023) Comment on Eva et al. laparoscopic surgery: Ann Med Surg (Lond). 2022;81:104562. Ann Med Surg (Lond) 85:1310–1311
Nadeau KC, Azuma H, Tilney NL (1995) Sequential cytokine dynamics in chronic rejection of rat renal allografts: roles for cytokines RANTES and MCP-1. Proc Natl Acad Sci USA 92:8729–8733

Naesens M, Roufosse C, Haas M, Lefaucheur C, Mannon RB, Adam BA, Aubert O, Bohmig GA, Callemeyn J, Clahsen-Van Groningen M, Cornell LD, Demetris AJ, Drachenberg CB, Einecke G, Fogo AB, Gibson IW, Halloran P, Hidalgo LG, Horsfield C et al (2024) The Banff 2022 Kidney Meeting Report: reappraisal of microvascular inflammation and the role of biopsy-based transcript diagnostics. Am J Transplant 24:338–349

Namen AE, Lupton S, Hjerrild K, Wignall J, Mochizuki DY, Schmierer A, Mosley B, March CJ, Urdal D, Gillis S (1988) Stimulation of B-cell progenitors by cloned murine interleukin-7. Nature 333:571–573

Nankivell BJ, Borrows RJ, Fung CL, O'Connell PJ, Allen RD, Chapman JR (2003) The natural history of chronic allograft nephropathy. N Engl J Med 349:2326–2333

Newell KA, Asare A, Kirk AD, Gisler TD, Bourcier K, Suthanthiran M, Burlingham WJ, Marks WH, Sanz I, Lechler RI, Hernandez-Fuentes MP, Turka LA, Seyfert-Margolis VL, Immune Tolerance Network, S. T. S. G (2010) Identification of a B cell signature associated with renal transplant tolerance in humans. J Clin Invest 120:1836–1847

Nickerson PW (2022) Rationale for the IMAGINE study for chronic active antibody-mediated rejection (caAMR) in kidney transplantation. Am J Transplant 22(Suppl 4):38–44. https://doi.org/10.1111/ajt.17210. PubMed PMID: 36453707

Nicosia M, Fairchild RL, Valujskikh A (2020) Memory T cells in transplantation: old challenges define new directions. Transplantation 104:2024–2034

Niedbala W, Wei XQ, Cai B, Hueber AJ, Leung BP, Mcinnes IB, Liew FY (2007) IL-35 is a novel cytokine with therapeutic effects against collagen-induced arthritis through the expansion of regulatory T cells and suppression of Th17 cells. Eur J Immunol 37:3021–3029

Nurieva R, Yang XXO, Martinez G, Zhang YL, Panopoulos AD, Ma L, Schluns K, Tian Q, Watowich SS, Jetten AM, Dong C (2007) Essential autocrine regulation by IL-21 in the generation of inflammatory T cells. Nature 448:480–4U8

Park SK, Kim SK, Kang SW, Kim BK, Chung BK, Park MS (2016) Association between tumor necrosis factor-alpha (TNF-α) polymorphism (−308, G/a) and acute rejection of solid organ allograft: a meta-analysis. Int J Clin Exp Med 9(9):17060–17068

Pedersen BK, Steensberg A, Schjerling P (2001) Muscle-derived interleukin-6: possible biological effects. J Physiol 536:329–337

Pethig K, Heublein B, Kutschka I, Haverich A (2000) Systemic inflammatory response in cardiac allograft vasculopathy: high-sensitive C-reactive protein is associated with progressive luminal obstruction. Circulation 102:III233–III236

Pflanz S, Timans JC, Cheung J, Rosales R, Kanzler H, Gilbert J, Hibbert L, Churakova T, Travis M, Vaisberg E, Blumenschein WM, Mattson JD, Wagner JL, To, W, Zurawski S, Mcclanahan TK, Gorman DM, Bazan JF, De Waal Malefyt R et al (2002) IL-27, a heterodimeric cytokine composed of EBI3 and p28 protein, induces proliferation of naive CD4+ T cells. Immunity 16:779–790

Pober JS, Sessa WC (2007) Evolving functions of endothelial cells in inflammation. Nat Rev Immunol 7:803–815

Pober JS, Chih S, Kobashigawa J, Madsen JC, Tellides G (2021) Cardiac allograft vasculopathy: current review and future research directions. Cardiovasc Res 117:2624–2638

Poggio ED, Augustine JJ, Arrigain S, Brennan DC, Schold JD (2021) Long-term kidney transplant graft survival-making progress when most needed. Am J Transplant 21:2824–2832

Qin Z, Wang R, Hou P, Zhang Y, Yuan Q, Wang Y, Yang Y, Xu T (2024) TCR signaling induces STAT3 phosphorylation to promote TH17 cell differentiation. J Exp Med 221

Raeber ME, Zurbuchen Y, Impellizzieri D, Boyman O (2018) The role of cytokines in T-cell memory in health and disease. Immunol Rev 283:176–193

Rauch MC, San Martin A, Ojeda D, Quezada C, Salas M, Carcamo JG, Yanez AJ, Slebe JC, Claude A (2009) Tacrolimus causes a blockage of protein secretion which reinforces its immunosuppressive activity and also explains some of its toxic side-effects. Transpl Immunol 22:72–81

Regateiro FS, Howie D, Cobbold SP, Waldmann H (2011) TGF-beta in transplantation tolerance. Curr Opin Immunol 23:660–669

Reinhardt RL, Khoruts A, Merica R, Zell T, Jenkins MK (2001) Visualizing the generation of memory CD4 T cells in the whole body. Nature 410:101–105

Rivino L, Messi M, Jarrossay D, Lanzavecchia A, Sallusto F, Geginat J (2004) Chemokine receptor expression identifies Pre-T helper (Th)1, Pre-Th2, and nonpolarized cells among human CD4+ central memory T cells. J Exp Med 200:725–735

Rodriguez G, Martinez GS, Negrete OD, Sun S, Guo W, Xie Y, Li L, Xiao C, Ross JA, Kirken RA (2023) JAK3 Y841 autophosphorylation is critical for STAT5B activation, kinase domain stability and dimer formation. Int J Mol Sci 24(15):11928

Roufosse C, Simmonds N, Clahsen-Van Groningen M, Haas M, Henriksen KJ, Horsfield C, Loupy A, Mengel M, Perkowska-Ptasinska A, Rabant M, Racusen LC, Solez K, Becker JU (2018) A 2018 reference guide to the Banff classification of renal allograft pathology. Transplantation 102:1795–1814

Rozman P, Svajger U (2018) The tolerogenic role of IFN-gamma. Cytokine Growth Factor Rev 41:40–53. https://doi.org/10.1016/j.cytogfr.2018.04.001. Epub 20180407. PubMed PMID: 29655565

Sajiir H, Ramm GA, Macdonald GA, McGuckin MA, Prins JB, Hasnain SZ (2025) Harnessing IL-22 for metabolic health: promise and pitfalls. Trends Mol Med 31(6):574–584. https://doi.org/10.1016/j.molmed.2024.10.016. Epub 20241121. PubMed PMID: 39578121

Sallusto F, Lenig D, Forster R, Lipp M, Lanzavecchia A (1999) Two subsets of memory T lymphocytes with distinct homing potentials and effector functions. Nature 401:708–712

Salomon RN, Hughes CC, Schoen FJ, Payne DD, Pober JS, Libby P (1991) Human coronary transplantation-associated arteriosclerosis. Evidence for a chronic immune reaction to activated graft endothelial cells. Am J Pathol 138:791–798

Sawant DV, Hamilton K, Vignali DA (2015) Interleukin-35: expanding its job profile. J Interf Cytokine Res 35(7):499–512. https://doi.org/10.1089/jir.2015.0015. Epub 20150428. PubMed PMID: 25919641; PMCID: PMC4507123

Sawitzki B, Harden PN, Reinke P, Moreau A, Hutchinson JA, Game DS, Tang Q, Guinan EC, Battaglia M, Burlingham WJ, Roberts ISD, Streitz M, Josien R, Boger CA, Scotta C, Markmann JF, Hester JL, Juerchott K, Braudeau C et al (2020) Regulatory cell therapy in kidney transplantation (The ONE Study): a harmonised design and analysis of seven non-randomised, single-arm, phase 1/2A trials. Lancet 395:1627–1639

Schenkel JM, Masopust D (2014) Tissue-resident memory T cells. Immunity 41:886–897

Schmitt E, Van Brandwijk R, Van Snick J, Siebold B, Rude E (1989) TCGF III/P40 is produced by naive murine CD4+ T cells but is not a general T cell growth factor. Eur J Immunol 19:2167–2170

Schraml BU, Hildner K, Ise W, Lee WL, Smith WA, Solomon B, Sahota G, Sim J, Mukasa R, Cemerski S, Hatton RD, Stormo GD, Weaver CT, Russell JH, Murphy TL, Murphy KM (2009) The AP-1 transcription factor Batf controls T(H)17 differentiation. Nature 460:405–409

Sharma VK, Bologa RM, Xu GP, Li B, Mouradian J, Wang J, Serur D, Rao V, Suthanthiran M (1996) Intragraft TGF-beta 1 mRNA: a correlate of interstitial fibrosis and chronic allograft nephropathy. Kidney Int 49(5):1297–1303. https://doi.org/10.1038/ki.1996.185. PubMed PMID: 8731094

Shen P, Roch T, Lampropoulou V, O'Connor RA, Stervbo U, Hilgenberg E, Ries S, Dang VD, Jaimes Y, Daridon C, Li R, Jouneau L, Boudinot P, Wilantri S, Sakwa I, Miyazaki Y, Leech MD, Mcpherson RC, Wirtz S et al (2014) IL-35-producing B cells are critical regulators of immunity during autoimmune and infectious diseases. Nature 507:366–370

Shi X, Que R, Liu B, Li M, Cai J, Shou D, Wen L, Liu D, Chen L, Liang T, Gong W (2016) Role of IL-21 signaling pathway in transplant-related biology. Transplant Rev (Orlando) 30:27–30

Shilling RA, Wilkes DS (2011) Role of Th17 cells and IL-17 in lung transplant rejection. Semin Immunopathol 33(2):129–134. https://doi.org/10.1007/s00281-011-0257-9. Epub 20110201. PubMed PMID: 21279808; PMCID:PMC3945469

Shimizu T, Hirai T, Unagam K, Takagi T, Ishida H (2025) Banff 2022 updated classification of renal allograft pathology. Nephron:1

Shiu KY, McLaughlin L, Rebollo-Mesa I, Zhao J, Burton H, Douthwaite H, Wilkinson H, Semik V, Dodd PC, Brookes P, Lechler RI, Hernandez-Fuentes MP, Kemper C, Dorling A (2017) Graft dysfunction in chronic antibody-mediated rejection correlates with B-cell-dependent indirect antidonor alloresponses and autocrine regulation of interferongamma production by Th1 cells. Kidney Int 91(2):477–492. https://doi.org/10.1016/j.kint.2016.10.009. Epub 20161215. PubMed PMID: 27988211; PMCID: PMC5258815

Shuai K (2006) Regulation of cytokine signaling pathways by PIAS proteins. Cell Res 16:196–202

Shuai K, Liu B (2005) Regulation of gene-activation pathways by PIAS proteins in the immune system. Nat Rev Immunol 5:593–605

Souffrant A, Wilson A, Hansen B, Tomofuji K, Otsuka R, Benichou G, Spitzer T, Kawai T (2025) Cytokines in transplantation tolerance. Best Pract Res Clin Haematol 38:101627

Soulillou JP, Giral M, Brouard S (2013) Operational tolerance in kidney transplantation-improved terminology may enable more precise investigation. Transplantation 96:e36–e38

Steensberg A, Van Hall G, Osada T, Sacchetti M, Saltin B, Klarlund Pedersen B (2000) Production of interleukin-6 in contracting human skeletal muscles can account for the exercise-induced increase in plasma interleukin-6. J Physiol 529(Pt 1):237–242

Stepkowski SM, Bitter-Suermann H, Duncan WR (1987) Induction of transplantation tolerance in rats by spleen allografts. II. Evidence that W3/25+ T suppressor/inducer and OX8+ T suppressor/effector cells are required to mediate specific unresponsiveness. Transplantation 44:443–448

Stepkowski SM, Erwin-Cohen RA, Behbod F, Wang ME, Qu X, Tejpal N, Nagy ZS, Kahan BD, Kirken RA (2002) Selective inhibitor of Janus tyrosine kinase 3, PNU156804, prolongs allograft survival and acts synergistically with cyclosporine but additively with rapamycin. Blood 99:680–689

Stepkowski SM, Kao J, Wang ME, Tejpal N, Podder H, Furian L, Dimmock J, Jha A, Das U, Kahan BD, Kirken RA (2005) The Mannich base NC1153 promotes long-term allograft survival and spares the recipient from multiple toxicities. J Immunol 175:4236–4246

Stoica SC, Cafferty F, Pauriah M, Taylor CJ, Sharples LD, Wallwork J, Large SR, Parameshwar J (2006) The cumulative effect of acute rejection on development of cardiac allograft vasculopathy. J Heart Lung Transplant 25:420–425

Sullivan JA, Tomita Y, Jankowska-Gan E, Lema DA, Arvedson MP, Nair A, Bracamonte-Baran W, Zhou Y, Meyer KK, Zhong W, Sawant DV, Szymczak-Workman AL, Zhang Q, Workman CJ, Hong S, Vignali DAA, Burlingham WJ (2020) Treg-cell-derived IL-35-coated extracellular vesicles promote infectious tolerance. Cell Rep 30:1039–1051.e5

Sutherland AP, Joller N, Michaud M, Liu SM, Kuchroo VK, Grusby MJ (2013) IL-21 promotes CD8+ CTL activity via the transcription factor T-bet. J Immunol 190:3977–3984

Swain SL, Weinberg AD, English M, Huston G (1990) IL-4 directs the development of Th2-like helper effectors. J Immunol 145:3796–3806

Tamargo CL, Kant S (2023) Pathophysiology of rejection in kidney transplantation. J Clin Med 12(12):4130

Tan JT, Ernst B, Kieper WC, Leroy E, Sprent J, Surh CD (2002) Interleukin (IL)-15 and IL-7 jointly regulate homeostatic proliferation of memory phenotype CD8+ cells but are not required for memory phenotype CD4+ cells. J Exp Med 195:1523–1532

Tay SS, Plain KM, Bishop GA (2009) Role of IL-4 and Th2 responses in allograft rejection and tolerance. Curr Opin Organ Transplant 14:16–22

Tejani A, Cortes L, Stablein D (1996) Clinical correlates of chronic rejection in pediatric renal transplantation. A report of the North American Pediatric Renal Transplant Cooperative Study. Transplantation 61:1054–1058

Thaunat O, Field AC, Dai J, Louedec L, Patey N, Bloch MF, Mandet C, Belair MF, Bruneval P, Meilhac O, Bellon B, Joly E, Michel JB, Nicoletti A (2005) Lymphoid neogenesis in chronic rejection: evidence for a local humoral alloimmune response. Proc Natl Acad Sci USA 102:14723–14728

Thomson AW, Bonham CA, Zeevi A (1995) Mode of action of tacrolimus (FK506): molecular and cellular mechanisms. Ther Drug Monit 17:584–591

Tian Y, Zajac AJ (2016) IL-21 and T cell differentiation: consider the context. Trends Immunol 37:557–568

Tough DF, Sun S, Zhang X, Sprent J (2000) Stimulation of memory T cells by cytokines. Vaccine 18:1642–1648

Tse GH, Johnston CJ, Kluth D, Gray M, Gray D, Hughes J, Marson LP (2015) Intrarenal B cell cytokines promote transplant fibrosis and tubular atrophy. Am J Transplant 15:3067–3080

Tyler JR, Robertson H, Booth TA, Burt AD, Kirby JA (2006) Chronic allograft nephropathy: intraepithelial signals generated by transforming growth factor-beta and bone morphogenetic protein-7. Am J Transplant 6:1367–1376

Uyttenhove C, Simpson RJ, Van Snick J (1988) Functional and structural characterization of P40, a mouse glycoprotein with T-cell growth factor activity. Proc Natl Acad Sci USA 85:6934–6938

Van Dyken SJ, Locksley RM (2013) Interleukin-4- and interleukin-13-mediated alternatively activated macrophages: roles in homeostasis and disease. Annu Rev Immunol 31:317–343

Vassalli G, Gallino A, Weis M, Von Scheidt W, Kappenberger L, Von Segesser LK, Goy JJ, Working Group Microcirculation of the Eurpean Society of, C (2003) Alloimmunity and nonimmunologic risk factors in cardiac allograft vasculopathy. Eur Heart J 24:1180–1188

Velleca A, Shullo MA, Dhital K, Azeka E, Colvin M, Depasquale E, Farrero M, Garcia-Guereta L, Jamero G, Khush K, Lavee J, Pouch S, Patel J, Michaud CJ, Shullo MA, Schubert S, Angelini A, Carlos L, Mirabet S, Patel J, Pham M, Urschel S, Kim KH, Miyamoto S, Chih S, Daly K, Grossi P, Jennings DL, Kim IC, Lim HS, Miller T, Potena L, Velleca A, Eisen H, Bellumkonda L, Danziger-Isakov L, Dobbels F, Harkess M, Kim D, Lyster H, Peled Y, Reinhardt Z (2023) The International Society for Heart and Lung Transplantation (ISHLT) guidelines for the care of heart transplant recipients. J Heart Lung Transplant 42:e1–e141

Vincenti F, Kirkman R, Light S, Bumgardner G, Pescovitz M, Halloran P, Neylan J, Wilkinson A, Ekberg H, Gaston R, Backman L, Burdick J (1998) Interleukin-2-receptor blockade with daclizumab to prevent acute rejection in renal transplantation. Daclizumab Triple Therapy Study Group. N Engl J Med 338:161–165

Vitetta ES, Ohara J, Myers CD, Layton JE, Krammer PH, Paul WE (1985) Serological, biochemical, and functional identity of B cell-stimulatory factor 1 and B cell differentiation factor for IgG1. J Exp Med 162:1726–1731

Waldmann TA (1989) The multi-subunit interleukin-2 receptor. Annu Rev Biochem 58:875–911

Waldmann H, Hilbrands R, Howie D, Cobbold S (2014) Harnessing FOXP3+ regulatory T cells for transplantation tolerance. J Clin Invest 124:1439–1445

Walsh PT, Strom TB, Turka LA (2004) Routes to transplant tolerance versus rejection; the role of cytokines. Immunity 20:121–131

Wang Y, Dai H, Liu Z, Cheng X, Tellides G, Dai Z (2006) Neutralizing IL-7 promotes long-term allograft survival induced by CD40/CD40L costimulatory blockade. Am J Transplant 6(12):2851–2860. https://doi.org/10.1111/j.1600-6143.2006.01550.x. PubMed PMID: 17062001

Wang J, Zhao X, Wan YY (2023) Intricacies of TGF-beta signaling in Treg and Th17 cell biology. Cell Mol Immunol 20(9):1002–1022. https://doi.org/10.1038/s41423-023-01036-7. Epub 20230523. PubMed PMID: 37217798; PMCID:PMC10468540

Wang X, Hong Y, Zou J, Zhu B, Jiang C, Lu L, Tian J, Yang J, Rui K (2025) The role of BATF in immune cell differentiation and autoimmune diseases. Biomark Res 13:22

Wei L, Laurence A, Elias KM, O'Shea JJ (2007) IL-21 is produced by Th17 cells and drives IL-17 production in a STAT3-dependent manner. J Biol Chem 282:34605–34610

Wilhelm MJ (2015) Long-term outcome following heart transplantation: current perspective. J Thorac Dis 7:549–551

Wollin M, Abele S, Bruns H, Weyand M, Kalden JR, Ensminger SM, Spriewald BM (2009) Inhibition of TNF-alpha reduces transplant arteriosclerosis in a murine aortic transplant model. Transpl Int 22:342–349

Xin G, Schauder DM, Lainez B, Weinstein JS, Dai Z, Chen Y, Esplugues E, Wen R, Wang D, Parish IA, Zajac AJ, Craft J, Cui W (2015) A critical role of IL-21-induced BATF in sustaining CD8-T-cell-mediated chronic viral control. Cell Rep 13:1118–1124

Xin PL, Jie LF, Cheng Q, Bin DY, Dan CW (2021) Pathogenesis and function of Interleukin-35 in rheumatoid arthritis. Front Pharmacol 12:655114

Xu X, Huang H, Cai M, Qian Y, Li Z, Bai H, Han Y, Xiao L, Zhou W, Wang X, Shi B (2013) Combination of IL-1 receptor antagonist, IL-20 and CD40 ligand for the prediction of acute cellular renal allograft rejection. J Clin Immunol 33(1):280–287. https://doi.org/10.1007/s10875-012-9777-x. Epub 20120905. PubMed PMID: 22948742

Xu H, Zhou R, Chen Z (2023) Tissue-resident memory T cell: ontogenetic cellular mechanism and clinical translation. Clin Exp Immunol 214:249–259

Yamani MH, Yousufuddin M, Starling RC, Tuzcu M, Ratliff NB, Cook DJ, Abdo A, Crowe T, Hobbs R, Rincon G, Bott-Silverman C, Mccarthy PM, Young JB (2004) Does acute cellular rejection correlate with cardiac allograft vasculopathy? J Heart Lung Transplant 23:272–276

Yang XO, Panopoulos AD, Nurieva R, Chang SH, Wang D, Watowich SS, Dong C (2007) STAT3 regulates cytokine-mediated generation of inflammatory helper T cells. J Biol Chem 282:9358–9363

Yapici U, Kers J, Bemelman FJ, Roelofs JJ, Groothoff JW, Van Der Loos CM, Van Donselaar-Van Der Pant KA, Idu MM, Claessen N, Ten Berge IJ, Florquin S (2011) Interleukin-17 positive cells accumulate in renal allografts during acute rejection and are independent predictors of worse graft outcome. Transpl Int 24:1008–1017

Ye P, Rodriguez FH, Kanaly S, Stocking KL, Schurr J, Schwarzenberger P, Oliver P, Huang W, Zhang P, Zhang J, Shellito JE, Bagby GJ, Nelson S, Charrier K, Peschon JJ, Kolls JK (2001) Requirement of interleukin 17 receptor signaling for lung CXC chemokine and granulocyte colony-stimulating factor expression, neutrophil recruitment, and host defense. J Exp Med 194:519–527

Ye C, Yano H, Workman CJ, Vignali DAA (2021) Interleukin-35: structure, function and its impact on immune-related diseases. J Interf Cytokine Res 41:391–406

Young DA, Hegen M, Ma HL, Whitters MJ, Albert LM, Lowe L, Senices M, Wu PW, Sibley B, Leathurby Y, Brown TP, Nickerson-Nutter C, Keith JC Jr, Collins M (2007) Blockade of the interleukin-21/interleukin-21 receptor pathway ameliorates disease in animal models of rheumatoid arthritis. Arthritis Rheum 56:1152–1163

Yuan X, Paez-Cortez J, Schmitt-Knosalla I, D'addio F, Mfarrej B, Donnarumma M, Habicht A, Clarkson MR, Iacomini J, Glimcher LH, Sayegh MH, Ansari MJ (2008) A novel role of CD4 Th17 cells in mediating cardiac allograft rejection and vasculopathy. J Exp Med 205:3133–3144

Zeng X, Pan Y, Lin J, Zheng Z, Wu H, Wang Y, Wu Y, Shen Y, Chen Y, Zhao Y, Xia Q, Duan Y, He K (2025) IL-21RTargeted Nano-immunosuppressant prevents acute rejection in allogeneic transplantation by blocking maturation of T follicular helper cells. Acta Biomater 199:346–360. https://doi.org/10.1016/j.actbio.2025.05.012. Epub 20250506. PubMed PMID: 40339970

Chapter 3
Exploring Chronic Rejection in Organ Transplantation Through Computational Modeling

Stefano Casarin and Elisa Serafini

Abstract Chronic rejection remains a significant challenge in solid organ transplantation, contributing to graft dysfunction and eventual failure despite advances in immunosuppressive therapies. Computational modeling has emerged as a powerful tool for understanding chronic rejection mechanisms, enhancing diagnostic precision, and identifying novel therapeutic targets. This chapter explores various computational approaches, including artificial intelligence, machine learning, ordinary differential equations, partial differential equations, agent-based models, and gene network analysis applied to solid organ transplantation: kidney, liver, heart, and lung. While computational models offer numerous advantages, including cost-effectiveness and the ability to integrate multi-omics data, challenges remain in terms of data quality, standardization, and clinical validation. Bridging these gaps will require comprehensive longitudinal studies and the development of hybrid models that combine diverse computational techniques. Emerging technologies such as single-cell transcriptomics and spatial genomics hold promise for enhancing predictive accuracy and understanding cellular heterogeneity. As computational methods evolve, their integration with experimental research will be essential for developing precision medicine strategies to improve long-term graft survival.

S. Casarin (✉)
Center for Precision Surgery, Houston Methodist Research Institute, Houston, TX, USA

LaSIE, UMR 7356 CNRS, La Rochelle Université, La Rochelle, France

Department of Surgery, Houston Methodist Hospital, Houston, TX, USA
e-mail: scasarin@houstonmethodist.org

E. Serafini
Center for Precision Surgery, Houston Methodist Research Institute, Houston, TX, USA

LaSIE, UMR 7356 CNRS, La Rochelle Université, La Rochelle, France

M. Kloc, R. M. Ghobrial (eds.), *Cellular and Molecular Mechanisms of Chronic Rejection in Organ Transplantation*, Results and Problems in Cell Differentiation 77, https://doi.org/10.1007/978-3-032-07686-1_3

3.1 Introduction

Organ transplantation offers a life-saving option for patients suffering from end-stage organ failure (Black et al. 2018). The number of organ transplants in the United States has steadily risen in the last decades, with over 48,000 transplants performed in 2024 alone. Kidney transplants remain the most common (~27,800 transplanted in 2024 (Lentine et al. 2023)), followed by liver (~11,500 (Kwong et al. 2024)), heart (~4500 (Colvin et al. 2024)), and lung (3400 (Perch et al. 2022)). However, the organ demand continues to overcome availability, with a new person being added to the transplant waiting list every 10 minutes. In 2024, over 103,000 individuals were on the national transplant waiting list, while only about 7000 living organ donations were available (according to Health Resources & Service Administration—Organ Donation Statistics, https://www.organdonor.gov/learn/organ-donation-statistics). Waiting times for transplants vary by organ type, with the average being 3–5 years for kidney (Stewart et al. 2023), 30 days to 5 years for liver (Trieu et al. 2018), and 4 months for heart and lung (Goldstein et al. 2016). The shortage of available organs underscores the importance of preventing organ rejection, as re-entering the waiting list after transplant failure is extremely challenging. Consequently, preventing organ rejection is crucial to avoid further strain on an already overburdened system.

Organ rejection occurs when the recipient's immune system recognizes the transplanted organ as a foreign body and initiates an immune response against it (Wood and Goto 2012). There are three main types of rejection: hyperacute, acute, and chronic. *Hyperacute* rejection happens within minutes to hours after transplantation and is caused by pre-existing antibodies in the recipient's blood that recognize the donor organ as foreign. It leads to immediate failure of the transplanted organ and its incidence; but today, it is very rare, thanks to improved matching and cross-matching techniques. *Acute* rejection occurs within the first few months and is characterized by the activation of T-cells and other immune cells attacking the transplanted organ. It is more common than hyperacute rejection and can typically be managed with immunosuppressive therapies. *Chronic* rejection is a long-term process that develops over the years, leading to progressive damage to the transplanted organ. It involves both immune and nonimmune factors, gradually causing fibrosis and narrowing blood vessels that sustain the graft. Chronic rejection arises from an interplay between early graft injury and sustained immune activation. It typically begins with an initial graft injury, such as ischemia-reperfusion injury, infections, or fluctuations in immunosuppression, which provoke an initial inflammatory response, triggering immune system activation. Over time, T cells recognize donor antigens and foster B cell maturation and production of donor-specific antibodies. These antibodies target the graft's vascular endothelium, activating complement and promoting chronic inflammation and fibrosis. The culmination of these processes is a severe graft vasculopathy, characterized by arterial narrowing and tissue remodeling, which ultimately compromises graft function (Moreau et al. 2013; Actor 2019).

Chronic rejection often progresses asymptomatically over a prolonged period, with symptoms only becoming evident once significant organ damage has already occurred (Kim and Brennan 2021). This delayed onset makes early detection particularly challenging, and, as a result, by the time chronic rejection is diagnosed, the organ may have already sustained considerable injury, limiting the effectiveness of treatment. Current therapeutic strategies are very often ineffective in fighting the progressive damage caused by chronic rejection, leading to graft dysfunction and, in many cases, necessitating retransplantation.

Current research on antichronic rejection therapies has predominantly relied on in vivo animal models. While these models have provided valuable insights into the immune mechanisms driving the rejection, they come with several intrinsic limitations (Vandamme 2014). In vivo models, particularly those using rodents, often fail to fully replicate the complexity of human immune responses, as they do not accurately mimic the human organ microenvironment, the genetic variability of transplant recipients, or the long-term dynamics of organ rejection. Animal models do not capture the full spectrum of chronic rejection's slow, progressive nature, which makes it challenging to study the late stages of rejection or assess the efficacy of treatments over extended periods. Additionally, animal models are resource-demanding and come with several ethical concerns. These limitations show the need for alternative models and methods, such as computational simulations, to better understand and address chronic rejection in human organ transplants (Colquitt et al. 2011). Flanking these studies with computational modeling can offer an integrated approach, addressing many of the roadblocks of traditional animal models. A main advantage of computational modeling is its ability to simulate the long-term dynamics of chronic rejection, which is instead challenging for in vivo models due to the limited lifespan of animals. Additionally, computational models can account for the genetic diversity of transplant recipients, which is often a challenge in animal studies that typically use genetically homogenous strains. Furthermore, computational models allow the integration of data from various sources, such as clinical observations, genomic information, and in vitro experiments, creating a more comprehensive view of chronic rejection. Finally, computational models can reduce resource demand and ethical concerns typically associated with in vivo models.

Given the potential of computational modeling to complement in vivo research, this chapter aims to provide a broad overview of computational approaches used to study chronic rejection in solid organ transplantation, with a focus, where applicable, on the four major transplanted organs: heart, kidney, liver, and lung. It will offer a representative sample of different modeling strategies developed to simulate the complex immune interactions and long-term pathological processes underlying chronic rejection.

Specifically, Sect. 3.2 will explore the use of artificial intelligence and mechanistic models of the biological mechanisms driving chronic rejection. Section 3.3 will illustrate gene expression analysis and network-based approaches to gain knowledge on the molecular basis of chronic rejection processes. Section 3.4 will focus on the identification and exploration of computationally derived biomarkers that can support the early detection and monitoring of chronic rejection. Finally, Sect. 3.5 will discuss the current strengths, limitations, and future perspectives of computational modeling in this context, highlighting both opportunities and challenges for advancing the field.

By reviewing existing models and their applications across these domains, this chapter highlights the value of computational simulations in deepening our understanding of chronic rejection and informing the development of more effective therapeutic interventions.

3.2 Computational Pathology Models

Computational pathology models have emerged as powerful tools for advancing the diagnosis, monitoring, and understanding of chronic rejection. These models, presented in Fig. 3.1, include artificial intelligence (AI) and machine learning (ML) techniques, mechanistic models based on ordinary/partial differential equations (ODEs/PDEs), and agent-based models (ABMs). AI and ML typically leverage digital pathology data to analyze morphological patterns and histopathological features that characterize chronic rejection across various organs, while ODEs, PDEs, and ABMs are crucial to capture the pathology multiscale dynamics.

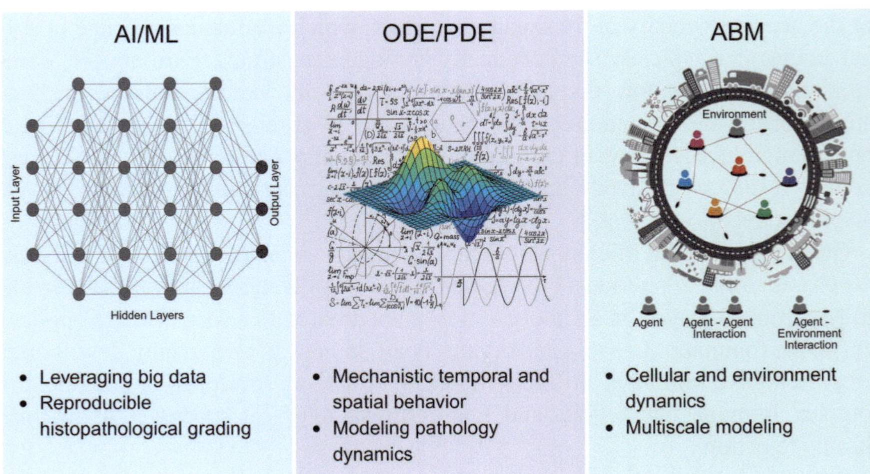

Fig. 3.1 Computational pathology modeling overview. (Created in BioRender. Casarin, S. (2025). https://BioRender.com/e6b12sw)

3.2.1 AI/ML Models: Leveraging Big Data

In the field of heart transplant (HTx), computational pathology models offer efficient analysis of complex biopsy data, particularly endomyocardial images, to assist in the histopathological grading of cardiac allograft vasculopathy (CAV)-mediated rejection, the most well-known manifestation of chronic rejection in HTx (Peyster et al. 2025). These methodologies enable pathologists to assess rejection severity with increased precision, reducing operator variability and subjectivity. Also, AI-based pipelines were developed to analyze stromal alterations in biopsies, demonstrating effective monitoring of CAV (Peyster et al. 2021). These studies highlight the power of AI models in detecting changes not easily observable through traditional approaches.

In renal pathology, AI-enhanced computational approaches have been applied to whole-slide images (WSIs) of kidney biopsies, facilitating the detection and subtyping of rejection events. A study used deep learning (DL) to automate the assessment of WSIs for kidney transplant rejection, with high precision in distinguishing borderline changes from outright tissue rejection, demonstrating the potential for algorithmic models to assist clinicians in identifying prognostic features such as fibrosis and interstitial inflammation (Fayzullin et al. 2024). Furthermore, convolutional neural networks (CNNs) have been employed to quantify tissue structures, including fibrosis extent and immune cell infiltration. These automated evaluations provide significant insights into the progression and severity of chronic kidney rejection (Fayzullin et al. 2024).

Importantly, privacy-preserving policies lately became a central topic in the development of AI models. As computational models become increasingly integrated into clinical and translational research settings, ensuring data privacy and maintaining model adaptability across institutions is becoming essential. Recent advances in privacy-preserving ML (e.g., federated learning, differential privacy) have opened new avenues for secure and collaborative modeling, particularly in high-dimensional domains such as digital pathology and multicenter omics integration, where data sensitivity is high and patient identifiers can persist even after anonymization. Importantly, research centers connected to hospital systems are facing escalating regulatory and ethical challenges in sharing patient-derived data, even when anonymized, with external collaborators. Concerns around data confidentiality, institutional compliance, and cross-jurisdictional privacy laws increasingly limit the feasibility of centralizing datasets, especially in studies involving longitudinal patient outcomes. In this context, the development of digital twin frameworks and computational models that can operate in decentralized or modular settings, without requiring raw data transfer, represents a promising path forward for multisite model validation and real-world translation. One key approach is federated learning, which enables AI models to be trained across decentralized datasets, such as biopsy scans or gene expression profiles from different transplant centers, without transferring patient data outside institutional firewalls. Instead, only model updates (not raw data) are shared and aggregated centrally, preserving patient confidentiality while benefiting from a broader data base (Hudaib et al.

2025; Zhang et al. 2024a; Teo et al. 2024). This strategy has shown promise for transplant applications, where data heterogeneity and strict regulatory barriers often limit the creation of centralized databases. Recent studies used federated learning in transplantation research. For example, a federated framework (*dGEM-disparity*) evaluates racial disparities in kidney graft failure, enabling collaborative model training across sites without sharing patient-level data and thus preserving privacy while assessing site-of-care effects on transplant outcomes (Tong et al. 2024). Additionally, the NephroCAGE consortium, a German-Canadian initiative, has applied federated learning to enhance the prediction and prevention of chronic kidney graft rejection using distributed datasets from international nephrology centers, while ensuring cross-border data confidentiality (Schapranow et al. 2023). These examples spotlight the transformative potential of federated modeling in supporting secure, multi-institutional collaboration in high-stakes clinical domains like transplantation.

Beyond privacy, continuous learning and monitoring represent another frontier for AI integration in post-transplant care. Unlike static models trained once and deployed indefinitely, continuous learning systems can adapt to new incoming data (updated histology reports, evolving biomarker panels, changes in immunosuppressive protocols, evolution of patient population demographic profile, etc.). This approach is often referred to an online learning, and it allows models to refine predictions over time, improving accuracy and resilience to data change. In the context of chronic rejection, which can evolve over months to years, continuously updated models could significantly enhance real-time risk stratification and reduce the need for repeat biopsies or invasive procedures, and they have been gaining tractions in the last few years. A review from Ramalhete et al. explores the application of AI and machine learning in kidney transplantation, emphasizing their role in enhancing donor-recipient matching, predicting complications, and personalizing post-transplant care, especially underscoring the potential of adaptive models to improve patient monitoring and clinical decision-making (Ramalhete et al. 2024). Another review article explores the use of AI and ML in organ transplantation, focusing on data-driven approaches to improve organ allocation, predict rejection risks, and support continuous, personalized post-transplant monitoring. Most interestingly, the study highlights the shift toward dynamic, real-time patient management systems (Olawade et al. 2025).

Together, privacy-preserving AI frameworks and adaptive learning models represent a tangible evolution in post-transplant care, enabling secure, real-time, and institutionally compliant clinical decision support across distributed health-care networks.

3.2.2 ODEs: Modelling Temporal Dynamics

Ordinary Differential Equations (ODEs) describe how variables change over time in a system of interest and have been crucial in modeling the temporal evolution of immune and inflammatory responses involved in chronic rejection. ODE-based models provide insights into the kinetics of immune cell proliferation, cytokine signaling, and graft damage progression (De Gaetano et al. 2012). For instance, an ODE model was developed to account for the proliferation of immune cells during the early immune reaction triggered by the transplanted graft (De Gaetano et al. 2012). This model captures the dynamic interplay between immune cell populations, such as T lymphocytes, and cytokines driving chronic inflammation, can quantify the temporal progression of immune activation, and predict outcomes such as graft fibrosis. Another ODE-based framework was used to evaluate five distinct stages of mouse HTx, spanning the contributions of innate and adaptive immunity in chronic rejection. The system tracked populations of immune cells and their interactions with the graft, offering a clearer understanding of rejection dynamics over time (Arciero et al. 2016). Additionally, ODEs have been employed to investigate vascular pathology, which often accompanies chronic rejection, including endothelial cell activation and vascular remodeling. These models provide a quantitative description of endothelial injury caused by immune activation and the subsequent release of signaling molecules, such as vascular endothelial growth factor (VEGF), which promote pathological remodeling of donor vessels (Bonin et al. 2017; Yen et al. 2011). In an ODE-based model, the proliferation and migration of vascular smooth muscle cells (VSMCs), driven by endothelial activation and immune responses during chronic rejection, was developed to replicate the signaling cascades involving growth factors like fibroblast growth factor (FGF) and transforming growth factor-beta (TGF-β), both pivotal to smooth muscle changes and extracellular matrix buildup (Bonin et al. 2017). Associated numerical simulations demonstrated the progression of luminal narrowing and suggested therapeutic strategies to mitigate smooth muscle cell (SMC) proliferation (Bonin et al. 2017). Furthermore, an ODE-based approach was employed to describe interactions between VEGF isoforms and their varying roles in vascular pathology. By encompassing VEGF receptor dynamics and tissue distribution rates, the system provided quantitative predictions regarding neointima development and specified conditions under which therapeutic VEGF inhibition might be effective (Yen et al. 2011).

The strength of ODE models is their capacity to compress multiple interdependent biological processes in a time-dependent manner. They uncover the progression of chronic rejection by providing predictions about immune responses, vascular changes, and therapeutic outcomes in a controlled and time-efficient in silico environment. Indeed, ODE models are computationally cost-effective, allowing researchers to run numerous simulations and test scenarios without the constraints faced in wet lab experiments. However, ODE models are based on the strong assumption of a stirred compartment, are intrinsically limited to temporal analyses, and may fail to capture spatial heterogeneity critical to understanding localized injury in vascularized transplanted organs. Incorporating spatial dimensions through PDEs or using ABMs would enhance the realism and predictive power of these simulations.

3.2.3 PDEs: Capturing Spatial and Temporal Phenomena

Partial differential equations (PDEs) extend the capabilities of ODEs by incorporating spatial dimensions, making them suitable for studying tissue-level processes and spatially heterogeneous immune responses in chronic rejection. PDE models are commonly used to simulate how various biological agents react and interact across the graft microenvironment (Southern et al. 2008). Examples of PDE applications in renal chronic rejection studied interstitial fibrosis and tubular atrophy. These models incorporate diffusion-reaction equations to simulate the movement of cytokines and chemokines through the graft tissue, allowing the delineation of the spatial distribution of immune mediators that influence tissue injury (De Gaetano et al. 2012). On the HTx side, a PDE approach has been used to model the movement and interactions of immune cells and soluble factors within CAV. The simulations revealed spatial patterns of graft vessel narrowing driven by localized immune cell infiltration and inflammatory signaling cascades (Schettini et al. 2023). PDEs have also been used to study the spatial aspects of allorecognition, wherein recipient T cells recognize donor antigens and instigate localized immune responses. By modeling the recruitment gradients of immune cells toward donor tissues, PDE frameworks help explain the spatial dynamics leading to rejection hotspots that exacerbate chronic damage (Ingulli 2010).

PDE models, although elegant in their formulation, are often constrained by their high computational complexity, challenges in solving high-dimensional parameter spaces, numerical stability issues, and difficulty in obtaining closed analytical solutions for their inherently complex structures. They somehow represent an intermediate solution between ODEs and ABMs, with the latter preferred to study the spatial detail of the events of interest.

3.2.4 ABMs: Simulating Complex Cellular Interactions

Agent-Based Simulation and Modeling (ABM) has emerged as a powerful method to simulate the emergent behavior of numerous interacting biological agents, such as immune and endothelial cells, and fibroblasts, within the context of chronic transplant rejection (An 2015). As a bottom-up approach, ABM simulates each agent's behavior based on rules that reflect biological dynamics. This allows for the study of collective behaviors arising from the agents' individual actions and their interactions with the surrounding environment (Tracy et al. 2018). Additionally, the utility of ABM lies in its ability to incorporate stochastic events and simulate individual heterogeneity often seen in transplant recipients. A notable ABM application involved simulating the progression of CAV (Serafini et al. 2023) using a bidimensional grid that replicates the arterial wall's response to two primary stimuli: inflammation and disrupted blood flow (low WSS). The model effectively captured key pathological features of CAV, including macrophage infiltration, chemoattractant release, SMC migration, and excessive extracellular matrix (ECM) production that led to intimal hyperplasia and luminal narrowing. Simulations of varying degrees of CAV (i.e., mild, moderate, and severe), under different levels of inflammatory and hemodynamic stimuli, along with potential artery remodeling scenarios post-HTx, demonstrated the ABM's flexibility in replicating diverse conditions. Sensitivity analysis of this model revealed that inflammation, rather than hemodynamic factors, predominantly drives the pathophysiology of CAV, aligning with clinical observations that chronic rejection is primarily mediated by immune mechanisms. This paper also offers a strong example of how an integrative approach (combining ODEs for molecular dynamics with ABMs for cellular behavior in a hybrid model) offers a more comprehensive view of chronic rejection, allowing detail of the multiscale nature of events within an ABM framework. Beyond CAV, new ABM-derived models of kidney allograft transplants have incorporated key pathological features such as interstitial fibrosis, capillary basement membrane multilayering, and glomerular injury (Baldwin 3rd et al. 2016). These simulations replicate the intricate interactions of immune cells, thereby advancing our knowledge of Antibody-Mediated Rejection (AMR) mechanisms (Baldwin 3rd et al. 2016).

In summary, ODEs, PDEs, and ABMs play pivotal roles in advancing our understanding of chronic rejection. Each approach contributes distinct but complementary insights: ODEs elucidate temporal kinetics of immune signaling; PDEs capture spatial dynamics of inflammation and tissue remodeling; and ABM brings emergent cellular behaviors to light. The integration of these methods into hybrid models holds immense potential for addressing the multiscale complexity of chronic rejection. As computational techniques continue to evolve, future research should focus on model calibration, validation against experimental and clinical data, and extending these frameworks to inform personalized medicine.

3.3 Gene Expression and Network Analysis

Gene expression profiling and network analysis have emerged as powerful approaches for elucidating the molecular foundations of chronic rejection, offering insights into the complex interplay of biological pathways involved. Via computational and molecular tools, as illustrated in Fig. 3.2, researchers have identified common molecular signatures, such as the common rejection module (CRM) (Sacreas et al. 2018), and have applied network analysis frameworks to decipher the biological networks underpinning chronic rejection.

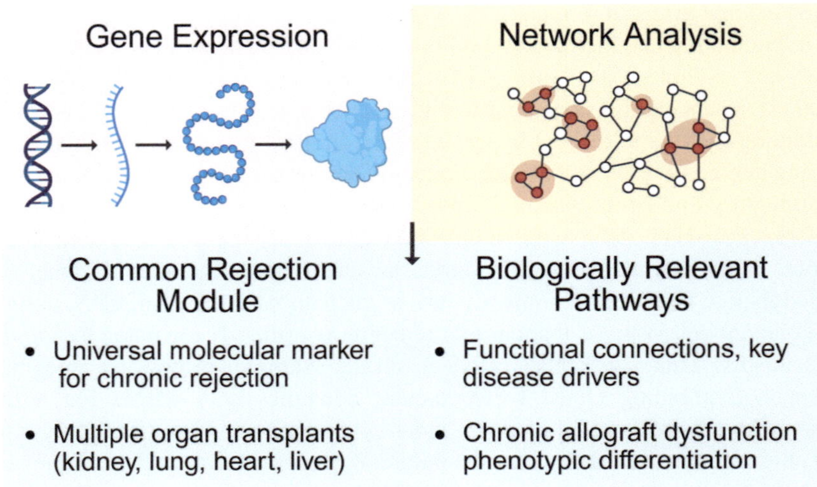

Fig. 3.2 Gene expression and network analysis overview. (Created in BioRender. Casarin, S. (2025). https://BioRender.com/lqwb6mq)

3.3.1 Molecular Signature of Rejection: The Common Rejection Module

The module consists of 11 genes: CD6 Molecule (CD6), Transporter 1 (TAP1), C-X-C Motif Chemokine Ligand 10 (CXCL10), C-X-C Motif Chemokine Ligand 9 (CXCL9), Inositol Polyphosphate-5-Phosphatase D (INPP5D), Interferon Stimulated Exonuclease Gene 20 (ISG20), Lymphocyte-Specific Protein Tyrosine Kinase (LCK), Natural Killer Cell Granule Protein 7 (NKG7), Proteasome Subunit Beta Type-9 (PSMB9), Runt-Related Transcription Factor 3 (RUNX3), and Brain Abundant Membrane Attached Signal Protein 1 (BASP1). The latter are overexpressed during acute rejection, with, however, varying degrees of expression in chronic rejection (Sacreas et al. 2018). Initial identification of the common rejection module (CRM) arose from a meta-analysis integrating gene expression data from biopsy-confirmed acute rejection across multiple organ transplants, including kidney, lung, heart, and liver. These findings suggested that a restricted set of genes drives inflammatory responses and tissue injury across various transplantation contexts. In lung transplantation, the CRM has been extensively validated to stratify patients based on phenotypes of chronic lung allograft dysfunction (CLAD), specifically bronchiolitis obliterans syndrome (BOS) and restrictive allograft syndrome (RAS). Gene expression profiling of tissue samples revealed elevated levels of CRM genes in both BOS and RAS, with significant distinctions in average scores of gene expression between these phenotypes (Sacreas et al. 2018).

In kidney transplants, the CRM genes' expression was found to be positively linked to the extent of graft injury. For instance, in studies analyzing renal allograft biopsies, the CRM score exhibited a high degree of specificity and sensitivity in distinguishing rejection cases from stable allografts (Khatri et al. 2013). Furthermore, this score was correlated with histological damage in chronic allograft injuries and predicted future graft injury with high precision. Assessments performed on independent datasets comprising hundreds of renal biopsy samples revealed that patients with higher CRM scores were more likely to experience progressive damage, as confirmed by longitudinal studies (Khatri et al. 2013). Additionally, CRM-based studies reinforced the role of these modules as universal biomarkers applicable to multiple organ types. In predictive modeling efforts, scores derived from the CRM framework were employed in six-month protocol biopsies and successfully anticipated severe histopathological damage within 2 years of transplantation (Khatri et al. 2013; Zarinsefat et al. 2020). These predictions were particularly useful in identifying subsets of patients at elevated risk despite their stable serum creatinine levels, underlining the clinical utility of CRM in guiding post-transplant monitoring and intervention strategies (Sigdel et al. 2019a).

Heart Transplant (HTx) studies have provided substantial evidence supporting the utility of CRM in preventing chronic graft dysfunction. Recent research has validated the 11 CRM genes as essential markers of rejection subtypes in heart allografts. Gene expression profiling in heart rejection demonstrated the CRM score's ability to differentiate between acute cellular rejection (ACR) and AMR. Studies leveraging CRM in HTx rejection have reduced the number of invasive biopsies required for post-transplant surveillance. In a cohort of HTx patients, CRM-based assays showed greater sensitivity and accuracy in diagnosing rejection compared to traditional histopathological methods, sparing patients from excessive biopsies during follow-up (Pham et al. 2010). Additionally, the tissue-specific implementation of CRM in HTx provided crucial insights into disease pathways, particularly those involved in T-cell-mediated rejection and antibody-mediated vascular injury. This level of specificity poses significant opportunities for tailoring immunosuppressive regimens to individual patients (Sigdel et al. 2019b).

CRM's universality extends to liver transplantation, where it has been validated as a reliable molecular marker for acute and chronic rejection. The overexpression of CRM genes was shown to correlate with measures of allograft injury, such as portal inflammation, bile duct damage, and hepatic venous outflow obstruction, all of which define liver graft rejection. These findings align closely with CRM gene roles observed in other organ systems, highlighting their participation in immune regulation and inflammatory response during rejection episodes. The molecular significance of CRM in liver graft rejection is underscored by experimental and clinical studies showing its association across phenotypes of immune-mediated allograft injury. For example, NKG7 and CXCL9 were significantly overexpressed in T-cell-mediated rejection (TCMR) of the liver, while TAP1 and PSMB9 have emerged as genetic markers associated with Antibody-Medicated Rejection (AMR). Importantly, computational simulations demonstrated that CRM-based interventions targeting inflammatory pathways might attenuate liver rejection without universally suppressing graft immunity, allowing for better therapeutic specificity (Ningappa et al. 2022). Moreover, CRM scores were able to predict long-term outcomes such as fibrosis progression and transplant loss, validating its use as a prognostic tool.

Blood-based assays detecting CRM gene expression in liver transplant recipients have shown promising results, enabling less invasive monitoring strategies and enhancing patient compliance (Levitsky et al. 2022).

The exploration of CRM applications to organs beyond lungs, kidneys, hearts, and livers is gaining momentum, though data remain relatively limited. Preliminary studies suggest that CRM could have significant implications for pancreas and small intestine transplants, where immune rejection remains the predominant cause of graft failure. By validating CRM expression profiles in these settings, researchers aim to develop noninvasive diagnostics and identify biomarkers predictive of AMR and chronic rejection.

3.3.2 Network Analysis as an Analytical Framework in Chronic Rejection

Network analysis is a method for studying how genes or proteins interact within interconnected systems to reveal underlying biological processes and disease mechanisms (Xia et al. 2015). Gene expression data can offer crucial insights into molecular changes, but it is through network analysis that researchers can really understand how these changes fit into broader biological systems. One popular approach is weighted gene co-expression network analysis (WGCNA), which helps identify clusters of co-regulated genes linked to traits like rejection phenotypes or fibrosis (Wang et al. 2023). By emphasizing functional connections, network analysis uncovers the underlying structure of gene regulation and highlights key drivers of disease progression. In kidney transplantation, WGCNA has been instrumental in identifying gene co-expression modules enriched with pathways related to immune activation and fibrosis in interstitial fibrosis and tubular atrophy, which is a central issue in chronic rejection. The pathways include inflammatory signaling, immune cell recruitment, and extracellular matrix remodeling, all of which are strongly associated with graft injury (Modena et al. 2016). These results are consistent with previous proteogenomic studies that identified fibrogenesis and chronic inflammation as major contributors to graft dysfunction (Nakorchevsky et al. 2010). Similarly, network analyses of chronic lung rejection have revealed specific inflammatory factors and molecular pathways that distinguish BOS from RAS phenotypes. Notably, modules linked to oxidative stress and T cell activation are more pronounced in RAS, reflecting heightened immune activation compared to BOS (Sacreas et al. 2018). This contrast highlights the value of network analysis for identifying pathways that could be targeted therapeutically or used as biomarkers for better phenotypic distinction. Additionally, multi-omics network approaches that integrate transcriptomic, proteomic, and epigenomic data have uncovered novel microRNA biomarkers involved in immune regulation and graft rejection (Lin et al. 2021). For example, some microRNAs identified through these methods are functionally related to gene modules connected to Th17 cell differentiation and macrophage polarization—both of which are pathways known to impact rejection outcomes.

Despite their potential, gene expression and network analysis techniques still face notable limitations. Chronic rejection phenotypes are often heterogeneous, making it difficult to generalize findings across different patients and organ systems. Moreover, relying solely on transcriptomic data overlooks post-transcriptional regulation and protein interactions that play significant roles in disease progression. Addressing these challenges will require more integrated multi-omics studies and longitudinal research that can track how gene expression and network structures evolve. Emerging technologies like single-cell transcriptomics and spatial genomics are also promising, offering a finer resolution for understanding cellular diversity and microenvironmental influences during chronic rejection.

3.4 Biomarker Discovery and Therapeutic Targeting Through Computational Predictions

Computational modeling transforms biomarker discovery and therapeutic target identification for chronic graft rejection, by uncovering immune mechanisms and tissue damage, and advancing precision medicine for improved patient outcomes (see Fig. 3.3). Omics-driven approaches are central to this effort. Platforms like Gene Expression Omnibus (GEO) and WEB-based GEne SeT AnaLysis Tool kit (WebGestalt) have been used to detect differentially expressed genes (DEGs) associated with AMR in kidney transplants. These DEGs include genes tied to antigen presentation, cytokine signaling, and leukocyte migration.

Protein–protein interaction networks have further clarified the molecular landscape of kidney graft rejection. Network-analysis-based tools have also identified proteins such as Signal Transducer and Activator of Transcription 1 (STAT1), Interferon Regulatory Factor 1 (IRF1), and LYN Proto-Oncogene (LYN), which are pivotal in immune cell signaling and cytotoxicity (Alfaro et al. 2021). These proteins are involved in NK cell cytotoxicity and leukocyte migration, making them attractive targets for therapeutic intervention. Additionally, computational drug prediction tools like Gene2Drug have identified potential small molecules that could modulate gene expression to mitigate graft rejection (Chen et al. 2024). Among these, imatinib, a tyrosine kinase inhibitor targeting Platelet-Derived Growth Factor Receptor Alpha (PDGFRA), has emerged as a promising candidate for managing AMR, demonstrating how computational methods can inform therapy design. Chronic liver rejection is driven by metabolic and immune dysregulation, making computational approaches essential for revealing its underlying mechanisms. The liver's unique immunological environment makes it particularly susceptible to immune-mediated injury post-transplantation. Metabolomic profiling using high-throughput techniques has shown suppressed fatty acid oxidation pathways in cases of chronic liver rejection. These metabolic shifts provide valuable clues about disease progression and suggest potential targets for metabolic reprogramming-based therapies (Zhang et al. 2024b). Computational immune-cell profiling algorithms like xCell and CIBERSORT have been essential in characterizing leukocyte

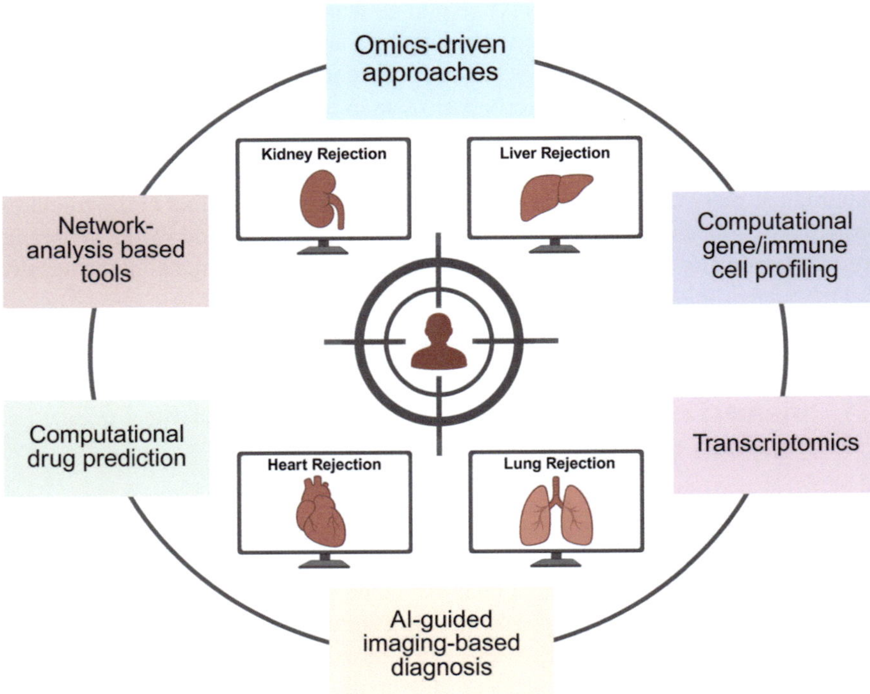

Fig. 3.3 Computational predictive model in solid organ rejection. (Created in BioRender. Casarin, S. (2025). https://BioRender.com/y5j6525)

infiltration patterns within liver grafts. These tools have revealed the prominent roles of T and B cells in driving rejection, pointing to new possibilities for immune modulation therapies (Xu et al. 2024). Additionally, protein–gene interaction networks constructed from transcriptomic data have spotlighted genes like Spi-1 Proto-Oncogene (SPI1) and Mitogen-Activated Protein Kinase (MAPK), which are linked to immune-mediated cellular dysfunction (Ningappa et al. 2022).

In HTx, chronic rejection primarily manifests as CAV, as seen above. Computational modeling has greatly improved diagnostic and predictive capabilities, allowing for earlier intervention. AI and ML have refined imaging-based diagnostics, enabling the detection of subtle morphological and vascular changes that signal CAV progression (Kveton et al. 2024; Peyster et al. 2018). These tools enhance precision in monitoring rejection, allowing for more timely clinical interventions. Gene expression profiling has further improved risk assessment in HTx. Studies like the CARGO gene-expression profiling initiative have identified critical molecular markers, such as NT-proBNP and Interleukin-6 (IL-6), that assist in early diagnosis and monitoring transplant outcomes. These insights are essential for precision medicine, enabling therapeutic strategies to be tailored to individual patients. Furthermore, computational protein interaction networks have shed light on pathways linked to cytokine signaling and endothelial damage in CAV. By identifying molecular targets like Lymphocyte-specific protein tyrosine kinase (LCK), researchers have explored pharmacological modulation using drugs such as Dasatinib to reduce vascular injury and improve graft survival (Wang and Sarwal 2015). In vivo imaging integrated with CFD-based tools has also helped link coronary artery morphology to altered hemodynamics and vascular remodeling, suggesting potential biomarkers for predicting CAV progression (Serafini et al. 2025).

The lung's constant exposure to environmental antigens presents unique challenges in chronic rejection, which is clinically recognized as Chronic Lung Allograft Dysfunction (CLAD). Computational approaches have been critical for tackling these complexities. Advanced imaging techniques combined with AI have improved the differentiation of CLAD phenotypes, particularly between BOS and RAS (Verleden et al. 2023). AI-assisted imaging has revealed strong correlations between pleuroparenchymal changes and fibrosis progression, facilitating early detection and improving therapeutic decision-making. Additionally, microbiome analysis has emerged as a valuable computational tool in lung transplantation. By integrating bronchoalveolar lavage molecular data, researchers have uncovered microbiota patterns associated with CLAD development. Notably, pathogen colonization, especially by *Pseudomonas aeruginosa*, has been linked to declining lung function, presenting potential opportunities for antimicrobial interventions. Pathway analyses have identified key roles for cytokine receptor signaling, alveolar macrophage activation, and interferon pathways like C-X-C Motif Chemokine Receptor 3 (CXCR3) in promoting lung transplant rejection. Targeting these immune pathways holds promise for reducing CLAD incidence and improving long-term transplant outcomes.

Despite the substantial progress made in computational modeling for chronic graft rejection biomarkers, several challenges remain. Integrating omics data from different platforms is complicated by the lack of standardized protocols, which limits cross-study comparisons and clinical applicability. Creating unified frameworks to bridge these gaps is essential for broader clinical adoption. Clinical translation remains another obstacle. While computational models have shown promise in theoretical and preclinical investigations, real-world validation across various solid organs is still lacking. Future research must focus on comprehensive clinical studies to confirm the efficacy of computational predictions. Expanding precision medicine applications through patient-specific models that integrate molecular, histological, and demographic data will be crucial for optimizing treatment strategies and enhancing transplant outcomes.

3.5 Discussion

While computational models offer transformative potential in transplant research, their infrastructure demands pose practical constraints, especially when considering widespread deployment across heterogeneous clinical environments. From

AI-driven histology analysis to hybrid ODE/ABM simulations, the models discussed in this chapter vary substantially in complexity and resource requirements.

DL models, such as CNNs used for whole-slide image analysis, are particularly compute-intensive, often requiring GPU acceleration and substantial memory. In resource-limited settings, this may impede clinical translation. To mitigate this, lightweight neural architectures, like MobileNet, and EfficientNet, have been explored to retain diagnostic accuracy while reducing computational load. These models are optimized for inference on CPUs or embedded systems and could power edge-deployed AI applications for biopsy scoring in low-resource environments.

In parallel, mechanistic models, particularly ABMs and PDE-based frameworks, are often constrained by long simulation times and high-dimensional parameter spaces. To address this, surrogate modeling techniques, such as reduced-order models or pretrained emulators, have gained traction. These approximations preserve the key dynamics of more complex simulations while enabling real-time or near-real-time evaluation, a capability essential for point-of-care decision support.

Moreover, the emergence of cloud-native infrastructure offers a viable path to scale these models without requiring extensive local resources. Containerized environments (e.g., using Docker or Kubernetes) allow transplant centers to deploy preconfigured AI/ML pipelines or simulation engines on demand, ensuring reproducibility and version control. When combined with interoperable data standards like FHIR and HL7, these tools can be integrated directly into electronic health records (EHRs), streamlining model access and automating feedback loops between predictions and clinical data.

Ultimately, balancing model fidelity with computational efficiency is key to enabling equitable access to advanced computational tools. Lightweight, scalable frameworks—coupled with thoughtful deployment strategies—will be essential to move from theoretical modeling toward broad clinical impact across diverse transplant ecosystems. The integration of computational modeling with experimental research has opened new avenues in understanding chronic graft rejection across various solid organ transplants. These models, ranging from AI and ML algorithms to differential equations and agent-based simulations, have provided crucial insights into the immune and molecular mechanisms underlying chronic rejection. By offering a flexible and cost-effective framework for analyzing complex biological systems, computational modeling has begun to fill gaps left by traditional experimental methods.

Recent advances have significantly contributed to the development of diagnostic tools and the identification of biomarkers and therapeutic targets. The use of AI and ML techniques has enhanced diagnostic precision by enabling automated analysis of biopsy samples from heart, kidney, liver, and lung transplants. The ability of these models to detect morphological changes associated with chronic rejection, particularly in conditions like CAV and CLAD, has improved both accuracy and efficiency. Computational network analyses, particularly through weighted gene coexpression network analysis and the CRM, have also uncovered gene signatures that appear consistent across different organ types. These studies helped clarify key molecular pathways related to inflammation, immune cell recruitment, and fibrosis.

Mechanistic modeling approaches employing ODEs and PDEs have provided valuable insights into the progression of immune responses, cytokine signaling, and tissue remodeling. ODEs have been particularly useful for capturing the temporal dynamics of immune activation, while PDEs have helped describe spatial aspects of inflammation and fibrosis. ABMs have further enhanced our understanding by allowing researchers to simulate interactions among individual cells and observe emergent behaviors that drive chronic rejection. Meanwhile, efforts to integrate multi-omics data, combining transcriptomic, proteomic, and epigenomic information, have helped researchers construct more comprehensive models capable of identifying new biomarkers and therapeutic targets with greater precision.

Despite these achievements, several limitations continue to be an obstacle to the clinical translation of computational models. The intrinsic complexity of chronic rejection poses a fundamental challenge, as current models often rely on simplified assumptions that may not fully capture the intricate interplay of immune and non-immune factors involved. These simplifications are necessary for practical reasons (computational power and interpretability among all) but can limit the accuracy of predictions, particularly when dealing with complex processes like post-transcriptional regulation or protein–protein interactions. Additionally, the heterogeneity of rejection phenotypes across different organs and patient populations complicates efforts to develop universally applicable models. For instance, pathways driving rejection in kidney transplants may differ significantly from those involved in heart or lung transplantation, making it difficult to generalize findings.

Data quality and availability remain an ongoing concern. Computational models are heavily dependent on the quality and completeness of the experimental data used to construct them. Limited access to large, high-quality datasets can result in models that are either biased or fail to capture the full spectrum of disease mechanisms. Rare rejection phenotypes often remain underrepresented, which restricts the applicability of findings to broader patient populations. Moreover, the integration of multi-omics data is still impeded by the absence of standardized protocols, making it difficult to compare results across studies or establish clinically applicable benchmarks. Clinical validation represents another critical challenge. While computational models have demonstrated impressive results in preclinical studies, their real-world utility remains limited without rigorous testing against clinical data. The complexity of human immune responses and the gradual progression of chronic rejection make it essential to validate predictive models through longitudinal studies that account for patient-specific variability and evolving clinical conditions. Without such validation, the transition from theoretical models to practical applications will continue to face significant barriers.

Moving forward, several strategies could enhance the clinical relevance and robustness of computational models for studying chronic rejection. Incorporating more comprehensive datasets that include post-transcriptional regulation, protein interactions, and spatial heterogeneity could improve model accuracy. Combining different modeling techniques, such as differential equations, ABMs, and ML algorithms, could also enhance predictive power by capturing both broad systemic trends and detailed cellular interactions. The development of patient-specific models that integrate molecular, histological, demographic, and clinical data offers a promising solution for improving risk stratification and tailoring immunosuppressive regimens to individual patients. Such models could prove invaluable for predicting long-term outcomes and guiding personalized treatment strategies.

The integration of diverse multi-omics data through advanced network analysis frameworks is another area ripe for further exploration. Standardizing protocols for data integration and interpretation would improve reproducibility and enhance the applicability of computational findings to clinical settings. Additionally, the application of emerging technologies like single-cell transcriptomics and spatial genomics could provide higher-resolution insights into cellular heterogeneity and microenvironmental influences during chronic rejection. Continued progress will depend on developing hybrid modeling frameworks that leverage the complementary strengths of different computational approaches. A unified platform incorporating all the aforementioned methods could offer a more comprehensive understanding of chronic rejection mechanisms. Such models could accommodate both large-scale systemic analyses and detailed investigations of cellular interactions, making them particularly valuable for understanding the complex pathways that drive graft dysfunction.

3.6 Conclusions

The combined use of computational models and experimental research offers significant potential for overcoming the challenges associated with chronic graft rejection. While recent advancements have contributed to enhanced diagnostic precision, biomarker discovery, and therapeutic target identification, several limitations remain. Improving model accuracy, standardizing data integration protocols, and conducting robust clinical validation will be essential to achieving widespread clinical adoption. As computational methods continue to advance and incorporate increasingly complex datasets, their role in informing precision medicine approaches and improving transplant outcomes is likely to grow.

Appendix

A list of the main abbreviations used in this chapter is offered below for reader convenience.

ABM	Agent-based model
ACR	Acute cellular rejection
AI	Artificial intelligence
AMR	Antibody-mediated rejection
BOS	Bronchiolitis obliterans syndrome

CAV	Cardiac allograft vasculopathy
CLAD	Chronic lung allograft dysfunction
CNN	Convolutional neural network
CRM	Common rejection module
DEG	Differentially expressed gene
DL	Deep learning
ECM	Extra cellular matrix
FGF	Fibroblast growth factor
HTx	Heart transplant
ML	Machine learning
ODE	Ordinary differential equation
PDE	Partial differential equation
PDGFRA	Platelet-derived growth factor receptor alpha
RAS	Restrictive allograft syndrome
SMC	Smooth muscle cell
TCMR	T-cell-mediated rejection
TGF-β	Transforming growth factor-β
VEGF	Vascular endothelial growth factor
VSMC	Vascular smooth muscle cells
WGCNA	Weighted gene coexpression network analysis
WSI	Whole slide image

References

Actor JK (2019) Chapter 11 – Transplantation immunology. In: Actor JK (ed) Introductory immunology, 2nd edn. Academic Press, pp 133–142

Alfaro R et al (2021) Computational prediction of biomarkers, pathways, and new target drugs in the pathogenesis of immune-based diseases regarding kidney transplantation rejection. Front Immunol 12:800968

An G (2015) Introduction of a framework for dynamic knowledge representation of the control structure of transplant immunology: employing the power of abstraction with a solid organ transplant agent-based model. Front Immunol 6:561

Arciero JC et al (2016) Combining theoretical and experimental techniques to study murine heart transplant rejection. Front Immunol 7:448

Baldwin WM 3rd, Valujskikh A, Fairchild RL (2016) Mechanisms of antibody-mediated acute and chronic rejection of kidney allografts. Curr Opin Organ Transplant 21(1):7–14

Black CK et al (2018) Solid organ transplantation in the 21(st) century. Ann Transl Med 6(20):409

Bonin CR et al (2017) Mathematical modeling based on ordinary differential equations: a promising approach to vaccinology. Hum Vaccin Immunother 13(2):484–489

Chen S, Zhu H, Jounaidi Y (2024) Comprehensive snapshots of natural killer cells functions, signaling, molecular mechanisms and clinical utilization. Signal Transduct Target Ther 9(1):302

Colquitt RB, Colquhoun DA, Thiele RH (2011) In silico modelling of physiologic systems. Best Pract Res Clin Anaesthesiol 25(4):499–510

Colvin MM et al (2024) OPTN/SRTR 2022 annual data report: heart. Am J Transplant 24(2s1):S305–S393

De Gaetano A et al (2012) Modeling rejection immunity. Theor Biol Med Model 9:18

Fayzullin A et al (2024) Towards accurate and efficient diagnoses in nephropathology: an AI-based approach for assessing kidney transplant rejection. Comput Struct Biotechnol J 24:571–582
Goldstein BA et al (2016) Assessment of heart transplant waitlist time and pre- and post-transplant failure: a mixed methods approach. Epidemiology 27(4):469–476
Hudaib A et al (2025) Exploring the implementation of federated learning in healthcare: a comprehensive review. Clust Comput 28(5):302
Ingulli E (2010) Mechanism of cellular rejection in transplantation. Pediatr Nephrol 25(1):61–74
Khatri P et al (2013) A common rejection module (CRM) for acute rejection across multiple organs identifies novel therapeutics for organ transplantation. J Exp Med 210(11):2205–2221
Kim MY, Brennan DC (2021) Therapies for chronic allograft rejection. Front Pharmacol 12:651222
Kveton M et al (2024) Digital pathology in cardiac transplant diagnostics: from biopsies to algorithms. Cardiovasc Pathol 68:107587
Kwong AJ et al (2024) OPTN/SRTR 2022 annual data report: liver. Am J Transplant 24(2s1):S176–S265
Lentine KL et al (2023) OPTN/SRTR 2021 annual data report: kidney. Am J Transplant 23(2 Suppl 1):S21–S120
Levitsky J et al (2022) Prediction of liver transplant rejection with a biologically relevant gene expression signature. Transplantation 106(5):1004–1011
Lin Y et al (2021) Multi-omics network characterization reveals novel microRNA biomarkers and mechanisms for diagnosis and subtyping of kidney transplant rejection. J Transl Med 19(1):346
Modena BD et al (2016) Gene expression in biopsies of acute rejection and interstitial fibrosis/tubular atrophy reveals highly shared mechanisms that correlate with worse long-term outcomes. Am J Transplant 16(7):1982–1998
Moreau, A., et al., Effector mechanisms of rejection. Cold Spring Harb Perspect Med, 2013. 3(11) a015461
Nakorchevsky A et al (2010) Molecular mechanisms of chronic kidney transplant rejection via large-scale proteogenomic analysis of tissue biopsies. J Am Soc Nephrol 21(2):362–373
Ningappa M et al (2022) A network-based approach to identify expression modules underlying rejection in pediatric liver transplantation. Cell Rep Med 3(4):100605
Olawade DB et al (2025) The impact of artificial intelligence and machine learning in organ retrieval and transplantation: a comprehensive review. Curr Res Transl Med 73(2):103493
Perch M et al (2022) The International Thoracic Organ Transplant Registry of the International Society for Heart and Lung Transplantation: thirty-ninth adult lung transplantation report-2022; focus on lung transplant recipients with chronic obstructive pulmonary disease. J Heart Lung Transplant 41(10):1335–1347
Peyster EG, Madabhushi A, Margulies KB (2018) Advanced morphologic analysis for diagnosing allograft rejection: the case of cardiac transplant rejection. Transplantation 102(8):1230
Peyster EG et al (2021) An automated computational image analysis pipeline for histological grading of cardiac allograft rejection. Eur Heart J 42(24):2356–2369
Peyster EG et al (2025) Computational pathology assessments of cardiac stromal remodeling: clinical correlates and prognostic implications in heart transplantation. JHLT Open 7:100202
Pham MX et al (2010) Gene-expression profiling for rejection surveillance after cardiac transplantation. N Engl J Med 362(20):1890–1900
Ramalhete L et al (2024) Revolutionizing kidney transplantation: connecting machine learning and artificial intelligence with next-generation healthcare—from algorithms to allografts. BioMedInformatics 4(1):673–689
Sacreas A et al (2018) The common rejection module in chronic rejection post lung transplantation. PLoS One 13(10):e0205107
Schapranow MP et al (2023) NephroCAGE-German-Canadian Consortium on AI for improved kidney transplantation outcome: protocol for an algorithm development and validation study. JMIR Res Protoc 12:e48892
Schettini F et al (2023) Computational reactive–diffusive modeling for stratification and prognosis determination of patients with breast cancer receiving Olaparib. Sci Rep 13(1):11951

Serafini E et al (2023) An agent-based model of cardiac allograft vasculopathy: toward a better understanding of chronic rejection dynamics. Front Bioeng Biotechnol 11:1190409
Serafini E et al (2025) Investigating the relationship between geometry and hemodynamics in an experimentally derived murine coronary computational model. Comput Biol Med 187:109793
Sigdel TK et al (2019a) A urinary Common Rejection Module (uCRM) score for non-invasive kidney transplant monitoring. PLoS One 14(7):e0220052
Sigdel T et al (2019b) Assessment of 19 genes and validation of CRM gene panel for quantitative transcriptional analysis of molecular rejection and inflammation in archival kidney transplant biopsies. Front Med (Lausanne) 6:213
Southern J et al (2008) Multi-scale computational modelling in biology and physiology. Prog Biophys Mol Biol 96(1):60–89
Stewart D, Mupfudze T, Klassen D (2023) Does anybody really know what (the kidney median waiting) time is? Am J Transplant 23(2):223–231
Teo ZL et al (2024) Federated machine learning in healthcare: a systematic review on clinical applications and technical architecture. Cell Rep Med 5(2):101419
Tong J et al (2024) Evaluating site-of-care-related racial disparities in kidney graft failure using a novel federated learning framework. J Am Med Inform Assoc 31(6):1303–1312
Tracy M, Cerdá M, Keyes KM (2018) Agent-based modeling in public health: current applications and future directions. Annu Rev Public Health 39:77–94
Trieu JA, Bilal M, Hmoud B (2018) Factors associated with waiting time on the liver transplant list: an analysis of the United Network for Organ Sharing (UNOS) database. Ann Gastroenterol 31(1):84–89
Vandamme TF (2014) Use of rodents as models of human diseases. J Pharm Bioallied Sci 6(1):2–9
Verleden SE et al (2023) Biomarkers for chronic lung allograft dysfunction: ready for prime time? Transplantation 107(2):341–350
Wang A, Sarwal MM (2015) Computational models for transplant biomarker discovery. Front Immunol 6:458
Wang Y et al (2023) Identification of PDCD1 as a potential biomarker in acute rejection after kidney transplantation via comprehensive bioinformatic analysis. Front Immunol 13:1076546
Wood KJ, Goto R (2012) Mechanisms of rejection: current perspectives. Transplantation 93(1):1–10
Xia J, Gill EE, Hancock REW (2015) NetworkAnalyst for statistical, visual and network-based meta-analysis of gene expression data. Nat Protoc 10(6):823–844
Xu Z et al (2024) Landscape of the immune infiltration and identification of molecular diagnostic markers associated with immune cells in patients with kidney transplantation. Sci Rep 14(1):24770
Yen P et al (2011) A two-compartment model of VEGF distribution in the mouse. PLoS One 6(11):e27514
Zarinsefat A et al (2020) Use of the tissue common rejection module score in kidney transplant as an objective measure of allograft inflammation. Front Immunol 11:614343
Zhang F et al (2024a) Recent methodological advances in federated learning for healthcare. Patterns 5(6):101006
Zhang C et al (2024b) Computational identification of novel potential genetic pathogenesis and otherwise biomarkers in acute liver allograft rejection. Heliyon 10(15):e33359

Part II
Immune Cells, Non-immune Cells, and Extracellular Molecules in Chronic Rejection

Chapter 4
The Role of the Endothelium as a Key Regulator of Chronic Transplant Rejection

Wolfgang Jungraithmayr, Birte Ohm, and Shampa Chatterjee

Abstract The vascular endothelium serves as an indispensable border for the interaction and exchange of molecules and cells between the blood flow and tissues. Upon transplantation, the endothelium of the allograft is the first to come into contact with the recipient's blood and becomes injured by alloreactive and nonalloreactive stimuli. Repetitive injuries lead to severe damage of the endothelium resulting in the development of chronic rejection over time.

The objective of this chapter is to highlight recent insights into the most relevant endothelial molecules, mediators, and pathways crucial in the evolvement of chronic organ rejection. We also discuss new avenues of drugs that can diminish or prevent damage to the endothelium leading to an otherwise irreversible development of chronic organ rejection.

4.1 Introduction

The vascular endothelium is a dynamic and interactive tissue that senses environmental cues and stimuli and initiates biochemical signaling cascades that play a crucial role in mechanotransduction (stop and restart of blood flow, blood pressure, and shear stress), metabolism, signaling, and aging. Upon transplantation (Tx) of a solid organ, the endothelium is the first site of contact between the recipient's blood

W. Jungraithmayr (✉)
Division of Thoracic Surgery, Rostock University Medical Center, Rostock, Germany
e-mail: wolfgang.jungraithmayr@med.uni-rostock.de

B. Ohm
Department of Thoracic Surgery, Medical Center – University of Freiburg, Faculty of Medicine, University of Freiburg, Freiburg, Germany

S. Chatterjee
Department of Physiology and Institute for Environmental Medicine, University of Pennsylvania Perelman School of Medicine, Philadelphia, PA, USA

M. Kloc, R. M. Ghobrial (eds.), *Cellular and Molecular Mechanisms of Chronic Rejection in Organ Transplantation*, Results and Problems in Cell Differentiation 77, https://doi.org/10.1007/978-3-032-07686-1_4

circulation and the donor organ. Indeed, the endothelium of the graft is the only obligate cell type of the donor with which circulating recipient immune cells interact directly.

Under physiologic conditions, the endothelium serves as an important barrier regulating the exchange of oxygen and nutrients and provides a well-organized influx and outflux of cells for organ supply. After Tx and reperfusion by the recipient's circulation, however, the endothelium becomes highly alloreactive, and immune cells influx through the injured endothelium resulting in allograft rejection.

The process of endothelial damage starts from the temporary absence of circulation causing hypoxia and leads to activation and injury even before transplantation. Subsequent organ reperfusion results in the production of reactive oxygen species (ROS) as a second step of endothelial injury. Endothelial cells are here both target and stimulator of the rejection response at the same time. The endothelium reacts to the injury with the production of specific membrane-bound molecules such as integrins and adhesion molecules along with the release of proinflammatory cytokines and chemokines while losing their intercellular junctions.

As semiprofessional antigen-presenting cells, the endothelium mediates both, cellular and humoral rejection. As an early event after Tx, preformed donor-specific antibodies (DSAs) recognize human leukocyte antigen (HLA) or ABO antigens on endothelium and thus damage endothelial cells and the graft. At later stages, so-called de novo anti-HLA DSA or non-HLA can activate and destroy the endothelium. On a cellular level, cytotoxic lymphocytes recognize HLA class I on endothelium with subsequent damage or, the endothelium itself initiates alloreactive T cell responses. Unlike these acute rejection events, the development of chronic rejection (CR) in a transplanted organ is believed to evolve from subclinical activation processes of the endothelial layer. As chronic rejection is still the major hurdle for long-term success after solid organ transplantation, the emphasis of transplant research today should be focused on pathological events that start at the endothelium as the first and most important target structure. We therefore point out and summarize in this chapter the most recent and future-orientated molecules, mediators, and pathways related to the endothelium that are crucial in the evolvement of CR. We also highlight new avenues of drugs that can diminish or prevent damage to the endothelium leading to an otherwise irreversible development of CR.

4.2 Endothelial Molecules Triggering or Protecting the Endothelium from Transplant Injury

Directly after transplantation, danger signals such as reactive oxygen species (ROS), proinflammatory cytokines, and complement activation products act on the endothelium, triggering a shift toward a proinflammatory and prothrombotic phenotype. Pattern-recognition receptors such as toll-like receptors (TLRs) detect damage-associated molecular patterns (DAMPs) released during ischemia-reperfusion injury, amplifying innate immune responses at the endothelial surface. Furthermore, the expression of both major histocompatibility complex (MHC) molecules and the costimulatory CD40 are upregulated, enhancing recognition by alloreactive T cells.

To protect their integrity, endothelia possess intrinsic protective structures, receptors, and pathways. In the following paragraphs, we discuss several of these protectors and their implications in the context of solid organ transplantation and associated chronic graft rejection.

4.2.1 Glycocalyx Degeneration Perpetuates Endothelial Dysfunction

The luminal surface of endothelia is covered by a 0.1–1 μm thick carbohydrate-rich matrix consisting of proteoglycans, glycosaminoglycans, and soluble proteins. The backbone of this structure is formed by proteoglycans, such as syndecans, and glycoproteins, which are anchored in the endothelial cells. These structures are then covalently bound by heparin and chondroitin glycosaminoglycans to create a negatively charged surface, which attracts blood plasma components, such as albumin. Under physiological conditions, the integrity of the glycocalyx is well maintained. Under stress, the glycocalyx is degraded, resulting in endothelial barrier dysfunction and increased cellular invasion. Thus, glycocalyx shedding is a major contributor to microvascular damage. In solid organ transplantation, glycocalyx shedding is associated with poor organ function. Clinically, elevated plasma levels of syndecan-1 correlate with poor graft function after liver transplantation (Schiefer et al. 2020). Similarly, glycocalyx shedding has been observed in kidney (Snoeijs et al. 2010) and lung transplantation, where it is associated with primary graft dysfunction (Sladden et al. 2019).

In summary, although a direct causal link between glycocalyx integrity and the prevention of chronic rejection is yet to be conclusively established, degradation of the endothelial glycocalyx is strongly associated with barrier dysfunction, cellular infiltration, and microvascular injury. These changes impair organ function and contribute to primary graft dysfunction, likely playing a role in the development of chronic rejection. Moreover, glycocalyx damage exacerbates endothelial dysfunction by downregulating endothelial nitric oxide synthase (eNOS), thereby reducing nitric oxide production, increasing reactive oxygen species generation, and further promoting graft injury.

4.2.2 Nitric Oxide Synthases Play a Differential Role in Ischemia-Reperfusion Injury and Chronic Rejection

Endothelial nitric oxide synthase (eNOS) generates nitric oxide (NO), a key molecule in preventing leukocyte adhesion, platelet aggregation, and vascular tone dysregulation. Endothelial dysfunction is marked by a decrease in NO production as the expression of eNOS is downregulated upon damage to the glycocalyx. Additionally, eNOS uncoupling occurs, further contributing to ROS generation (Gielis et al. 2017). The protective effect of NO on solid organ grafts has been demonstrated in preclinical studies. In murine heart transplantation, the activity of inducible nitric oxide synthase (iNOS) has been associated with enhanced allograft survival and a reduction of endothelial cell activation (Qian et al. 2001). In a porcine ex vivo model of lung xenotransplantation, supplementation of the perfusion fluid with an NO donor achieved a decrease in von Willebrand factor release, inhibiting vasoconstriction, thrombin generation, and complement activation (Park et al. 2015). While the beneficial effect of NO on ischemia-reperfusion injury (IRI) is well established across different solid organ grafts, its direct influence on chronic graft rejection is poorly studied. In a murine model of tracheal transplantation, it was demonstrated that the activity of the NOS isoforms may have specific implications for the graft's fate. While eNOS deficiency had no significant effect on airway obliteration, allografts in iNOS-deficient recipients showed reduced inflammation and less luminal occlusion (Minamoto and Pinsky 2002). In order to provide protection for solid organ grafts, maintaining eNOS function in the early phase after transplantation can protect from damage during ischemia reperfusion injury. However, to protect the graft from long-term damage, inhibiting iNOS may be a more promising strategy.

4.2.3 Complement Regulatory Proteins on Endothelial Cells Protect from Chronic Graft Rejection

The complement system plays a critical role in transplantation and modulates inflammation and immune responses in the graft. During IRI, all three pathways of complement activation are involved. Independent of the mode of activation, all pathways converge in the activation of the complement component C3 and the subsequent formation of the membrane attack complex (MAC) from the recruitment of complement proteins C5b-C9, resulting in cell lysis. The deposition of the complement protein C4d on endothelial cells is an established marker for graft injury in solid organs, especially in kidneys and hearts (Murata and Baldwin 3rd. 2009).

Endothelia are equipped with complement-regulatory surface molecules that protect them from complement-mediated damage. CD59 on the endothelial surface prevents the assembly of the MAC by binding to C8 and C9. Endothelial damage during IRI of the liver is exacerbated in CD59-deficient rodents (Zhang et al. 2011). Promotor polymorphisms in the CD59 gene that correlate with reduced expression in lung grafts have been associated with a higher incidence of chronic lung allograft rejection (Budding et al. 2016).

CD55, also known as decay-accelerating factor (DAF), is another surface protein expressed in endothelial cells that mediates protection from complement-mediated damage by inhibiting the formation of C3 convertase. In a mouse model of tracheal transplantation, loss of CD55 has been linked to microvascular injury, which promotes leukocyte infiltration and graft vasculopathy (Khan et al. 2020). Overexpression of CD55 in kidney organoids results in decreased activation of the complement system upon stimulation with anti-HLA antibodies, underlining the significance of the protein in protection from humoral graft rejection (Gaykema et al. 2022).

In summary, CD55 and CD59 limit complement-mediated damage in IRI, which plays a major role in early graft dysfunction following organ transplantation. CD55 inhibits the formation of C3 and C5 convertase, thereby protecting cells from complement-mediated damage, whereas CD59 prevents the formation of the membrane attack complex (MAC), protecting transplanted tissue from lysis.

4.2.4 Modulation of Purinergic Signaling Can Protect the Endothelia

After transplantation, extracellular ATP (eATP) is released from damaged or stressed cells, acting as a proinflammatory danger signal (Zeiser et al. 2016). eATP then binds to P2X and P2Y purinergic receptors on immune and endothelial cells, promoting inflammation and endothelial cell activation. Protection from eATP signaling is provided by the ectonucleotidases CD39 (NTPDase1) and CD73 (Ecto-5′-nucleotidase), which are highly expressed in endothelial cells under physiological conditions and degrade eATP to adenosine. In preclinical models, extracellular adenosine has been demonstrated to exert immunosuppressive effects on multiple different immune cells, such as dendritic cells, T effector cells, and regulatory T cells (Wang et al. 2024). Indeed, adenosinergic signaling protected solid organ grafts from ischemia-reperfusion injury in murine models. In a model of renal transplantation with prolonged cold storage, the transgenic overexpression of human CD39 in the graft resulted in improved graft function and reduced histological tissue injury (Crikis et al. 2010). In adoptive transfer experiments, the authors demonstrated that the expression of CD39 on either the vasculature or circulating cells mitigated ischemia-reperfusion injury. In a model of chronic airway rejection using tracheal allografts, CD73-deficient recipient animals showed a significant increase in luminal obstruction of the airway graft when compared to their wild-type counterparts (Ohtsuka et al. 2010). Treatment with an adenosine receptor agonist resulted in reduced leukocyte infiltration and airway obliteration. In a phase I clinical trial, it was demonstrated that the application of the adenosine receptor agonist regadenoson in lung transplant recipients is feasible without dose-limiting toxicity or drug-related mortality (Lau et al. 2020).

In clinical liver transplantation, contrasting findings have been reported regarding the role of extracellular adenosine. In contrast to their nontolerant counterparts, operationally tolerant liver allograft recipients exhibit a higher expression of the adenosine-degrading enzyme adenosine deaminase (Baroja-Mazo et al. 2019). Moreover, the authors report that the presence of extracellular adenosine inhibits both human T effector cells and human regulatory T cells alike in vitro. These findings point towards either the organ-specific effects of extracellular adenosine or the need for more complex strategies to adequately regulate purinergic signaling during solid organ transplantation.

4.3 Signaling Pathways of the Endothelial Cell Promoting the Development of Chronic Rejection

Chronic rejection of a transplanted graft involves the thickening of blood vessels or vascular remodeling; this is called transplant vasculopathy (Pilmore et al. 2000; Pober et al. 2021) (TV) and is characterized by the thickening of the neointima that in turn arises from proliferation of smooth muscle cell layer below the endothelium (Onuta et al. 2010). TV eventually contributes to bronchiolitis obliterans syndrome (BOS) in lung transplantation, cardiac allograft vasculopathy in cardiac transplantation, and arteriosclerosis after renal transplantation (Kitai et al. 2019).

The signaling that eventually leads to TV and BOS is initiated at the allograft endothelium which by virtue of being in direct contact with the host circulation participates in the recruitment of the host cells into the graft. This process requires a "bidirectional activation" of both the allograft endothelial layer and host immune cells (leukocytes, neutrophils, etc.) (Al-Lamki et al. 2008). Such activation leads to a chronic inflammatory state. Chronic allograft rejection develops over an extended period (months to years) and occurs largely through the chronic inflammatory state. The hallmark of this state is the activation of the endothelium via immunological and nonimmunological signaling.

4.3.1 Immunological Factors or Antibody-Mediated Rejection (AMR)

Endothelial cells express several antigens and are therefore a target for the host immune response (Piotti et al. 2014; Amersfoort et al. 2022). Donor-specific antibodies (DSAs) are directed against both major histocompatibility antigens (HLA-1 and non-HLA antigens) of the donor (Tikkanen et al. 2016; Brandon et al. 2023). The binding of DSA to HLA molecules on the donor endothelial cells triggers endothelial activation that in turn leads to the development of a proinflammatory endothelial phenotype, that is, increased expression of inducible proinflammatory adhesion molecules, such as intercellular adhesion molecule-1 (ICAM-1), vascular cell adhesion molecule-1 (VCAM-1) on the endothelial surface; the upregulation of chemokines monocyte chemoattractant protein-1 (MCP-1) and growth factors such as Platelet-Derived Growth Factor (PDGF) and transforming growth factor (TGF-β) (Kummer et al. 2020). These chemokines and growth factors provide a chemotactic gradient for effective recruitment and transmigration of immune cells. Furthermore, several signaling cascades viz. mitogen activated protein (MAP) kinase pathway, the extracellular regulated kinase (ERK) pathway, and the nuclear factor (NF)-kappa B and fibroblast growth factor (FGF) pathways are also activated. Collectively, all these facilitate adherence of immune cells to the endothelial layer; this is followed by their transmigration across the endothelial lining into the tissue (Cook-Mills and Deem 2005). The continued production of antibodies against the graft endothelium that results in an activated endothelium results in chronic antibody-mediated rejection.

4.3.2 *Endothelial Cells as Antigen-Producing Cells*

The host immune system distinguishes the self from the "nonself" by recognizing the graft as foreign. This process is facilitated by antigen-presenting cells (APCs) that bear antigens of MHC class I and II (HLA system for humans). In general, APCs are bone marrow-derived monocytes or dendritic cells that present HLA class I (HLA-A, -B, and -C) and II (HLA-DR, -DP, and –DQ). Upon transplant, the (adaptive) immune responses are initiated when foreign antigens are presented (to naïve T and B cells) by traditional APCs (monocytes, dendritic cells etc.). T- and B- lymphocytes have corresponding receptors (T-cell receptors and B-cell receptors) that then "recognize" the antigen as foreign, leading to the activation of T and B lymphocyte-mediated rejection. Thus, antigen presentation via the MHC or HLA system is a pivotal event in graft rejection (Mahdi 2013; Alegre et al. 2016). Activated endothelial cells have been reported to exhibit features that are typical of immune cells, such as the presentation of antigen-MHC complex to T- and B-cells (Pober et al. 2017). There is evidence suggesting that the release of MHC and HLA from endothelial cells stimulates an autoimmune response leading to chronic rejection (Al-Lamki et al. 2008; Rose 1998). Thus, the endothelium is the first target of alloimmune responses in transplanted organs.

4.3.3 *Nonimmunological Factors*

The ischemia-reperfusion (I/R) associated with Tx leads to endothelial cell activation (Chatterjee et al. 2014; Chatterjee 2004). The initial activation occurs primarily because of the cessation of blood flow and leads to NADPH oxidase 2 assembly and ROS production (Tao et al. 2016; Browning et al. 2014). The restoration of flow further increases ROS production. ROS leads to endothelial oxidative injury. ROS produced upon I/R plays a crucial role in molecular, structural, and functional properties of endothelial cells and tissues either directly (oxidative damage) or via various signaling pathways, which either activate or inhibit numerous transcription factors (TFs) (Kummer et al. 2020; Verrier 2004). These regulators/TFs include

Nuclear erythroid-2 related factors (Nrf-2), CCAAT/enhancer-binding protein delta (CEBPD), Activator Protein-1 (AP-1), Hypoxia-inducible factor 1(HIF-1), Nuclear factor κB (NF-κB), Specificity Protein-1 (SP-1), and Forkhead Box class O (FoxO) transcription factors. These play central roles in cellular processes, such as inflammation, innate immune response, cell proliferation, cell survival, and programmed cell death. Endothelial ROS production and TF activation drive an inflammatory phenotype that is characterized by CAMs; interactions between CAMs on the surface of neutrophils and ECs, for example, P-selectin, E-selectin, and ICAM-1. There are also procoagulatory and prothrombotic changes on the surface of the endothelium, resulting in vascular occlusion (Nordling et al. 2015).

The onset of inflammation on the endothelium is also characterized by changes in endothelial cell-cell junctional structure that can lead to increased permeability. As it is the interface between the circulation and tissues, endothelial permeability plays a major role in the trafficking of cells in and out of the circulation and into the site of inflammation within various organs (Augustin and Koh 2024). Besides facilitating trafficking, activated endothelial cells also drive migration and proliferation of vascular smooth muscle cells that in turn leads to neointima formation followed by a reduction in organ perfusion and eventually results in tissue injury (Kummer et al. 2020; Hellenthal et al. 2022).

Overall, the endothelium signaling cascades serve to initiate rejection and can therefore serve as targets to prevent or attenuate interactions between immune cells and endothelium, thus controlling chronic rejection processes.

4.4 Emerging Drugs that Preserve the Endothelial Barrier from Chronic Transplant Rejection

4.4.1 Complement Inhibition

It is believed that chronic rejection arises from persistent inflammation of the endothelium. This inflammation is multifactorial and is described as transplant vasculopathy that forms the base for bronchiolitis obliterans syndrome, the hallmark of CR in lung Tx. A major risk factor for the development of CR is the occurrence of donor-specific anti-HLA antibodies (DSAs). The binding of these DSAs on alloantigens at the endothelial surface activates the complement cascade mainly via the classical and alternative pathway that ultimately results in endothelial damage by the production of anaphylatoxins. One of these anaphylatoxins is C5a, a split product of C5 that is, along with C5b, degraded by C5 convertase. C5a increases the permeability of the endothelium and attracts inflammatory cells by chemotaxis, particularly neutrophils. The endothelium can be protected against DSA toxicity by employing potent inhibitors. For example, inhibitors of the C5a receptor were recently introduced, among them Avacopan (van Leeuwen et al. 2025). According to recent EULAR guidelines, Avacopan is approved and recommended as a steroid-sparing agent for the treatment of ANCA-positive vasculitis (Hellmich et al. 2024). A series of clinical trials have proven Avacopan as an effective treatment option in the management of vasculitis (van Leeuwen et al. 2025; Jayne et al. 2021; Merkel et al. 2020).

In the context of transplantation, Eculizumab (E.) is one of the most commonly used inhibitors of the complement cascade. E. is an anti-C5 monoclonal antibody that binds to the complement C5 protein and inhibits the cleavage of C5 into C5a and C5b and thus inhibits the forming of the membrane attack complex (MAC), the final effector pathway of complement activation. E. has its primary indication for the treatment of paroxysmal nocturnal hemoglobinuria and atypical hemolytic uremic syndrome in which aberrant activation of the complement system occurs during disease progression. In order to reduce chronic antibody-mediated rejection damaging the endothelium by complement activation, E. has been employed in patients treated for severe chronic antibody-mediated rejection (AMR) after lung (Mallet et al. 2025) and heart Tx (Kearney et al. 2021) in order to bridge for re-Tx. A clinical trial was recently performed applying E. in high-risk heart transplant recipients showing a lower incidence of AMR or death (Coutance et al. 2023). Moreover, E. seems to have a role in the prophylactic use of complement inhibition for heart transplantation: Patek and colleagues observed a dramatic reduction in the risk of biopsy-proven AMR in patients treated with E. (Patel et al. 2021). In another phase 2, randomized, multicenter clinical trial, a potential benefit for E. could also be observed in preventing acute AMR in recipients sensitized to their living-donor kidney transplants (Marks et al. 2019). E. was also considered as a promising agent to reduce thrombotic microangiopathy after kidney (Avila et al. 2021) or lung Tx given as a short-term cycle (Trujillo et al. 2024). Also, E. was successfully employed in late AMR after the failure of intensified conventional immunosuppressive treatment in pediatric renal Tx (Siddiqui et al. 2022). While E. therapy appears to have the ability to stabilize renal function in the short term, the expression of endothelial cell-associated transcripts predicting an ongoing humoral injury was not shown to be reduced (Kulkarni et al. 2017). Several other monoclonal antibodies, such as crovalimab, pozelimab, tesidolumab, and ravulizumab that inhibit complement activation through targeting C5, have been introduced and although there is robust data showing encouraging results in the prophylaxis or treatment of AMR, the beneficial effect on outcome is uncertain, particularly for long-term outcome (Schroppel et al. 2020).

4.4.2 *IL 6 inhibition*

Upon allogeneic transplantation, inflamed endothelial cells enhance the production of IL 6 after HLA-DR antibody binding. IL 6 blockade is therefore increasingly investigated as a therapeutic target. Earlier investigations on coronary allograft injury by Fogal et al. reported significantly increased IL 6 levels in endothelial cells when human coronary arteries were transplanted into immunodeficient mice.

Treatment with anti-IL 6 antibody showed significant reductions in endothelial cell IL 6 expression, limited T cell infiltration, and significant reductions in intimal proliferation (Fogal et al. 2011). Also, the Th17+ cell population was reduced and the proportion of FoxP3 T cells was increased. Lion and colleagues reported that in cultures treated with clazakizumab, another monoclonal antibody against IL 6, endothelial cells did not activate T effector functions, decreasing Th17 and Th1 cell populations. This was additionally associated with a decline in complement activation reflected by reduced C5b-C9 MACs. In a follow-up study, Lion and colleagues tested humanized antibody blockade of IL 6 on endothelial cell interactions with allogeneic peripheral blood mononuclear cells, and after anti-HLA or DSA binding to endothelial cells in an in vitro human experimental model, they found that, upon IL 6 blockade, CD4+, T cell activation was impaired; however, in this case, complement activation remained unchanged (Lion et al. 2023). Although these results are generated in vitro, they shed light on IL 6 as a promising target to attenuate endothelial injury apart from its role in the recruitment of immune cells against the endothelium. The IMAGINE trial (Interleukin-6 Blockade Modifying Antibody-mediated Graft Injury and Estimated Glomerular Filtration Rate Decline) (NCT03744910), employing clazakizumab against AMR in kidney Tx, is currently the only clinical trial for the treatment of chronic rejection in this regard (Berger et al. 2024).

These observations have led to the application of the monoclonal IL 6 receptor antibody Tocilizumab in, for example, the setting of lung Tx. January and colleagues analyzed in a case-control study the effect of Tocilizumab on DSA and found that the treatment with T. resulted in a higher clearance of DSA, a lower recurrence of DSA, a lower incidence of new DSA, and lower rates of graft failure when compared to those treated for AMR without Tocilizumab. The incidence of side effects was similar between groups (January et al. 2023). A prospective, multicenter phase 2 clinical trial was done in which 200 primary heart transplant recipients were randomized to receive either Tocilizumab or placebo plus standard triple maintenance immunosuppression (NCT03644667). This trial was just completed in March 2025 and results are pending. Interestingly, in porcine xenotransplantation, human anti-IL 6 antibody binds to human IL 6 and prevents IL 6R activation but does not bind to porcine IL 6 and thus does not prevent inflammation. In contrast, IL 6R inhibitors such as Tocilizumab block IL 6 activity through binding of human and porcine IL-6 to human IL-6R (Iwanczyk et al. 2025).

4.4.3 Extracellular Vesicles

Extracellular vesicles (EVs) are particles that are enclosed by a lipid double layer and can be secreted by virtually all cells (Kalluri and LeBleu 2020). These membranous structures can contain nucleic acids, proteins, ligands, and molecules and serve either as biomarkers for predicting disease outcomes, but they can function also as therapeutic carriers for injured tissue and cells, restoring physiological conditions in a regenerative or anti-inflammatory manner. Brugiere and colleagues analyzed the potential of EV-bound HLA-G plasma levels as liquid biopsies in an exploratory study, suggesting that CLAD onset and subsequent graft failure can be potentially predicted (Brugiere et al. 2025). To exert therapeutic effects, EVs need to be customized to become specific to target cells. In an approach to target and potentially protect activated and inflamed endothelium, Zheng and colleagues genetically engineered EVs that display the tetrasaccharide glycan Sialyl Lewis-X on their surface. Sialyl Lewis-X is the ligand for E-selectin that is expressed on endothelial cells during inflammation and responsible for leukocyte adhesion (Zheng et al. 2022). Authors here presented a novel glycoengineering strategy for surface display of Sialyl Lewis-X on EVs showing that endothelial repair with EVs from mesenchymal stem cells is possible. Another approach of tailoring EVs to treat endothelial damage was presented by He and colleagues: they applied a three-dimensional protocol for hiPSC differentiation into endothelial cells (ECs) with isolation of EVs from the derived hiPSC-EC validating the therapeutic effect in intimal hyperplasia models. They found that intravenously injected EVs accumulate on the carotid artery endothelium and significantly alleviate the intimal thickening. Here, miR-126 was the most abundant in hiPSC-EC-produced EVs. miR-126 was able to prevent endothelial apoptosis upon EV inflammation, promoting EC migration and tube formation through autophagy upregulation (He et al. 2022). A prominent role of miR-126 was similarly assigned before, showing that extracellular vesicles released from endothelial progenitor cells stimulate angiogenesis through the transfer of miR-126 in endothelial cell culture systems (Deregibus et al. 2007). As an alternative to EVs, so-called membrane particles (MPs) can fulfill major tasks otherwise done by EVs. MPs are nanovesicles that are artificially generated from the membranes of mesenchymal stem cells (MSCs) and possess some of the therapeutic properties of MSCs. Merino and colleagues could convincingly show that small circular fragments of cell membranes from MSC can ameliorate TNFα-induced endothelial injury by restoring endothelium integrity, blocking cell transmigration, and stimulating angiogenesis (Merino et al. 2021).

EVs have gained increased attention in transplant surveillance in different organs and they have started to be extensively studied for their utilization as therapeutics in transplantation. Even if EV therapies seem to have little risk in terms of engrafting in vitro engineered cells, clinical trials providing long term-data in solid organ transplanted patients are still pending.

4.4.4 *Modulators of Chemokine-Glycosaminoglycan (GAG) Interactions and the Endothelial Glycocalyx*

Immune cell invasion after transplantation of solid organs is directed by chemokines that bind to glycosaminoglycans (GAGs). GAG-bound chemokines on the luminal side of vascular endothelial cells are key for the initiation of immune responses and for the forming of directional gradients allowing immune cells to traverse the vascular endothelium and to enter the engrafted organ. Zanetti and colleagues reported that pretransplant treatment with the Myxoma virus-derived chemokine modulating protein, M-T7, a chemokine-modulating protein that blocks chemokine–GAG interactions before renal transplantation, led to a disruption of this chemokine–GAG interactions with a significant reduction in endothelial inflammation (Zanetti et al. 2022). This approach is an example of how viruses can be a potential source of new immune-modulating protein therapeutics for the prevention of the endothelial injury of the transplant organ. Crijns and colleagues in a comprehensive review a wide range of experimental approaches and emerging strategies including clinical trials that target the chemokine–GAG interactions in endothelial inflammation (Crijns et al. 2020).

Another approach to maintaining the endothelial integrity has been proposed by Siren and colleagues. They rebuild the glycocalyx structure via cell-surface engineering by attaching functionalized polymers to the endothelial surface through an enzyme-mediated conjugation strategy using guinea pig liver tissue transglutaminase (Siren et al. 2021). The subsequent strong binding between lysin residues of the endothelium and the polymer led to less leukocyte attachment. Furthermore, authors integrated sialic acid into the polymers in order to additionally create an immunosuppressive environment, thus further attenuating the attachment of immune cells onto the endothelial surface. This approach is an elegant way to shield the entry of immune cells through injured endothelium and to circumvent systemic immunosuppression. Moreover, this clinically translatable strategy can be combined with other ex vivo organ preservation techniques to optimize the prevention of chronic organ rejection.

4.5 Summary and Conclusion

Chronic graft rejection after organ transplantation comprises the major challenge for graft survival and a successful clinical outcome. The signaling that leads to rejection is a complex process and involves interactions between several cell types: the critical players being the endothelium of the graft and the immune cell subsets and soluble factors from the recipient.

Immediately after transplantation, the endothelium of the graft vessel wall is activated by virtue of its encounters with the immune system of the host and by I/R injury. Activation is manifested by the expression of several surface molecules on the endothelial cells, which collectively promote transmigration of immune cells (leukocytes) from the recipient's bloodstream across the endothelial monolayer into the vessel wall and tissue. Cytokines released from activated endothelial cells as well as leukocytes facilitate a proinflammatory microenvironment that contributes to the development of CR.

Thus, therapeutic strategies to improve the outcome of transplantation need to be focused on improving endothelial health and function. There are promising strategies and compounds on the way such as inhibitors of interleukin 6 or complement, but also therapeutic carriers such as extracellular vesicles and glycocalyx polymers that can effectively protect the endothelial integrity that can attenuate chronic allograft rejection and improve patient outcome after transplantation.

References

Alegre ML, Lakkis FG, Morelli AE (2016) Antigen presentation in transplantation. Trends Immunol 37:831–843

Al-Lamki RS, Bradley JR, Pober JS (2008) Endothelial cells in allograft rejection. Transplantation 86:1340–1348

Amersfoort J, Eelen G, Carmeliet P (2022) Immunomodulation by endothelial cells – partnering up with the immune system? Nat Rev Immunol 22:576–588

Augustin HG, Koh GY (2024) A systems view of the vascular endothelium in health and disease. Cell 187:4833–4858

Avila A, Gavela E, Sancho A (2021) Thrombotic microangiopathy after kidney transplantation: an underdiagnosed and potentially reversible entity. Front Med (Lausanne) 8:642864

Baroja-Mazo A, Revilla-Nuin B, de Bejar A, Martinez-Alarcon L, Herrero JI, El-Tayeb A, Muller CE, Aparicio P, Pelegrin P, Pons JA (2019) Extracellular adenosine reversibly inhibits the activation of human regulatory T cells and negatively influences the achievement of the operational tolerance in liver transplantation. Am J Transplant 19:48–61

Berger M, Baliker M, Van Gelder T, Bohmig GA, Mannon RB, Kumar D, Chadban S, Nickerson P, Lee LA, Djamali A (2024) Chronic active antibody-mediated rejection: opportunity to determine the role of interleukin-6 blockade. Transplantation 108:1109–1114

Brandon W, Dunn C, Bollineni S, Joerns J, Lawrence A, Mohanka M, Timofte I, Torres F, Kaza V (2023) Management of donor-specific antibodies in lung transplantation. Front Transplant 2:1248284

Browning E, Wang H, Hong N, Yu K, Buerk DG, DeBolt K, Gonder D, Sorokina EM, Patel P, De Leon DD, Feinstein SI, Fisher AB, Chatterjee S (2014) Mechanotransduction drives post ischemic revascularization through K(ATP) channel closure and production of reactive oxygen species. Antioxid Redox Signal 20:872–886

Brugiere O, Dreyfuss D, Guilet R, Rong S, Hirschi S, Renaud-Picard B, Reynaud-Gaubert M, Coiffard B, Bunel V, Messika J, Demant X, Le Pavec J, Dauriat G, Saint Raymond C, Falque L, Mornex JF, Tissot A, Lair D, Le Borgne Krams A, Bousseau V, Magnan A, Picard C, Roux A, Glorion M, Carmagnat M, Gazeau F, Aubertin K, Carosella E, Vallee A, Landais C, Rouas-Freiss N, LeMaoult J, the CC (2025) Circulating vesicular-bound HLA-G as noninvasive predictive biomarker of CLAD after lung transplantation. Transplantation 109:736–745

Budding K, van de Graaf EA, Kardol-Hoefnagel T, Broen JC, Kwakkel-van Erp JM, Oudijk EJ, van Kessel DA, Hack CE, Otten HG (2016) A promoter polymorphism in the CD59 complement regulatory protein gene in donor lungs correlates with a higher risk for chronic rejection after lung transplantation. Am J Transplant 16:987–998

Chatterjee SFA (2004) The role of ischemia-reperfusion injury in graft rejection. In: Wilkes DS, Burlingham WJ (eds) Immunobiology of organ transplantation. Kluwer/Academic, New York, pp 545–563

Chatterjee S, Nieman GF, Christie JD, Fisher AB (2014) Shear stress-related mechanosignaling with lung ischemia: lessons from basic research can inform lung transplantation. Am J Physiol Lung Cell Mol Physiol 307:L668–L680

Cook-Mills JM, Deem TL (2005) Active participation of endothelial cells in inflammation. J Leukoc Biol 77:487–495
Coutance G, Kobashigawa JA, Kransdorf E, Loupy A, Desire E, Kittleson M, Patel JK (2023) Intermediate-term outcomes of complement inhibition for prevention of antibody-mediated rejection in immunologically high-risk heart allograft recipients. J Heart Lung Transplant 42:1464–1468
Crijns H, Vanheule V, Proost P (2020) Targeting chemokine-glycosaminoglycan interactions to inhibit inflammation. Front Immunol 11:483
Crikis S, Lu B, Murray-Segal LM, Selan C, Robson SC, D'Apice AJ, Nandurkar HH, Cowan PJ, Dwyer KM (2010) Transgenic overexpression of CD39 protects against renal ischemia-reperfusion and transplant vascular injury. Am J Transplant 10:2586–2595
Deregibus MC, Cantaluppi V, Calogero R, Lo Iacono M, Tetta C, Biancone L, Bruno S, Bussolati B, Camussi G (2007) Endothelial progenitor cell derived microvesicles activate an angiogenic program in endothelial cells by a horizontal transfer of mRNA. Blood 110:2440–2448
Fogal B, Yi T, Wang C, Rao DA, Lebastchi A, Kulkarni S, Tellides G, Pober JS (2011) Neutralizing IL-6 reduces human arterial allograft rejection by allowing emergence of CD161+ CD4+ regulatory T cells. J Immunol 187:6268–6280
Gaykema LH, van Nieuwland RY, Dekkers MC, van Essen MF, Heidt S, Zaldumbide A, van den Berg CW, Rabelink TJ, van Kooten C (2022) Inhibition of complement activation by CD55 overexpression in human induced pluripotent stem cell derived kidney organoids. Front Immunol 13:1058763
Gielis JF, Quirynen L, Briede JJ, Roelant E, Cos P, Van Schil PEY (2017) Pathogenetic role of endothelial nitric oxide synthase uncoupling during lung ischaemia-reperfusion injury. Eur J Cardiothorac Surg 52:256–263
He Y, Li Q, Feng F, Gao R, Li H, Chu Y, Li S, Wang Y, Mao R, Ji Z, Hua Y, Shen J, Wang Z, Zhao M, Yao Q (2022) Extracellular vesicles produced by human-induced pluripotent stem cell-derived endothelial cells can prevent arterial stenosis in mice via autophagy regulation. Front Cardiovasc Med 9:922790
Hellenthal KEM, Brabenec L, Wagner NM (2022) Regulation and dysregulation of endothelial permeability during systemic inflammation. Cells 11:1935
Hellmich B, Sanchez-Alamo B, Schirmer JH, Berti A, Blockmans D, Cid MC, Holle JU, Hollinger N, Karadag O, Kronbichler A, Little MA, Luqmani RA, Mahr A, Merkel PA, Mohammad AJ, Monti S, Mukhtyar CB, Musial J, Price-Kuehne F, Segelmark M, Teng YKO, Terrier B, Tomasson G, Vaglio A, Vassilopoulos D, Verhoeven P, Jayne D (2024) EULAR recommendations for the management of ANCA-associated vasculitis: 2022 update. Ann Rheum Dis 83:30–47
Iwanczyk Z, Hara H, Cooper DKC, Maenaka A (2025) Inhibition of inflammation by IL-6 blockade in xenotransplantation. Cytokine 189:156897
January SE, Fester KA, Halverson LP, Witt CA, Byers DE, Vazquez-Guillamet R, Alexander-Brett J, Tague LK, Kreisel D, Gelman A, Puri V, Bahena RN, Takahashi T, Hachem RR, Kulkarni HS (2023) Tocilizumab for antibody-mediated rejection treatment in lung transplantation. J Heart Lung Transplant 42:1353–1357
Jayne DRW, Merkel PA, Schall TJ, Bekker P, Group AS (2021) Avacopan for the treatment of ANCA-associated vasculitis. N Engl J Med 384:599–609
Kalluri R, LeBleu VS (2020) The biology, function, and biomedical applications of exosomes. Science 367:eaau6977
Kearney K, Macdonald P, Hayward C, Muthiah K (2021) The use of eculizumab as a bridge to retransplantation for chronic antibody-mediated rejection in a heart transplant recipient: a case report. Eur Heart J Case Rep 5:ytab180
Khan MA, Shamma T, Kazmi S, Altuhami A, Ahmed HA, Assiri AM, Broering DC (2020) Hypoxia-induced complement dysregulation is associated with microvascular impairments in mouse tracheal transplants. J Transl Med 18:147

Kitai T, Okamoto T, Miyakoshi C, Niikawa H, Alvarez PA, Krittanawong C, Xanthopoulos A, McCurry KR (2019) Impact of combined heart and lung transplantation on bronchiolitis obliterans syndrome, cardiac allograft vasculopathy, and long-term survival. J Heart Lung Transplant 38:1170–1177

Kulkarni S, Kirkiles-Smith NC, Deng YH, Formica RN, Moeckel G, Broecker V, Bow L, Tomlin R, Pober JS (2017) Eculizumab therapy for chronic antibody-mediated injury in kidney transplant recipients: a pilot randomized controlled trial. Am J Transplant 17:682–691

Kummer L, Zaradzki M, Vijayan V, Arif R, Weigand MA, Immenschuh S, Wagner AH, Larmann J (2020) Vascular signaling in allogenic solid organ transplantation – the role of endothelial cells. Front Physiol 11:443

Lau CL, Beller JP, Boys JA, Zhao Y, Phillips J, Cosner M, Conaway MR, Petroni G, Charles EJ, Mehaffey JH, Mannem HC, Kron IL, Krupnick AS, Linden J (2020) Adenosine A2A receptor agonist (regadenoson) in human lung transplantation. J Heart Lung Transplant 39:563–570

Lion J, Maitre ML, de Truchis C, Taupin JL, Poussin K, Haziot A, Chong E, Glotz D, Mooney N (2023) Restriction of interleukin-6 alters endothelial cell immunogenicity in an allogenic environment. Clin Transpl 37:e14851

Mahdi BM (2013) A glow of HLA typing in organ transplantation. Clin Transl Med 2:6

Mallet H, Razat L, Perrier Q, Briault A, Raymond CS, Falque L, Rostaing L, Degano B, Bedouch P (2025) Effectiveness of complement inhibitors against refractory antibody-mediated rejection of lung transplantation: two clinical cases. Transpl Immunol 89:102174

Marks WH, Mamode N, Montgomery RA, Stegall MD, Ratner LE, Cornell LD, Rowshani AT, Colvin RB, Dain B, Boice JA, Glotz D, Group CS (2019) Safety and efficacy of eculizumab in the prevention of antibody-mediated rejection in living-donor kidney transplant recipients requiring desensitization therapy: a randomized trial. Am J Transplant 19:2876–2888

Merino A, Sablik M, Korevaar SS, Lopez-Iglesias C, Ortiz-Virumbrales M, Baan CC, Lombardo E, Hoogduijn MJ (2021) Membrane particles derived from adipose tissue mesenchymal stromal cells improve endothelial cell barrier integrity. Front Immunol 12:650522

Merkel PA, Niles J, Jimenez R, Spiera RF, Rovin BH, Bomback A, Pagnoux C, Potarca A, Schall TJ, Bekker P, Investigators C (2020) Adjunctive treatment with avacopan, an Oral C5a receptor inhibitor, in patients with antineutrophil cytoplasmic antibody-associated vasculitis. ACR Open Rheumatol 2:662–671

Minamoto K, Pinsky DJ (2002) Recipient iNOS but not eNOS deficiency reduces luminal narrowing in tracheal allografts. J Exp Med 196:1321–1333

Murata K, Baldwin WM 3rd. (2009) Mechanisms of complement activation, C4d deposition, and their contribution to the pathogenesis of antibody-mediated rejection. Transplant Rev (Orlando) 23:139–150

Nordling S, Hong J, Fromell K, Edin F, Brannstrom J, Larsson R, Nilsson B, Magnusson PU (2015) Vascular repair utilising immobilised heparin conjugate for protection against early activation of inflammation and coagulation. Thromb Haemost 113:1312–1322

Ohtsuka T, Changelian PS, Bouis D, Noon K, Harada H, Lama VN, Pinsky DJ (2010) Ecto-5′-nucleotidase (CD73) attenuates allograft airway rejection through adenosine 2A receptor stimulation. J Immunol 185:1321–1329

Onuta G, van Ark J, Rienstra H, Boer MW, Klatter FA, Bruggeman CA, Zeebregts CJ, Rozing J, Hillebrands JL (2010) Development of transplant vasculopathy in aortic allografts correlates with neointimal smooth muscle cell proliferative capacity and fibrocyte frequency. Atherosclerosis 209:393–402

Park HS, Kim JE, You HJ, Gu J, Yoo B, Lee S, Lee HJ, Hwang HY, Hwang Y, Kim HK, Kim YT (2015) Beneficial effect of a nitric oxide donor in an ex vivo model of pig-to-human pulmonary xenotransplantation. Xenotransplantation 22:391–398

Patel JK, Coutance G, Loupy A, Dilibero D, Hamilton M, Kittleson M, Kransdorf E, Azarbal B, Seguchi O, Zhang X, Chang D, Geft D, Czer L, Varnous S, Kobashigawa JA (2021) Complement inhibition for prevention of antibody-mediated rejection in immunologically high-risk heart allograft recipients. Am J Transplant 21:2479–2488

Pilmore HL, Painter DM, Bishop GA, McCaughan GW, Eris JM (2000) Early up-regulation of macrophages and myofibroblasts: a new marker for development of chronic renal allograft rejection. Transplantation 69:2658–2662

Piotti G, Palmisano A, Maggiore U, Buzio C (2014) Vascular endothelium as a target of immune response in renal transplant rejection. Front Immunol 5:505

Pober JS, Merola J, Liu R, Manes TD (2017) Antigen presentation by vascular cells. Front Immunol 8:1907

Pober JS, Chih S, Kobashigawa J, Madsen JC, Tellides G (2021) Cardiac allograft vasculopathy: current review and future research directions. Cardiovasc Res 117:2624–2638

Qian Z, Gelzer-Bell R, Yang Sx SX, Cao W, Ohnishi T, Wasowska BA, Hruban RH, Rodriguez ER, Baldwin WM 3rd, Lowenstein CJ (2001) Inducible nitric oxide synthase inhibition of weibel-palade body release in cardiac transplant rejection. Circulation 104:2369–2375

Rose ML (1998) Endothelial cells as antigen-presenting cells: role in human transplant rejection. Cell Mol Life Sci 54:965–978

Schiefer J, Faybik P, Koch S, Tudor B, Kollmann D, Kuessel L, Krenn CG, Berlakovich G, Baron DM, Baron-Stefaniak J (2020) Glycocalyx damage within human liver grafts correlates with graft injury and postoperative graft function after orthotopic liver transplantation. Transplantation 104:72–78

Schroppel B, Akalin E, Baweja M, Bloom RD, Florman S, Goldstein M, Haydel B, Hricik DE, Kulkarni S, Levine M, Mehrotra A, Patel A, Poggio ED, Ratner L, Shapiro R, Heeger PS (2020) Peritransplant eculizumab does not prevent delayed graft function in deceased donor kidney transplant recipients: results of two randomized controlled pilot trials. Am J Transplant 20:564–572

Siddiqui MA, Baskin E, Yarbug Karakayali F, Gemici A, Gulleroglu K, Caltik Yilmaz A, Moray G, Haberal M (2022) Use of eculizumab in pediatric patients with late antibody-mediated rejection after kidney transplantation. Exp Clin Transplant 20:49–52

Siren EMJ, Luo HD, Tam F, Montgomery A, Enns W, Moon H, Sim L, Rey K, Guan Q, Wang JJ, Wardell CM, Monajemi M, Mojibian M, Levings MK, Zhang ZJ, Du C, Withers SG, Choy JC, Kizhakkedathu JN (2021) Prevention of vascular-allograft rejection by protecting the endothelial glycocalyx with immunosuppressive polymers. Nat Biomed Eng 5:1202–1216

Sladden TM, Yerkovich S, Grant M, Zhang F, Liu X, Trotter M, Hopkins P, Linhardt RJ, Chambers DC (2019) Endothelial glycocalyx shedding predicts donor organ acceptability and is associated with primary graft dysfunction in lung transplant recipients. Transplantation 103:1277–1285

Snoeijs MG, Vink H, Voesten N, Christiaans MH, Daemen JW, Peppelenbosch AG, Tordoir JH, Peutz-Kootstra CJ, Buurman WA, Schurink GW, van Heurn LW (2010) Acute ischemic injury to the renal microvasculature in human kidney transplantation. Am J Physiol Renal Physiol 299:F1134–F1140

Tao JQ, Sorokina EM, Vazquez Medina JP, Mishra MK, Yamada Y, Satalin J, Nieman GF, Nellen JR, Beduhn B, Cantu E, Habashi NM, Jungraithmayr W, Christie JD, Chatterjee S (2016) Onset of inflammation with ischemia: implications for donor lung preservation and transplant survival. Am J Transplant 16:2598–2611

Tikkanen JM, Singer LG, Kim SJ, Li Y, Binnie M, Chaparro C, Chow CW, Martinu T, Azad S, Keshavjee S, Tinckam K (2016) De novo DQ donor-specific antibodies are associated with chronic lung allograft dysfunction after lung transplantation. Am J Respir Crit Care Med 194:596–606

Trujillo H, Huerta A, Alonso R, Serrano ML, Aguilar M, Morales E, Cavero T (2024) Eculizumab as salvage treatment for thrombotic microangiopathy after lung transplantation. Clin Transpl 38:e15443

van Leeuwen JR, Quartuccio L, Draibe JB, Gunnarson I, Sprangers B, Teng YKO (2025) Evaluating avacopan in the treatment of ANCA-associated vasculitis: design, development and positioning of therapy. Drug Des Devel Ther 19:23–37

Verrier ED (2004) Activation of the endothelium in cardiac allografts. J Heart Lung Transplant 23:S229–S233

Wang B, Zhou A, Pan Q, Li Y, Xi Z, He K, Li D, Li B, Liu Y, Liu Y, Xia Q (2024) Adenosinergic metabolism pathway: an emerging target for improving outcomes of solid organ transplantation. Transl Res 263:93–101

Zanetti IR, Burgin M, Zhang L, Yeh ST, Ambadapadi S, Kilbourne J, Yaron JR, Lowe KM, Daggett-Vondras J, Fonseca D, Boyd R, Wakefield D, Clapp W, Lim E, Chen H, Lucas A (2022) Virus-derived chemokine modulating protein pre-treatment blocks chemokine-glycosaminoglycan interactions and significantly reduces transplant immune damage. Pathogens 11:588

Zeiser R, Robson SC, Vaikunthanathan T, Dworak M, Burnstock G (2016) Unlocking the potential of purinergic signaling in transplantation. Am J Transplant 16:2781–2794

Zhang J, Hu W, Xing W, You T, Xu J, Qin X, Peng Z (2011) The protective role of CD59 and pathogenic role of complement in hepatic ischemia and reperfusion injury. Am J Pathol 179:2876–2884

Zheng W, He R, Liang X, Roudi S, Bost J, Coly PM, van Niel G, Andaloussi SEL (2022) Cell-specific targeting of extracellular vesicles through engineering the glycocalyx. J Extracell Vesicles 11:e12290

Chapter 5
Pathogenesis of Chronic Rejection: Graft Endothelial Cell Trans-presentation of IL-15 Connects Alloantibody- and Cell-Mediated Steps in Allograft Vasculopathy

Clancy W. Mullan and Jordan S. Pober

Abstract Chronic rejection of allografts appears mechanistically distinct from acute antibody-mediated and or acute cell-mediated rejection. Histological features point to roles for alloantibody and lymphocytic infiltrates, and allograft vasculopathy is a common feature in chronic rejection of several different solid organs. Vasculopathy, in turn, can cause late graft failure due to chronic ischemia, parenchymal cell loss, and replacement fibrosis. Molecular analyses implicate graft vascular endothelial cells in this process. Here, we review evidence favoring a hypothesis linking alloantibody and complement to induction of endothelial cell surface expression and trans-presentation of IL-15 as a critical signal that enhances lymphocyte activation during antigen-mediated recruitment of alloreactive T cells. These T cells trans-migrate into the intima and secrete interferon-γ, which acts both on vascular smooth muscle cells within the arterial wall to stimulate their proliferation, resulting in vasculopathy, and back on the endothelial cells to provide positive feedback for this sequence of events.

Supported by Yale Department of Surgery Scientist Training Award (C.W.M.) and NIH grants R21AI180187 and R01AI175206 (J.S.P.)

C. W. Mullan
Department of Surgery, Yale School of Medicine, New Haven, CT, USA

J. S. Pober (✉)
Departments of Immunobiology, Pathology and Dermatology, Yale School of Medicine, New Haven, CT, USA
e-mail: jordan.pober@yale.edu

M. Kloc, R. M. Ghobrial (eds.), *Cellular and Molecular Mechanisms of Chronic Rejection in Organ Transplantation*, Results and Problems in Cell Differentiation 77, https://doi.org/10.1007/978-3-032-07686-1_5

5.1 Introduction

Solid vascularized organ allografts can trigger potent host immunological responses that lead to graft cell destruction over weeks or months, most often associated with infiltrating cytotoxic T lymphocytes (CTLs) and/or antibodies reactive with donor alloantigens. Both CTL and antidonor antibodies largely target non-self-allelic forms of major histocompatibility complex gene-encoded proteins (human leukocyte antigens [HLA]) expressed on graft cells. Over the past three to four decades, patient screening and selection to limit the extent of HLA mismatch and to avoid the presence of high-titer preformed antidonor antibodies combined with regimens of effective and tolerated immunosuppressive agents have significantly reduced graft loss due to these forms of acute rejection during the first post-transplant year. Despite these significant decreases in the incidence, severity, and graft loss due to acute rejection, there remains a significant rate (typically 3–10%, depending on the organ) of graft loss per year over the subsequent decade that has not much changed despite maintenance of immunosuppression. Although the role of the immune system in this process was once questioned, it is now evident that most cases are associated with the development of a de novo donor-specific antibody and graft tissue alterations in transcripts that implicate the immune system (Loupy and Lefaucheur 2018; Halloran et al. 2010). The underlying process has common features across several different organ types (kidney, heart, and lung) and is now best described as "chronic rejection," involving immune mechanisms that are not well targeted by the immunosuppressive agents used to control acute rejection. The shift of the nature of the immune response is not surprising in retrospect, as the immune system's response to an allograft strongly resembles the immune response to viral infections and viruses that cannot be acutely eliminated by the immune system-mediated cytolysis of infected cells will activate alternative immune mechanisms more focused on fibrotic "containment" than eradication. Here, we will begin with a brief description of the histological and transcriptomic patterns associated with chronic rejection in kidney, heart, and lung allografts and then propose a hypothesis regarding a central role in this process played by a unique cytokine, interleukin-15, that is induced in graft endothelial cells (ECs), linking donor-specific antibodies to alloreactive T cell responses (Fig. 5.1).

5.2 Histopathologic Changes in Chronic Rejection

Chronic rejection has three common histological features in diverse organs: loss of organ parenchymal cells, interstitial fibrosis, and alterations in graft vasculature.

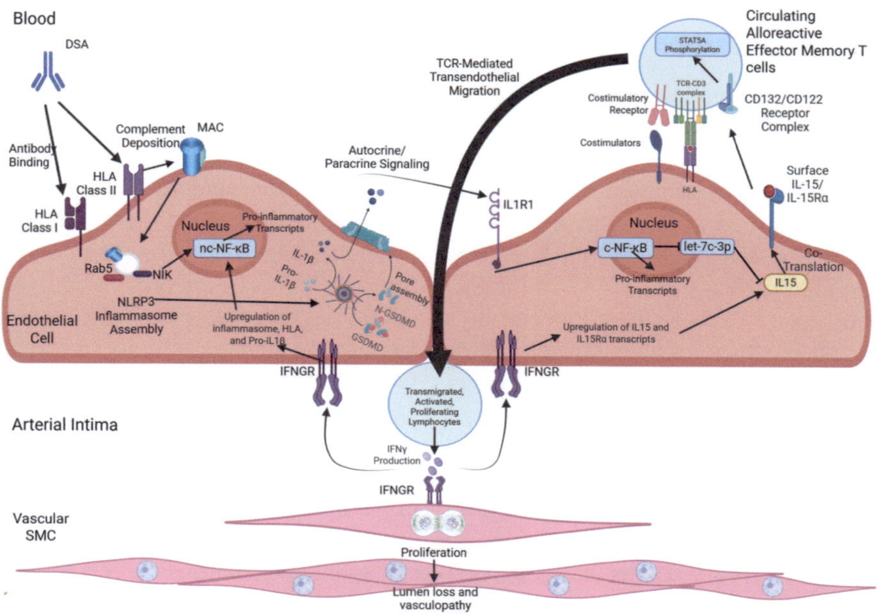

Fig. 5.1 A hypothesis linking alloantibody, graft endothelial immunogenicity, and chronic rejection. Donor-specific antibody (DSA) binds to endothelial cell human leukocyte antigen (HLA) class I and II complexes, resulting in classical complement activation and membrane attack complex (MAC) formation. MAC is internalized on Rab5⁺ endosomes and results in NIK stabilization with subsequent noncanonical NF-κB activation and nuclear localization of p52 with induction of proinflammatory transcripts. Simultaneously, NIK catalyzes the assembly of an NLRP3 inflammasome that proteolytically processes inactive pro-IL-1β to active IL-1β and inactive gasdermin D membrane pore-forming N-gasdermin D, allowing active IL-1β release. Autocrine/paracrine signaling via IL1R1 activates canonical NF-κB signaling in ECs, resulting in additional inflammatory transcripts including ICOSL and downregulation of let-7c-3p miRNA, the latter releasing a blockade on the translation of IL-15 mRNA to protein, surface mobilization in a complex with IL-15Rα, and trans-presentation to circulating alloreactive T cells and (not shown) NK cells. Increased costimulatory molecules support the T cell receptor-mediated transendothelial migration to the intima, where these highly activated alloreactive T cells display enhanced effector molecule production, proliferation, and IFNγ production. IFNγ acts on both nearby vascular smooth muscle cells to stimulate proliferation and vasculopathy and endothelial cells to increase expression of numerous players in this pathway including HLA, inflammasome components, pro-IL-1β, IL-15, and IL-15Rα. Abbreviations: c-NF-κB—canonical nuclear factor kappa-light-chain-enhancer of activated B cells. DSA—donor-specific antibody. GSDMD—Gasdermin D. HLA- human leukocyte antigen. IFNγ—Interferon γ. IFNGR—Interferon γ Receptor. IL-15Rα—interleukin 15 receptor α chain (CD215). IL-15—interleukin 15. IL-1β—Interleukin 1β. IL1R1—Interleukin 1 Receptor 1. MAC- membrane attack complex. nc-NF-κB—noncanonical nuclear factor kappa-light-chain-enhancer of activated B cells. NIK- Nuclear factor-κB inducing kinase. NLRP3—NOD-, LRR-, and pyrin domain-containing protein 3. SMC—Smooth muscle cell. STAT5A—Signal transducer and activator of transcription 5A. TCR—T cell receptor

5.2.1 *Kidney*

In kidneys, the primary parenchymal cell loss involves the epithelial lining of the tubules, and the primary site of fibrosis is the interstitium surrounding the tubules. The diffuse nature of the process differs from an acute infarct in that the fibrosis does not condense into a scar and the tubular loss is widespread. Microvascular changes involve both glomerular and peritubular capillaries, which show increased thickening and layering of the basement membrane, suggestive of cycles of injury and repair (Haas et al. 2018). Infiltrates of mononuclear cells are present but typically sparse except where they are organized into aggregates resembling tertiary lymphoid structures within which microvessels develop markers of high endothelial venules found in lymph nodes. Indeed, microvascular inflammation has been recognized as a criterion for rejection diagnosis (Naesens et al. 2024). Large vessels may show intimal hyperplasia with infiltration of mononuclear leukocytes (Busch et al. 1971). Intimal expansion of graft arteries results in loss of lumen and reduced perfusion. Importantly, the endothelial cell (EC) lining of the renal vascular tree remains largely intact and, except in cases of severe acute rejection, remains of donor origin in the kidneys (Lagaaij et al. 2001). Despite morphological normality, ECs may be coated by donor-specific antibodies and consequent deposition of both early (C3d, C4d) and late (polyC9) complement components. C4d deposition on the endothelium is a widely accepted biomarker of chronic rejection of kidney allografts (Wallace et al. 2005). A substantial percentage of kidney transplant recipients currently not diagnosed with rejection have molecular evidence of rejection that results in a higher risk of graft loss compared to true nonrejecting controls (Halloran et al. 2025).

5.2.2 *Heart*

In the heart, the primary cell type lost is the cardiac myocyte and fibrosis occurs diffusely around residual myocytes (i.e., “replacement fibrosis”). As in the kidney, this pattern differs from an acute infarct in that the fibrotic areas do not coalesce into defined scars. The entire coronary vasculature may be altered, but changes are most pronounced in the arterial tree, affecting both extra and intramyocardial vessels (Billingham 1994). As in the kidney, the primary change in the arterial wall is infiltration of mononuclear leukocytes and accumulation of cells bearing markers of vascular smooth muscle cells (SMCs) that expand the intima and compromise the lumen without adequate compensatory outward remodeling. Multiple terms have been used to describe these changes, but they are now commonly referred to as cardiac allograft vasculopathy (CAV) and allograft vasculopathy may be applied to describe similar changes seen in other allograft organ types. CAV lesions may start focally and form eccentrically, much like atheromas, but unlike atheromas, the lesions of allograft vasculopathy eventually expand to be continuous and concentric. Ultimately, clear evidence of vascular damage may develop, including hemorrhagic, necrotic, and thrombotic changes (Berry et al. 2013). In donor hearts in which arteries contain atheromas, the changes of allograft vasculopathy may develop on top of pre-existing lesions and are morphologically distinct. The mononuclear cell infiltrates found within the neointima consist of both lymphocytes and myeloid cells and are localized primarily subjacent to the endothelial lining. Tertiary lymphoid structures may form around the adventitia of the extramyocardial arterial tree. Interestingly, the endothelium of the arterial tree remains of donor origin although about 20% of the lining of microvessels within the myocardium may be replaced by cells of recipient origin; this microvascular replacement does not correlate with rejection (Minami et al. 2005).

5.2.3 Lung

In the lung, the airspaces and vascular tree may be lost to fibrotic and destructive changes, diminishing the volume of functional airspace and compromising ventilation, perfusion, and gas exchange (von der Thüsen et al. 2018). The most striking abnormality is in the walls of the bronchi, which, like the large vessels of the heart and kidney, show smooth muscle cell (SMC) expansion of the wall that impinges on the airway lumen (bronchiolitis obliterans). Similar to other organs, there are also changes in the vasculature. The ECs that form the inner layer of the vasculature may be also decorated by donor-specific antibody and complement, but this finding is less consistent in the lung than in hearts or kidneys (Cone et al. 2021).

In summary, across several organ types, the common histopathologic themes of chronic rejection are invasion of the allograft by recipient immune cells, expansion of SMCs (typically within the intima of large vessels or, in the case of lung, additionally within the airways), and replacement fibrosis of the parenchyma. The arterial changes in the heart and the kidney produce graft hypoperfusion that may be responsible for parenchymal cell loss and replacement fibrosis, but it is also possible, especially in the lung, that vascular and parenchymal changes are caused by distinct processes. The changes to graft parenchyma result in functional loss over time and, ultimately, in graft failure. It is important to emphasize that chronic rejection typically lacks evidence of extensive graft cell injury and death. However, around 10% of transplant recipients still suffer an acute rejection episode, which despite reversal, increases the risk of chronic rejection and late graft loss (Khush et al. 2018; Matas et al. 1994). Thus, although acute and chronic rejection are readily distinguished by histology, there is a link between them, and clues have been revealed by transcriptomic analyses of biopsies.

5.3 Molecular Profile of Chronic Rejection

Early transcriptional analysis of acutely rejecting organs focused primarily on known inflammatory mediators such as T cell-derived cytokine and effector molecules such as interferon gamma (IFNγ), perforin, Fas-L and granzyme B consistent with CTL-mediated rejection (Strehlau et al. 1997). Acute antibody-mediated rejection was less common and differed in that there were both prominent NK cell and EC transcripts (Baan et al. 1996; Krams et al. 1992). Importantly, transcriptional profiles of chronically rejecting organs resemble those of antibody-mediated rejection and correlate well with clinical variables (Holweg et al. 2011; Mengel et al. 2010). However, despite this commonality, chronic rejection typically lacks histological evidence of extensive graft cell injury and death noted in acute T cell-mediated rejection or fibrinoid necrosis of the vasculature seen in acute antibody-mediated rejection. The unexpected prominence of EC-specific transcripts increased focus on what roles this cell type might be playing in rejection. We will consider this in the next section.

5.4 Endothelial Cells in Allograft Rejection

Graft vessels are lined by a contiguous monolayer of ECs, which although exhibiting some variation among organs and vascular segments within the same organ, have many common features that identify them as a recognizable cell type. Unbiased clustering of single-cell RNA sequencing (scRNAseq) data from essentially every organ confirms the shared features and identity of ECs although there are differences within cells classified as ECs between organs and among different vascular segments within the same organ. Under healthy conditions, ECs inhibit coagulation and thrombosis, maintain a permselective barrier between blood and tissue, and minimize the activation of inflammatory leukocytes (Abrahimi et al. 2015). When activated, most often by inflammatory cytokines such as tumor necrosis factor (TNF) or interleukin 1 (IL-1), ECs lining arterioles cause medial SMCs to relax and increase local perfusion while ECs lining postcapillary venules upregulate adhesion molecules that bind leukocytes and display activating chemokines that induce leukocytes to undergo transendothelial migration, recruiting them into the graft parenchyma. A reduction in permselectivity in venules and capillaries allows circulating macromolecules such as antibodies, fibrinogen, and fibronectin to enter the tissue. The latter two proteins can create a provisional matrix that supports the infiltrating leukocytes. These functions of activated ECs all contribute to changes observed in acute T cell-mediated rejection, but ECs also appear to play a more specific role in adaptive immune responses as antigen-presenting cells.

5.4.1 Origins of Appreciation for an EC Role in Alloimmunity

The adaptive immune functions of ECs were initially discovered in the context of organ transplantation. Rat models of allograft rejection had suggested that the activation of a host immune response depended on the presence within the graft of "passenger leukocytes," more specifically donor dendritic cells, which were presumed to act by migrating to lymphoid organs and presenting graft MHC molecules to activate host T cells (Lechler and Batchelor 1982; Lechler et al. 2001). Rat kidney grafts depleted of dendritic cells were not rejected, but a subsequent infusion of donor strain dendritic cells could initiate the rejection of a stable graft. However, attempts to deplete passenger leukocytes from mouse organs or human organs failed to prevent rejection (Brewer et al. 1989; Kreisel et al. 2002). This led to the notion that cells other than passenger leukocytes could have the capacity to trigger a rejection response, and this capacity could be species-specific. Histological analyses of human organs showed that ECs appeared to be the only nonhematopoietic cell population within the graft that expressed high basal levels of HLA class I and class II molecules (Baldwin 3rd et al. 1981; Lautenschlager et al. 1984). Mouse ECs typically express class I but not class II HLA molecules at baseline. Furthermore, cultured human umbilical vein ECs shared with leukocytes the capacity to activate allogeneic T cells in a mixed leukocyte-endothelial reaction, a property not seen with cultured fibroblasts (Hirschberg et al. 1980). Since donor-derived dendritic cells rapidly (within 2–3 days) disappear from the graft and are rapidly destroyed within the spleen (there is typically no lymphatic drainage of a graft for the first few weeks) and ECs remain of donor origin (Minami et al. 2005), transplant immunologists concluded that ECs are the most-likely tissue cell candidate responsible for initiating rejection.

While Class I HLA molecules are constitutively produced by most nucleated cells, human ECs are essentially the only tissue resident structural cell type (i.e. other than resident leukocytes) that basally (in the absence of inflammation) expresses class II HLA in situ; human ECs also express significantly higher levels of class I HLA molecules than most other tissue-resident structural cell types. Expression levels of class I HLA molecules fall when ECs are cultured and expression of Class II HLA molecules is completely lost (Daar et al. 1984). Expression of both Class I and Class II HLA molecules in cultured human ECs can be restored after stimulation by interferon-γ (IFNγ) (Pober et al. 1983). Expression levels of HLA molecules on most structural cell types can be increased in culture or in vivo by high-dose IFNγ injection. However, in vivo experiments can be complicated by the participation of endogenously generated cytokines made in response to the exogenous IFNγ. This confounder can be eliminated by implantation of human coronary artery segments as aortic interposition grafts in immunodeficient mice. Because mouse IFNγ is unable to engage human IFNγ receptors and vice versa, the normal basal expression levels of class I and II HLA molecules present prior to the implantation are lost on the human ECs of the implanted vessel segment but restored when the animals are given human IFNγ. Because of species-dependent specificities of the cytokine receptor, none of the mouse cells respond to this treatment.

Cytolysis is a response mediated by differentiated and activated CTL and involves engagement of the T cell receptor for antigen (TCR) by a peptide from a non-self-protein bound to and displayed on the target cell surface by a self-class I HLA moleculee. Most cell types expressing relevant HLA protein targets can be killed by CTL. Although normally restricted to peptides displayed by self-allelic forms of HLA molecules, many T cells will cross-react with a variety of peptides (self and non-self protein-derived) displayed by non-self-allelic forms of HLA molecules (Lakkis and Lechler 2013). The functionality of EC HLA proteins, initially detected by antibody binding, in allogeneic recognition by T cells was established by showing both CD8+ and CD4+ CTL respond to allogeneic ECs in a manner blocked by anticlass I and anticlass II HLA antibodies respectively (Epperson and Pober 1994; Salomon et al. 1991).

5.4.2 EC Activation of T Lymphocytes

The activation of resting T cells depends upon HLA recognition but also requires additional antigen-independent signals referred to as "costimulators." Unlike most tissue cells, cultured human ECs can directly activate resting allogeneic $CD4^+$ and CD8 T memory but not naïve T cells (Shiao et al. 2007) although the $CD4^+$ T cell response is dependent upon prior reinduction of class II HLA protein expression. Prolonged culture of human CD8 memory T cells with allogeneic ECs will stimulate differentiation into functional cytotoxic T lymphocytes (CTLs) capable of killing (Biedermann and Pober 1998; Dengler and Pober 2000). While both central and effector memory T cells will proliferate and produce cytokines in culture with allogeneic ECs, effector memory T cells are preferentially activated (Shiao et al. 2007) and additionally are activated to undergo transendothelial migration independent of chemokines, an in vitro assay of diapedesis modeling tissue entry in vivo (Manes and Pober 2008, 2016). Recognition of HLA by the T cell in the absence of costimulation can prevent that T cell from being subsequently activated, a state referred to as clonal anergy (Boussiotis et al. 1994; Gimmi et al. 1993).

Many of the costimulators on ECs that contribute to these processes have been identified. Although human ECs appear to lack both B7-1 [CD80] and B7-2 [CD86] (the best-studied costimulators expressed by dendritic cells which engage T cell CD28 and CTLA-4), human ECs do express LFA-3 [CD58] (which engages T cell CD2), ICOSL (which in humans, but not mice, engages T cell CD28 and CTLA4 as well as ICOS), 4-1BBL (which engages T cell 4-1BB) and Ox40L (which engages T cell Ox40). LFA-3, which is not expressed in mice, may be the most important in that elevated levels of CD2 on T cells is a characteristic of memory T cells and may explain the difference between naïve and memory T cell responses. In almost all assays of blocking T cell activation (except transmigration (Manes et al. 2022)), blocking LFA-3 interactions with CD2 is the most inhibitory of T cell activation (Shiao et al. 2007). Human ECs also can express PD-L1 and PD-L2, inhibitory ligands that engage T cell PD-1 (Rodig et al. 2003). While these properties of EC have largely been primarily discovered in the context of transplantation and alloimmunity, antigen presentation by ECs is critical for antitumor immunity (Schaaf et al. 2018), autoimmunity (Murdaca et al. 2012), and responses to infection (Raia and Zafrani 2022).

5.5 Complement Deposition on Graft Endothelium Leads to Pro-inflammatory Activation

5.5.1 Historical Perspective

While early transplantation science was concerned with hyperacute and acute rejection, the evolution of the field has resulted in greater attention being focused on chronic allograft changes. As we have noted, antidonor antibodies are a characteristic component of chronic rejection. The vast majority of donor-reactive antibodies recognize non-self HLA antigens, whether existing prior to transplantation as broadly recognizing non-self HLA [panel reactive antibodies (PRA)] in presensitized individuals or arising after transplantation as de novo donor-specific antibodies (DSA). Since ECs express the highest levels of these molecules in the graft, such antibodies primarily bind to the graft EC surface (Lee et al. 2002; Loupy et al. 2011). Such antibodies may directly alter ECs by cross-linking of class I or II HLA molecules (Jin et al. 2018; Valenzuela et al. 2013), but they are almost always of the IgG1 isotype and can activate complement by the classical pathway. Consequently, deposition of C4d in an established graft is a diagnostic marker of AMR (Murata and Baldwin III 2009). The effects on the EC transcriptome directly induced by the binding of anti-HLA antibodies are much more limited than the transcriptional changes that occur when bound anti-HLA antibodies activate complement (Jane-Wit et al. 2013).

5.5.2 Complement Deposited on Graft ECs Is Nonlytic and Inflammatory

Complement activation on the ECs appears to be a contributor to rejection pathology. This is obvious in acute antibody-mediated rejection where the consequence is EC loss and fibrinoid necrosis of the vessel wall (Haas et al. 2018). These effects are attributed to both the production of anaphylatoxins that activate leukocytes (C3a and C5a) as well as the assembly of membrane attack complexes (MACs) on the ECs. MAC is a complex hetero-multimer of the five terminal complement proteins:

C5b, C6, C7, C8, and C9. The poly-C9 structure forms a membrane-embedded pore that canonically is believed to lyse the cell upon which it is deposited by inducing cell swelling to the point of lysis driven by an osmotic influx of water. However, antibody and complement deposition with subsequent assembly of the MAC is generally nonlytic in nucleated cells, including ECs, unless there is extremely heavy deposition. However, in chronic rejection, both the antibody and MACs can be detected on the EC surface of ECs that have a normal morphological appearance. Nevertheless, such ECs do show an altered phenotype, displaying proteins previously described to be expressed as a result of activation by IL-1 or TNF. Mechanisms underlying these changes have been revealed by in vitro and in vivo experiments as well as analyses of human tissues.

Studies in cultured human ECs have shown that MACs are internalized by clathrin-mediated endocytosis and result in the formation of Rab5+MAC+ early endosomes (Jane-Wit et al. 2015). Unlike typical Rab5+ endosomes, the presence of MAC leads to lipid remodeling, recruitment of Akt, and recruitment and stabilization of nuclear factor-κB (NF-kB)-inducing kinase (NIK). NIK stabilization (i.e., prevention of ubiquitinylation and degradation by the proteosome) by this mechanism is independent of engagement of a tumor necrosis factor (TNF) superfamily receptor such as CD40, occurs rapidly compared to these usual mechanisms, and leads to activation of noncanonical NF-kB transcription factor (Jane-Wit et al. 2013). Noncanonical NF-κB-dependent activation of ECs, in turn, was found to result in the upregulation of pro-IL1β transcription leading to the production of biologically inactive pro-IL1β protein that accumulates in the cytosol (Xie et al. 2019). Moreover, NIK stabilized on Rab5+ endosomes performs another function not previously associated with this enzyme, namely initiation of assembly of a NOD-, LRR-, and pyrin domain-containing protein 3 (NLRP3) inflammasome. The assembled NLRP3 inflammasome recruits and activates caspase 1, a protease that cleaves both pro-IL1β and gasdermin D. The amino terminal portion of gasdermin D forms a pore in the plasma membrane of the cell that allows cleaved (and now active) mature IL-1β to be secreted into the local milieu. The released IL-1β activates the ECs in an autocrine or paracrine fashion through binding to IL-1R1 on the EC cell surface (Xie et al. 2019). In other words, signaling initiated antibody binding and complement activation on the EC surface results in EC synthesis and release of active IL-1β which, acting through a cell surface receptor on the ECs, results in EC expression of adhesion molecules for leukocytes, increased costimulators for T cells (notably ICOSL), and cell junctional changes leading plasma protein entry into the tissue, all critical for T cell activation and recruitment. These IL-1β-induced changes all involve activation of canonical NF-κB signaling (Pober and Sessa 2007). Thus the elimination of canonical NF-κB signaling (i.e., IL-1β signaling) by targeted siRNA knockdown of protein NF-κB expression eliminates many of the changes defined as the inflammatory phenotype of alloantibody and complement-treated ECs. During acute rejection in vivo, this pathway may be redundant with the effects of leukocyte-derived IL-1 or TNF but may be a primary mechanism in chronic rejection. Interestingly, some heterogeneity to antibody/MAC-mediated EC activation may exist between vascular beds (Gunawardana et al. 2021). In cell culture, the response to alloantibody and complement is enhanced by pretreatment of ECs with IFNγ. While IFNγ does not on its own trigger inflammasome assembly, its action on ECs at typical circulating concentrations leads to transcriptional increases of machinery critical to recognition by circulating DSA (namely class I and II HLA molecules) and for activation by MAC, including NLRP3, pro-caspase 1, gasdermin D (Xie et al. 2019).

5.6 IL-1β Auto-activated Endothelial Cells Trans-present IL-15 to T Cells

5.6.1 IL-15 Functions Similar to Other Costimulatory Molecules

The requirement of resting T cells, including memory T cells, for costimulatory signals in addition to TCR recognition of HLA molecules to become activated and avoid anergy is known as the "two signal" hypothesis of T cell activation. It subsequently has been shown that cytokines provided by an antigen-presenting cell may also contribute to preventing clonal anergy and boosting T cell activation, leading to the "three signal" hypothesis (Curtsinger et al. 2003). Several different cytokines may play such a role, but IL-15 has emerged as a particularly important cytokine produced by ECs in this context. Specifically, it is an important component of the IL-1β-mediated EC inflammatory activation phenotype, and activated ECs express surface IL-15 in complex with the IL-15 receptor α (IL-15Rα; CD215) (Xie et al. 2020). IL-15 was initially termed interleukin-T for its capacity to stimulate T cell proliferation, which it does by engaging the IL-2Rβ chain (CD122) and the cytokine common γ chain (CD132) receptor complex. The same two cytokine receptor proteins, when combined on the T cell surface with the IL-2Rα chain (CD25), constitute the high affinity IL-2 receptor. In the case of IL-15, the third component of the high affinity receptor, namely, IL-15 Rα (CD215), resides on the surface of that antigen-presenting cell where it binds and displays IL-15 to the T cell. Both IL-15 and IL-2 can act on T cells (or NK cells) in the absence of their respective α chains, but the affinity for these cytokines is much reduced and requires pharmacological rather than physiological concentrations. The IL-15/IL-15Rα complex shows remarkably high affinity for the IL-2 receptor and can out-compete IL-2 for binding (Grabstein et al. 1994). IL-15 is basally expressed at the transcript level by numerous cell types, but its translation and subcellular localization is tightly regulated though incompletely understood. In ECs, the level of total IL-15 transcripts is upregulated by IFN-γ, and IFNγ-treated cultured human ECs express increased levels of IL-15/IL-15Rα protein complexes within the cell nucleus. However, IFNγ-treated human ECs in culture or ECs in situ in their basal state do not express IL-15 on their cell surface (Xie et al. 2020). IL-15 transcriptional upregulation has, in fact, been observed in rejection biopsies from kidney, lung, liver, and heart grafts

(Pavlakis et al. 1996; Conti et al. 2003; Shi et al. 2004; Baan et al. 1999), but protein-level detection has been limited by lack of validated antibodies, possibly due to homology between IL-15 and other 4 alpha helix cytokines (Mullan et al. 2024).

5.6.2 IL-15 Is Regulated at Multiple Levels of Expression

The complex regulation of IL-15 expression has been increasingly appreciated as relevant to EC in transplant immunology. IL-15 protein is intrinsically unstable unless it is bound to IL-15Rα; these complexes form during protein translation and both proteins must be coexpressed. The transcript structure of IL-15 is unusual. There are many splice variants; eight variants are detectable in human ECs (along with 22 variants for IL-15Rα) (Mullan et al. 2024) that primarily involve variations of the relatively long 5′ untranslated region (UTR) and two 3′ UTR variants (one long and one short) (Bamford et al. 1996). The 5′ UTR variants are associated with poor translation efficiency, and the wild-type IL-15 mRNA contains 10 interfering start codons that inhibit translation; interestingly, this regulatory mechanism is found in other four alpha helix cytokine family members such as IL-7 and IL-9 (Kozak 1987, 1991). The 3′ UTR has long been understood to repress the expression of IL-15, without a clear mechanistic understanding. Several binding sites for miRNAs have been reported that downregulate protein expression (Mullan et al. 2024). On the other hand, HuR protein may bind and stabilize message to increase protein expression; this interaction can be antagonized by a long noncoding RNA CLEC4GP1 (Yin et al. 2022).

Once translated, complexes of IL-15/IL-15Rα have been found in the nucleus, cytosol, and on the cell surface. Subcellular localization is largely regulated by two variant signal peptides generated by differential splicing, the 48-amino acid long-signal peptide (LSP) and the 21-amino acid short-signal peptide (SSP). Newly translated IL-15 proteins containing the IL-15-SSP are retained in the cytosol or targeted to the nucleus while proteins containing the IL-15-LSP are targeted to the endoplasmic reticulum for translocation to the cell surface. Critically, IL-15-LSP alone does not result in surface presentation but depends on coexpression with IL-15Rα. The two proteins are cotranslated and associate early in the endoplasmic reticulum post-translation, resulting in mutual stabilization and membrane-bound display rather than secretion, presumably because IL-15Rα is a transmembrane protein (Bergamaschi et al. 2008). Because the IL-15/IL-15Rα complex is on the surface of the cell presenting the antigen, the T cell must be in contact with the IL-15-expressing cell to receive a signal. This process, called *trans*-presentation, allows IL-15 to augment the proliferation and activation of T cells, especially memory T cells, which are primed to respond to IL-15 *trans*-presentation by increased expression of IL-2Rβ (Geginat et al. 2003). Furthermore, circulating effector memory T cells must encounter activated ECs displaying IL-15/IL-15Rα to enter the tissue, be it graft parenchyma or the intima of a graft artery (Xie et al. 2020). It is not known whether the intracellular localization of short signal peptide variants is relevant to the generation of an alloimmune response, but transcriptomic evidence that the long signal peptide IL-15 dominates the EC transcript pool suggests this is the primary relevant molecular form (Mullan et al. 2024).

5.6.3 IL-15 Is a Known Activator of Lymphocyte Populations Critically Implicated in the Rejection Response

5.6.3.1 The T Cell Response to IL-15

The principal targets of IL-15 trans-presentation are effector memory T cells and NK cells, and there is considerable evidence that IL-15 plays a role in allograft rejection. IL-15 blockade using an IL-15Rα fragment completely prevented chronic rejection in a minor histocompatibility mismatch heart transplantation mouse model (Smith et al. 2000). The principal response of lymphocytes to IL-15 signaling is tyrosine phosphorylation and nuclear translocation of STAT5 protein. Mismatched allografts into STAT5 knockout mouse recipients have very limited CD4 and CD8 T cell infiltrates (though, interestingly, increased evidence of antibody-mediated rejection) (Zhang et al. 2006). In a chronic rejection mouse model, eliminating IL-15 presentation by dendritic cells eliminates chronic rejection and decreases intragraft $CD8^+$ resident memory T cell populations (Tieu et al. 2023). Thus, while other costimulatory molecules are certainly relevant in rejection, IL-15 signaling is critical and, apparently, nonredundant for, at least, the $CD8^+$ T cell alloimmune response.

The role of IL-15 raises the question of what signals are responsible for inducing cell surface IL-15/IL-15Rα on graft cells, particularly on graft ECs. Previously, it had been shown that TNF, but not IL-1β, treatment of ECs resulted in increased IL-15 mRNA expression (Liu et al. 2009). Surprisingly, IL-1 receptor 1 knockout mice demonstrated a deficient IL-1β-augmented, IL-15-mediated antitumor lymphocyte response (Lee et al. 2019). In fact, the surface expression of IL-15/IL-15Rα heterodimers in the setting of EC IL-1β activation appears to be significantly responsible for the recruitment, activation, and proliferation of alloimmune T cells (Xie et al. 2020, 2021). The answer to this seeming paradox is that IL-1 induces expression of IL-15/IL-15Rα surface complexes on human ECs in a manner that is independent of transcription or changes in mRNA levels. Specifically, ECs basally express an IL-15 mRNA that includes the LSP, but it is not translated, because Let7c3P miRNA is bound to the 3′ UTR. Activation of canonical NF-κB by either TNF or IL-1 represses transcription of the noncoding RNA that gives rise to Let7c, and the decline of Let7c within a few hours enables translation to proceed and IL-15/IL-15Rα complexes to appear on the cell surface. ECs generally do not synthesize TNF, but they will synthesize and secrete IL-1β above as we have described in response to alloantibody and complement signaling.

5.6.3.2 NK Cells, Rejection, and IL-15

Although less numerous than T cells by histology, Natural killer (NK) cells were also identified in the transcriptional analysis of late-rejecting grafts, along with ECs, as a feature to distinguish chronic from acute rejection. NK cells constitute a minority of the circulating immune cell population (approximately 5% of lymphocytes) and are naturally cytotoxic cells that do not depend on antigen presentation for activity (Sun and Lanier 2011). Populations of NK cells are highly heterogeneous in their expression of both activating and inhibitory receptors, and these cells will become activated when encountering graft cells that express more activating than inhibitory ligands. Some inhibitory killer immunoglobulin-like receptors (KIRs) recognize self-allelic forms of HLA-C or B, allowing them to be more easily activated by HLA-mismatched graft cells. CD16 is an important activating receptor that binds to antibody Fc domains Thus, graft cells that are mismatched at the HLA C or B loci and have bound donor-specific antibodies of the IgG1 isotype are capable of acting NK cells and are subject to antibody-dependent cell-mediated cytotoxicity (ADCC). Additional activating signals may include certain microbial productions, damage-associated molecular patterns (DAMPs), and importantly trans presented IL-15 (Li and Lan 2023; Parkes et al. 2017; Koenig et al. 2019). Regardless of the activating mechanism, NK cells are profoundly cytotoxic cells with high expression of granzymes, perforin, FasL, and stimuli of other pathways capable of killing graft cells. Minor histocompatibility proteins such as MICA/MICB (ligand of NKG2D on NK cells), and antibodies against these targets, are now recognized as significant contributors to the alloimmune response, although MICA/B are not able to trigger full activation via NKG2D because human NKG2D, unlike the mouse version, cannot engage the adapter protein DAP12, (Benichou et al. 2011). Instead, the human version of NKG2D signals through a different adaptor protein, DAP10, which boosts the cell response but cannot replace engagement of an activating receptor. NK cells in rejection are commonly CD16 (FcγRIII) positive, which is an activating receptor that recognizes the Fc protein of isotype IgG1 antibodies bound to the surface of a target cell. In this way, NK cells can mediate antibody-dependent cell-mediated cytotoxicity. NK cell infiltrates in rejecting allografts are correlated with the severity of inflammation, and are prominent in the vascular spaces of specimens with mixed and antibody-mediated rejection (Diebold et al. 2024; Hidalgo et al. 2012; Lamarthée et al. 2023). In addition to cytotoxic effects, activation of NK cells leads to IFNγ expression, increasing EC expression of MHC, IL15, IL15RA, and other pro-rejection molecules as described earlier (Pontrelli et al. 2020).

The relevance of NK cells to the alloresponse supports the critical role of IL-15 in the generation and propagation of rejection. IL-15 or IL-15Rα knockout mice have dramatic reductions in NK cell populations (Lodolce et al. 1998; Kennedy et al. 2000). In RAG gene knockout mice otherwise incapable of rejecting skin allografts, treatment with an IL-15/IL-15Rα fusion is sufficient to stimulate NK cells to reject allografts. These activated and expanded NK populations are $CD44^+$, $NKG2D^+$, and $CD16^+$, mimicking the highly activated phenotype found in NK cells from rejecting human allografts (Kroemer et al. 2008). Similar to T cells, IL-15 stimulated NK cells phosphorylate STAT5 and STAT3 via JAK1/JAK3 and also activate the PI3K-AKT pathway, resulting in both a pro-survival and a highly cytotoxic effector profile (Marçais et al. 2014). In response to IL-15 activation, NK cells copiously produce pro-inflammatory cytokines like IFNγ and TNFα, and, most importantly, IL-15 synergizes with other cytokines and signals capable of activating NK cells to further potentiate cytokine production, ADCC, and antibody-independent cytotoxicity (Carson et al. 1994, 1995; Waldmann and Tagaya 1999; Fehniger et al. 1999). This effect is especially pronounced with IL-12 and IL-18, which are known to be implicated in the IFNγ production of arterial wall-invading T cells (Ranjbaran et al. 2007; Fehniger et al. 1999). Thus, IL-15 production by ECs, known to be critical to the $CD8^+$ T_{EM} response, is a likely critical source of IL-15 for the generation and maintenance of the NK alloimmune response.

5.7 Activated Lymphocytes Produce IFNγ that Results in Allograft Vasculopathy

Histologically, we have noted that allograft vasculopathy is a common feature of chronic rejection in multiple different organs. In this section, we will discuss how IFNγ can drive the characteristic features of allograft vasculopathy. Multiomic analysis of renal transplant biopsies has identified graft IFNγ production and signaling as a determinant of clinical outcomes with both antibody-mediated and T cell-mediated rejection (Zhang et al. 2024). In mouse models, mismatched cardiac allografts transplanted into IFNγ knockout animals fail to develop graft vasculopathy despite acute rejection in the myocardium (Nagano et al. 1997). Transcript analysis of human artery specimens affected by CAV shows upregulation of IFNγ (Huibers et al. 2011). Since IFNγ signaling is species-restricted, human coronary arteries implanted into immunodeficient mice will develop arteriosclerosis with intimal expansion of SMCs even in the absence of human leukocytes (Tellides et al. 2000). Furthermore, blockade of IFNγ in a similar model that included human T cells allogeneic to the artery graft reduces pathologic remodeling of the arterial walls (Wang et al. 2004). Intimal expansion results from the vascular SMC response to IFNγ, resulting in proliferation mediated by PI3K/AKT/mTORC1 axis activity (Wang et al. 2007) and increased matrix production. Intimal SMCs are present in healthy human arteries, but it is unclear if there is additional recruitment of SMCs from the media. Intimal expansion is initially compensated by outward remodeling, but eventually, the latter process cannot keep up, resulting in progressive loss of vessel lumen.

The production of IFNγ by host T cells (and possibly NK cells) within the graft arterial intima likely depends upon the recognition of alloantigen. Human SMCs are unable to activate resting T cells, although they might support previously activated T cells. Spatially, the infiltrating T cells are in close contact with infiltrating myeloid cells and with the ECs that line the graft vessels. In the setting of CAV, these ECs are often coated with alloantibody and MACs. Importantly, in the human xenograft coronary artery transplant model, there are no human myeloid cells, and in a version of this model incorporating alloantibody-enhanced T cell-mediated allograft vasculopathy, antibody blockade of EC IL-15 *trans*-presentation reduced the infiltration of T effector memory cells, especially by CD8+ T cells and reduced production of effector molecules including IFNγ (Xie et al. 2020). In vitro, IL-15 blockade reduces the polyfunctionality and cytokine production by both alloantigen-activated CD4+ and CD8+ T effector memory cells (Xie et al. 2021). In other words, the specific-capacity alloantibody-induced IL-15 to be transpresented to T cells (and perhaps NK cells) by graft ECs results in an increased production of IFNγ by those T cells that have transmigrated through the EC lining.

5.8 Summary and Speculation

We have summarized data supporting the hypothesis of how alloantibody binding to graft ECs accompanied by complement activation and MAC deposition activate ECs to synthesize and secrete IL-1β. We have further described how this pathway, which activates canonical NF-κB signaling, leads to EC surface expression of IL-15/IL-15Rα complexes on the EC that can be trans-presented to bound effector memory T cells. We propose that host T cells that recognize non-self HLA molecules as they undergo transendothelial migration into the graft artery wall receive a signal that boosts the production of IFNγ. IFNγ produced within the vessel intima will drive SMC proliferation, resulting in changes in allograft vasculature. As the vascular lumen is progressively narrowed, the graft will fail with changes characterized by parenchymal cell loss and replacement fibrosis. This process links DSA to the activation of infiltrating T cells that produce IFNγ, both of which are well-documented features of allograft vasculopathy, by complement MAC-mediated surface induction of IL-15/IL-15Rα complexes on the graft EC lining. The absence of overt cell death within the wall of arteries involved in CAV is unexplained. i.e. it is not entirely clear how or why the cytotoxic effects of alloantibody or alloreactive T cells are mitigated in chronic but not acute rejection. Possible explanations include the expression of coinhibitory molecules like PD-L1 on ECs by IFNγ that can "exhaust" T cells, allowing them to produce cytokines but not cytolysis (Rodig et al. 2003). Alternatively, or additionally, graft cells may develop mechanisms to resist lymphocyte-mediated cytolysis, just as they can resist complement-mediated lysis. Graft cell resistance to injury is broadly termed accommodation and may involve the expression of specific proteins that block cell death pathways (Soares et al. 1999). Of course, our "one-size fits all" hypothesis does not exclude considerations of alternative mechanisms of allograft vasculopathy or chronic rejection, but it describes a potential chain of therapeutic targets that could augment current regimens that are effective against acute but not chronic rejection. Our model places graft ECs as central players, and graft ECs are uniquely accessible to genetic manipulation during ex vivo transfer and (increasingly) perfusion prior to transplant, the risks of off-gene therapy to the recipient are minimized. More specifically, targeting the pathway leading to IL-15 transpresentation by ECs appears to have significant appeal.

References

Abrahimi P, Liu R, Pober JS (2015) Blood vessels in allotransplantation. Am J Transplant 15:1748–1754

Baan C, Niesters H, Balk A, Mochtar B, Zondervan P, Weimar W (1996) The intragraft cytokine mRNA pattern reflects the efficacy of steroid antirejection therapy. J Heart Lung Transplant 15:1184–1193

Baan CC, Knoop CJ, van Gelder T, Holweg CT, Niesters HG, Smeets TJ, van der Ham F, Zondervan PE, Maat LP, Balk AH (1999) Anti-CD25 therapy reveals the redundancy of the intragraft cytokine network after clinical heart transplantation. Transplantation 67:870–876

Baldwin WM 3rd, Claas FH, van ES LA, van Rood JJ (1981) Distribution of endothelial-monocyte and HLA antigens on renal vascular endothelium. Transplant Proc 13:103–107

Bamford RN, Battiata AP, Burton JD, Sharma H, Waldmann TA (1996) Interleukin (IL) 15/IL-T production by the adult T-cell leukemia cell line HuT-102 is associated with a human T-cell lymphotrophic virus type I region/IL-15 fusion message that lacks many upstream AUGs that normally attenuates IL-15 mRNA translation. Proc Natl Acad Sci 93:2897–2902

Benichou G, Yamada Y, Aoyama A, Madsen JC (2011) Natural killer cells in rejection and tolerance of solid organ allografts. Curr Opin Organ Transplant 16:47–53

Bergamaschi C, Rosati M, Jalah R, Valentin A, Kulkarni V, Alicea C, Zhang G-M, Patel V, Felber BK, Pavlakis GN (2008) Intracellular interaction of interleukin-15 with its receptor α during production leads to mutual stabilization and increased bioactivity. J Biol Chem 283:4189–4199

Berry GJ, Burke MM, Andersen C, Bruneval P, Fedrigo M, Fishbein MC, Goddard M, Hammond EH, Leone O, Marboe C (2013) The 2013 International Society for Heart and Lung Transplantation Working Formulation for the standardization of nomenclature in the pathologic diagnosis of antibody-mediated rejection in heart transplantation. J Heart Lung Transplant 32(12):1147–1162. Elsevier.

Biedermann BC, Pober JS (1998) Human endothelial cells induce and regulate cytolytic T cell differentiation. J Immunol 161:4679–4687

Billingham ME (1994) Pathology and etiology of chronic rejection of the heart. Clin Transpl 8:289–292

Boussiotis VA, Freeman GJ, Griffin JD, Gray GS, Gribben JG, Nadler LM (1994) CD2 is involved in maintenance and reversal of human alloantigen-specific clonal anergy. J Exp Med 180:1665–1673

Brewer Y, Palmer A, Taube D, Welsh K, Bewick M, Bindon C, Hale G, Waldmann H, Dische F, Parsons V et al (1989) Effect of graft perfusion with two CD45 monoclonal antibodies on incidence of kidney allograft rejection. Lancet 2:935–937

Busch GJ, Galvanek EG, Reynolds ES, JR. (1971) Human renal allografts. Analysis of lesions in long-term survivors. Hum Pathol 2:253–298

Carson WE, Giri JG, Lindemann MJ, Linett ML, Ahdieh M, Paxton R, Anderson D, Eisenmann J, Grabstein K, Caligiuri MA (1994) Interleukin (IL) 15 is a novel cytokine that activates human natural killer cells via components of the IL-2 receptor. J Exp Med 180:1395–1403

Carson WE, Ross ME, Baiocchi RA, Marien MJ, Boiani N, Grabstein K, Caligiuri MA (1995) Endogenous production of interleukin 15 by activated human monocytes is critical for optimal production of interferon-gamma by natural killer cells in vitro. J Clin Invest 96:2578–2582

Cone BD, Zhang JQ, Sosa RA, Calabrese F, Reed EF, Fishbein GA (2021) Phosphorylated S6 ribosomal protein expression by immunohistochemistry correlates with de novo donor-specific HLA antibodies in lung allograft recipients. J Heart Lung Transplant 40:1164–1171
Conti F, Frappier J, Dharancy S, Chereau C, Houssin D, Weill B, Calmus Y (2003) Interleukin-15 production during liver allograft rejection in humans. Transplantation 76:210–216
Curtsinger JM, Lins DC, Mescher MF (2003) Signal 3 determines tolerance versus full activation of naive CD8 T cells: dissociating proliferation and development of effector function. J Exp Med 197:1141–1151
Daar AS, Fuggle SV, Fabre JW, Ting A, Morris PJ (1984) The detailed distribution of MHC Class II antigens in normal human organs. Transplantation 38:293–298
Dengler TJ, Pober JS (2000) Human vascular endothelial cells stimulate memory but not naive CD8+ T cells to differentiate into CTL retaining an early activation phenotype. J Immunol 164:5146–5155
Diebold M, Farkash EA, Barnes J, Regele H, Kozakowski N, Schatzl M, Mayer KA, Haindl S, Vietzen H, Hidalgo LG (2024) Natural killer cell presence in antibody-mediated rejection. Transpl Int 37:13209
Epperson DE, Pober JS (1994) Antigen-presenting function of human endothelial cells. Direct activation of resting CD8 T cells. J Immunol 153:5402–5412
Fehniger TA, Shah MH, Turner MJ, Vandeusen JB, Whitman SP, Cooper MA, Suzuki K, Wechser M, Goodsaid F, Caligiuri MA (1999) Differential cytokine and chemokine gene expression by human NK cells following activation with IL-18 or IL-15 in combination with IL-12: implications for the innate immune response. J Immunol 162:4511–4520
Geginat J, Lanzavecchia A, Sallusto F (2003) Proliferation and differentiation potential of human CD8+ memory T-cell subsets in response to antigen or homeostatic cytokines. Blood 101:4260–4266
Gimmi CD, Freeman GJ, Gribben JG, Gray G, Nadler LM (1993) Human T-cell clonal anergy is induced by antigen presentation in the absence of B7 costimulation. Proc Natl Acad Sci USA 90:6586–6590
Grabstein KH, Eisenman J, Shanebeck K, Rauch C, Srinivasan S, Fung V, Beers C, Richardson J, Schoenborn MA, Ahdieh M (1994) Cloning of a T cell growth factor that interacts with the β chain of the interleukin-2 receptor. Science 264:965–968
Gunawardana H, Romero T, Yao N, Heidt S, Mulder A, Elashoff DA, Valenzuela NM (2021) Tissue-specific endothelial cell heterogeneity contributes to unequal inflammatory responses. Sci Rep 11:1949
Haas M, Loupy A, Lefaucheur C, Roufosse C, Glotz D, Seron DA, Nankivell B, Halloran P, Colvin R, Akalin E (2018) The Banff 2017 Kidney Meeting Report: revised diagnostic criteria for chronic active T cell–mediated rejection, antibody-mediated rejection, and prospects for integrative endpoints for next-generation clinical trials. Am J Transplant 18(2):293–307. Wiley Online Library.
Halloran PF, de Freitas DG, Einecke G, Famulski KS, Hidalgo LG, Mengel M, Reeve J, Sellares J, Sis B (2010) The molecular phenotype of kidney transplants. Am J Transplant 10:2215–2222
Halloran PF, Madill-Thomsen KS, Böhmig G, Bromberg J, Budde K, Barner M, Mackova M, Chang J, Einecke G, Eskandary F (2025) Subthreshold rejection activity in many kidney transplants currently classified as having no rejection. Am J Transplant 25:72–87
Hidalgo L, Sellares J, Sis B, Mengel M, Chang J, Halloran P (2012) Interpreting NK cell transcripts versus T cell transcripts in renal transplant biopsies. Am J Transplant 12:1180–1191
Hirschberg H, Bergh O, Thorsby E (1980) Antigen-presenting properties of human vascular endothelial cells. J Exp Med 152:249s–255s
Holweg CT, Potena L, Luikart H, Yu T, Berry GJ, Cooke JP, Valantine HA, Mocarski ES (2011) Identification and classification of acute cardiac rejection by intragraft transcriptional profiling. Circulation 123:2236–2243

Huibers M, De Jonge N, Van Kuik J, Koning ES, Van Wichen D, Dullens H, Schipper M, De Weger R (2011) Intimal fibrosis in human cardiac allograft vasculopathy. Transpl Immunol 25:124–132

Jane-Wit D, Manes TD, Yi T, Qin L, Clark P, Kirkiles-Smith NC, Abrahimi P, Devalliere J, Moeckel G, Kulkarni S (2013) Alloantibody and complement promote T cell–mediated cardiac allograft vasculopathy through noncanonical nuclear factor-κB signaling in endothelial cells. Circulation 128:2504–2516

Jane-Wit D, Surovtseva YV, Qin L, Li G, Liu R, Clark P, Manes TD, Wang C, Kashgarian M, Kirkiles-Smith NC (2015) Complement membrane attack complexes activate noncanonical NF-κB by forming an Akt+ NIK+ signalosome on Rab5+ endosomes. Proc Natl Acad Sci 112:9686–9691

Jin YP, Valenzuela NM, Zhang X, Rozengurt E, Reed EF (2018) HLA class II-triggered signaling cascades cause endothelial cell proliferation and migration: relevance to antibody-mediated transplant rejection. J Immunol 200:2372–2390

Kennedy MK, Glaccum M, Brown SN, Butz EA, Viney JL, Embers M, Matsuki N, Charrier K, Sedger L, Willis CR (2000) Reversible defects in natural killer and memory CD8 T cell lineages in interleukin 15–deficient mice. J Exp Med 191:771–780

Khush KK, Cherikh WS, Chambers DC, Goldfarb S, Hayes D Jr, Kucheryavaya AY, Levvey BJ, Meiser B, Rossano JW, Stehlik J (2018) The International Thoracic Organ Transplant Registry of the International Society for Heart and Lung Transplantation: thirty-fifth adult heart transplantation report-2018; focus theme: multiorgan transplantation. J Heart Lung Transplant 37:1155–1168

Koenig A, Chen C-C, Marçais A, Barba T, Mathias V, Sicard A, Rabeyrin M, Racapé M, Duong-Van-Huyen J-P, Bruneval P (2019) Missing self triggers NK cell-mediated chronic vascular rejection of solid organ transplants. Nat Commun 10:5350

Kozak M (1987) An analysis of 5′-noncoding sequences from 699 vertebrate messenger RNAs. Nucleic Acids Res 15:8125–8148

Kozak M (1991) An analysis of vertebrate mRNA sequences: intimations of translational control. J Cell Biol 115:887–903

Krams SM, Falco DA, Villanueva JC, Rabkin J, Tomlanovich SJ, Vincenti F, Amend WJ, Melzer J, Garovoy MR, Roberts JP et al (1992) Cytokine and T cell receptor gene expression at the site of allograft rejection. Transplantation 53:151–156

Kreisel D, Krupnick AS, Gelman AE, Engels FH, Popma SH, Krasinskas AM, Balsara KR, Szeto WY, Turka LA, Rosengard BR (2002) Non-hematopoietic allograft cells directly activate CD8+ T cells and trigger acute rejection: an alternative mechanism of allorecognition. Nat Med 8:233–239

Kroemer A, Xiao X, Degauque N, Edtinger K, Wei H, Demirci G, Li XC (2008) The innate NK cells, allograft rejection, and a key role for IL-15. J Immunol 180:7818–7826

Lagaaij EL, Cramer-Knijnenburg GF, van Kemenade FJ, van ES LA, Bruijn JA, van Krieken JH (2001) Endothelial cell chimerism after renal transplantation and vascular rejection. Lancet 357:33–37

Lakkis FG, Lechler RI (2013) Origin and biology of the allogeneic response. Cold Spring Harb Perspect Med 3(8):a014993

Lamarthée B, Callemeyn J, Van Herck Y, Antoranz A, Anglicheau D, Boada P, Becker JU, Debyser T, De Smet F, De Vusser K (2023) Transcriptional and spatial profiling of the kidney allograft unravels a central role for FcyRIII+ innate immune cells in rejection. Nat Commun 14:4359

Lautenschlager I, Taskinen E, Inkinen K, Lehto V-P, Virtanen I, Häyry P (1984) Distribution of the major histocompatibility complex antigens on different cellular components of human liver. Cell Immunol 85:191–200

Lechler RI, Batchelor JR (1982) Restoration of immunogenicity to passenger cell-depleted kidney allografts by the addition of donor strain dendritic cells. J Exp Med 155:31–41

Lechler R, Ng WF, Steinman RM (2001) Dendritic cells in transplantation—friend or foe? Immunity 14:357–368

Lee P-C, Terasaki PI, Takemoto SK, Lee P-H, Hung C-J, Chen Y-L, Tsai A, Lei H-Y (2002) All chronic rejection failures of kidney transplants were preceded by the development of HLA antibodies. Transplantation 74:1192–1194
Lee P-H, Yamamoto TN, Gurusamy D, Sukumar M, Yu Z, Hu-Li J, Kawabe T, Gangaplara A, Kishton RJ, Henning AN (2019) Host conditioning with IL-1β improves the antitumor function of adoptively transferred T cells. J Exp Med 216:2619–2634
Li Q, Lan P (2023) Activation of immune signals during organ transplantation. Signal Transduct Target Ther 8:110
Liu X, Zuo Y, Zhang W, Yang D, Xiong C, Zhang X (2009) Expression of interleukin-15 and its receptor on the surface of stimulated human umbilical vein endothelial cells. J Huazhong Univ Sci Technolog Med Sci 29:527–534
Lodolce JP, Boone DL, Chai S, Swain RE, Dassopoulos T, Trettin S, Ma A (1998) IL-15 receptor maintains lymphoid homeostasis by supporting lymphocyte homing and proliferation. Immunity 9:669–676
Loupy A, Lefaucheur C (2018) Antibody-mediated rejection of solid-organ allografts. N Engl J Med 379:1150–1160
Loupy A, Cazes A, Guillemain R, Amrein C, Hedjoudje A, Tible M, Pezzella V, Fabiani J, Suberbielle C, Nochy D (2011) Very late heart transplant rejection is associated with microvascular injury, complement deposition and progression to cardiac allograft vasculopathy. Am J Transplant 11:1478–1487
Manes TD, Pober JS (2008) Antigen presentation by human microvascular endothelial cells triggers ICAM-1-dependent transendothelial protrusion by, and fractalkine-dependent transendothelial migration of, effector memory CD4+ T cells. J Immunol 180:8386–8392
Manes TD, Pober JS (2016) Significant differences in antigen-induced transendothelial migration of human CD8 and CD4 T effector memory cells. Arterioscler Thromb Vasc Biol 36:1910–1918
Manes TD, Wang V, Pober JS (2022) Costimulators expressed on human endothelial cells modulate antigen-dependent recruitment of circulating T lymphocytes. Front Immunol:13:1016361
Marçais A, Cherfils-Vicini J, Viant C, Degouve S, Viel S, Fenis A, Rabilloud J, Mayol K, Tavares A, Bienvenu J (2014) The metabolic checkpoint kinase mTOR is essential for IL-15 signaling during the development and activation of NK cells. Nat Immunol 15:749–757
Matas AJ, Gillingham KJ, Payne WD, Najarian JS (1994) The impact of an acute rejection episode on long-term renal allograft survival (t1/2) 1, 2. Transplantation 57:857–859
Mengel M, Sis B, Kim D, Chang J, Famulski K, Hidalgo L, Einecke G, De Freitas D, Tymchak W, Burton J (2010) The molecular phenotype of heart transplant biopsies: relationship to histopathological and clinical variables. Am J Transplant 10:2105–2115
Minami E, Laflamme MA, Saffitz JE, Murry CE (2005) Extracardiac progenitor cells repopulate most major cell types in the transplanted human heart. Circulation 112:2951–2958
Mullan CW, Summer L, Lopez-Giraldez F, Tobiasova Z, Manes TD, Yasothan S, Song G, Jane-Wit D, Saltzman WM, Pober JS (2024) IL-1β induces human endothelial surface expression of IL-15 by relieving let-7c-3p suppression of protein translation. J Immunol 213:1338–1348
Murata K, Baldwin WM III (2009) Mechanisms of complement activation, C4d deposition, and their contribution to the pathogenesis of antibody-mediated rejection. Transplant Rev 23:139–150
Murdaca G, Colombo BM, Cagnati P, Gulli R, Spanò F, Puppo F (2012) Endothelial dysfunction in rheumatic autoimmune diseases. Atherosclerosis 224:309–317
Naesens M, Roufosse C, Haas M, Lefaucheur C, Mannon RB, Adam BA, Aubert O, Böhmig GA, Callemeyn J, Clahsen-van Groningen M (2024) The Banff 2022 Kidney Meeting Report: reappraisal of microvascular inflammation and the role of biopsy-based transcript diagnostics. Am J Transplant 24:338–349
Nagano H, Mitchell RN, Taylor MK, Hasegawa S, Tilney NL, Libby P (1997) Interferon-gamma deficiency prevents coronary arteriosclerosis but not myocardial rejection in transplanted mouse hearts. J Clin Invest 100:550–557

Parkes MD, Halloran PF, Hidalgo LG (2017) Evidence for CD16a-mediated NK cell stimulation in antibody-mediated kidney transplant rejection. Transplantation 101:e102–e111
Pavlakis M, Strehlau J, Lipman M, Shapiro M, Maslinski W, Strom T (1996) Intragraft IL-15 transcripts are increased in human renal allograft rejection. Transplantation 62:543–545
Pober JS, Sessa WC (2007) Evolving functions of endothelial cells in inflammation. Nat Rev Immunol 7:803–815
Pober JS, Collins T, Gimbrone MA Jr, Cotran RS, Gitlin JD, Fiers W, Clayberger C, Krensky AM, Burakoff SJ, Reiss CS (1983) Lymphocytes recognize human vascular endothelial and dermal fibroblast Ia antigens induced by recombinant immune interferon. Nature 305:726–729
Pontrelli P, Rascio F, Castellano G, Grandaliano G, Gesualdo L, Stallone G (2020) The role of natural killer cells in the immune response in kidney transplantation. Front Immunol 11:1454
Raia L, Zafrani L (2022) Endothelial activation and microcirculatory disorders in sepsis. Front Med 9:907992
Ranjbaran H, Sokol SI, Gallo A, Eid RE, Iakimov AO, D'alessio A, Kapoor JR, Akhtar S, Howes CJ, Aslan M, Pfau S, Pober JS, Tellides G (2007) An inflammatory pathway of IFN-gamma production in coronary atherosclerosis. J Immunol 178:592–604
Rodig N, Ryan T, Allen JA, Pang H, Grabie N, Chernova T, Greenfield EA, Liang SC, Sharpe AH, Lichtman AH, Freeman GJ (2003) Endothelial expression of PD-L1 and PD-L2 down-regulates CD8+ T cell activation and cytolysis. Eur J Immunol 33:3117–3126
Salomon RN, Hughes CC, Schoen FJ, Payne DD, Pober JS, Libby P (1991) Human coronary transplantation-associated arteriosclerosis. Evidence for a chronic immune reaction to activated graft endothelial cells. Am J Pathol 138:791–798
Schaaf MB, Garg AD, Agostinis P (2018) Defining the role of the tumor vasculature in antitumor immunity and immunotherapy. Cell Death Dis 9:115
Shi R, Yang J, Jaramillo A, Steward NS, Aloush A, Trulock EP, Patterson GA, Suthanthiran M, Mohanakumar T (2004) Correlation between interleukin-15 and granzyme B expression and acute lung allograft rejection. Transpl Immunol 12:103–108
Shiao SL, Kirkiles-Smith NC, Shepherd BR, Mcniff JM, Carr EJ, Pober JS (2007) Human effector memory CD4+ T cells directly recognize allogeneic endothelial cells in vitro and in vivo. J Immunol 179:4397–4404
Smith XG, Bolton EM, Ruchatz H, Wei X, Liew FY, Bradley JA (2000) Selective blockade of IL-15 by soluble IL-15 receptor alpha-chain enhances cardiac allograft survival. J Immunol 165:3444–3450
Soares MP, Lin Y, Sato K, Stuhlmeier KM, Bach FH (1999) Accommodation. Immunol Today 20:434–437
Strehlau J, Pavlakis M, Lipman M, Shapiro M, Vasconcellos L, Harmon W, Strom TB (1997) Quantitative detection of immune activation transcripts as a diagnostic tool in kidney transplantation. Proc Natl Acad Sci USA 94:695–700
Sun JC, Lanier LL (2011) NK cell development, homeostasis and function: parallels with CD8+ T cells. Nat Rev Immunol 11:645–657
Tellides G, Tereb DA, Kirkiles-Smith NC, Kim RW, Wilson JH, Schechner JS, Lorber MI, Pober JS (2000) Interferon-γ elicits arteriosclerosis in the absence of leukocytes. Nature 403:207–211
Tieu R, Zeng Q, Zhao D, Zhang G, Feizi N, Manandhar P, Williams AL, Popp B, Wood-Trageser MA, Demetris AJ, Tso JY, Johnson AJ, Kane LP, Abou-Daya KI, Shlomchik WD, Oberbarnscheidt MH, Lakkis FG (2023) Tissue-resident memory T cell maintenance during antigen persistence requires both cognate antigen and interleukin-15. Sci Immunol 8:eadd8454
Valenzuela NM, Mulder A, Reed EF (2013) HLA class I antibodies trigger increased adherence of monocytes to endothelial cells by eliciting an increase in endothelial P-selectin and, depending on subclass, by engaging FcγRs. J Immunol 190:6635–6650
von der Thüsen JH, Vandermeulen E, Vos R, Weynand B, Verbeken EK, Verleden SE (2018) The histomorphological spectrum of restrictive chronic lung allograft dysfunction and implications for prognosis. Mod Pathol 31:780–790

Waldmann T, Tagaya Y (1999) The multifaceted regulation of interleukin-15 expression and the role of this cytokine in NK cell differentiation and host response to intracellular pathogens. Annu Rev Immunol 17:19–49

Wallace WD, Reed EF, Ross D, Lassman CR, Fishbein MC (2005) C4d staining of pulmonary allograft biopsies: an immunoperoxidase study. J Heart Lung Transplant 24:1565–1570

Wang Y, Burns WR, Tang PC, Yi T, Schechner JS, Zerwes HG, Sessa WC, Lorber MI, Pober JS, Tellides G (2004) Interferon-γ plays a nonredundant role in mediating T-cell-dependent outward vascular remodeling of allogeneic human coronary arteries. FASEB J 18:606–608

Wang Y, Bai Y, Qin L, Zhang P, Yi T, Teesdale SA, Zhao L, Pober JS, Tellides G (2007) Interferon-gamma induces human vascular smooth muscle cell proliferation and intimal expansion by phosphatidylinositol 3-kinase dependent mammalian target of rapamycin raptor complex 1 activation. Circ Res 101:560–569

Xie CB, Qin L, Li G, Fang C, Kirkiles-Smith NC, Tellides G, Pober JS, Jane-Wit D (2019) Complement membrane attack complexes assemble NLRP3 inflammasomes triggering IL-1 activation of IFN-γ–primed human endothelium. Circ Res 124:1747–1759

Xie CB, Jiang B, Qin L, Tellides G, Kirkiles-Smith NC, Jane-Wit D, Pober JS (2020) Complement-activated interferon-γ–primed human endothelium transpresents interleukin-15 to CD8+ T cells. J Clin Invest 130:3437–3452

Xie CB, Zhou J, Mackay S, Pober JS (2021) Complement-activated human endothelial cells stimulate increased polyfunctionality in alloreactive T cells. Am J Transplant 21:1902–1909

Yin A, Chen W, Tang L, Zhong M, Jia B (2022) Pseudogene CLEC4GP1 modulates trophoblast cell apoptosis and invasion via IL-15 inhibition. Exp Cell Res 418:113215

Zhang Y, Kirken RA, Furian L, Janczewska S, Qu X, Hancock WW, Wang M, Tejpal N, Kerman R, Kahan BD (2006) Allograft rejection requires STAT5a/b-regulated antiapoptotic activity in T cells but not B cells. J Immunol 176:128–137

Zhang H, Zhang D, Xu Y, Zhang H, Zhang Z, Hu X (2024) Interferon-γ and its response are determinants of antibody-mediated rejection and clinical outcomes in patients after renal transplantation. Genes Immun 25:66–81

Chapter 6
Macrophage Actin Cytoskeleton in Chronic Rejection

Malgorzata Kloc, Marta Halasa, Souhail A. Thabet, Jarek Wosik, Ahmed Uosef, Arijita Subuddhi, and Rafik Mark Ghobrial

Abstract There is no cure for chronic rejection of transplanted organs. Macrophages are heavily involved in chronic rejection. Macrophages' movement into the graft depends on the actin cytoskeleton and its regulators the GTPase RhoA and its effector ROCK kinase. In this chapter, we describe how the interference with macrophage RhoA/ROCK pathway disrupts macrophage actin cytoskeleton, impairing their functions and migration to the allograft, and preventing the development of chronic rejection, and how such interference can be applied to clinical therapies.

M. Kloc (✉) · M. Halasa · A. Uosef · R. M. Ghobrial
The Houston Methodist Research Institute, Transplant Immunology, Houston, TX, USA

The Houston Methodist Hospital, Department of Surgery, Houston, TX, USA
e-mail: mkloc@houstonmethodist.org

S. A. Thabet
The Houston Methodist Hospital, Department of Surgery, Houston, TX, USA

J. Wosik
Electrical and Computer Engineering Department, University of Houston, Houston, TX, USA

Texas Center for Superconductivity, University of Houston, Houston, TX, USA

A. Subuddhi
The Houston Methodist Research Institute, Transplant Immunology, Houston, TX, USA

The Houston Methodist Hospital, Department of Surgery, Houston, TX, USA

Emory/GA Tuberculosis Research Advancement Center (TRAC), Emory Vaccine Center, Atlanta, GA, USA

M. Kloc, R. M. Ghobrial (eds.), *Cellular and Molecular Mechanisms of Chronic Rejection in Organ Transplantation*, Results and Problems in Cell Differentiation 77, https://doi.org/10.1007/978-3-032-07686-1_6

6.1 Introduction

Solid organ transplantation is a lifesaving procedure in patients with organ failure or end-stage dysfunction (Black et al. 2018). However, the graft survival above 10 years post-transplantation remains challenging. Therefore, adequate immunosuppression is crucial in preventing acute and chronic rejection (Small et al. 2022).

In heart transplant, patients are typically managed with triple therapy that incorporates an antimetabolite in addition to a calcineurin inhibitor and steroids that are gradually tapered in the first year after transplant (Kim et al. 2018). Induction with basiliximab or antithymocyte globulin is commonly indicated in sensitized, high-risk, or multiorgan transplant recipients. Given the high immunogenicity of cardiac allografts, intensive immunosuppression is essential to prevent cellular and antibody-mediated rejection (Chang et al. 2020).

Kidney transplant maintenance immunosuppression is relatively similar, consisting of a calcineurin inhibitor (tacrolimus), an antiproliferative agent (mycophenolate), and corticosteroids if needed. Induction therapy is tailored to immunologic risk. Antithymocyte globulin (rATG) is the standard induction therapy used at most centers and for high immune risk recipients (such as those with high panel reactive antibody titers or those with immune-mediated kidney disease), while IL-2 receptor antagonists such as basiliximab are recommended for lower immune risk recipients (older patients, those already on immunosuppression, or well-matched donor recipient HLA type). A third induction agent, alemtuzumab (Campath), is another alternative that provides longer-term depletion of lymphocytes for higher immune risk individuals or those undergoing steroid-sparing regimen. Even though triple-drug maintenance remains standard, steroid-sparing or early withdrawal strategies may be used in selected low-risk individuals to mitigate calcineurin inhibitor (CNI) nephrotoxicity, which is a key contributor to chronic allograft dysfunction. CNI-sparing regimens incorporating belatacept or mTOR inhibitors can also be employed in certain cases (Hardinger and Brennan 2013; Wojciechowski and Wiseman 2021).

Lung transplantation demands the most aggressive immunosuppression due to a high rejection risk. Maintenance therapy, like in heart and kidney transplants, typically includes tacrolimus, mycophenolate, and low-dose corticosteroids, following induction most commonly with an IL-2 antagonist or antithymocyte globulin. Even minor alloimmune responses can cause bronchiolitis obliterans, the hallmark of chronic lung rejection, so regimens may incorporate mTOR inhibitors or extracorporeal therapies to address late graft injury (Pecoraro et al. 2017; Small et al. 2022).

In contrast to heart, kidney, and lung transplantation, liver allografts are relatively tolerogenic; hence, standard immunosuppression can often be minimized. Due to the lower rejection risk and fragile clinical condition of end-stage liver disease patients, induction therapy is also less commonly used. Typical maintenance consists of a calcineurin inhibitor, such as tacrolimus, along with an antiproliferative agent combined with a short course of corticosteroids that are rapidly tapered during the first week, maintained, and eventually withdrawn by 3–6 months post-transplant. Azathioprine or Everolimus may be added only if needed (Di Maira et al. 2020; Moini et al. 2015; Safarini et al. 2025).

Calcineurin Inhibitors, such as tacrolimus and cyclosporine, bind to FK506-binding protein or cyclophilin, respectively, forming complexes that inhibit calcineurin activity. Calcineurin is a phosphatase essential for activating a transcription factor NFAT (nuclear factor of activated T cells), which drives interleukin-2 production. Inhibition of calcineurin prevents IL-2 transcription, thereby suppressing the activation, proliferation, and differentiation of T cells, primarily affecting helper in addition to cytotoxic and regulatory subsets (Safarini et al. 2025).

Antimetabolites: Mycophenolate mofetil (MMF) is a prodrug of mycophenolic acid. MMF inhibits inosine monophosphate dehydrogenase (IMPDH) type II, the rate-limiting enzyme in the synthesis of guanosine nucleotides. The type II isoform is expressed in activated lymphocytes and, by depletion of guanosine nucleotides, suppresses T and B cell proliferation. By comparison, azathioprine is converted to 6-mercaptopurine and eventually to active metabolites, such as 6-thioguanine nucleotides, which are integrated into DNA replication, thereby disrupting purine synthesis and ultimately arresting lymphocyte proliferation (Allison and Eugui 2000; Broen and van Laar 2020).

Corticosteroids diffuse through the cell membrane and bind to intracellular glucocorticoid receptors, which move to the nucleus and affect gene transcription. Receptor activation in immune cells downregulates proinflammatory genes and upregulates inhibitory proteins such as IκBα, which sequesters NF-κB in the cytoplasm. Preventing NF-κB activation decreases the expression of cytokines and adhesion molecules, which, in turn, suppresses acute and chronic rejection (Auphan et al. 1995).

Mammalian target of rapamycin (mTOR) inhibitors, sirolimus and Everolimus, binds to the intracellular protein FKBP12 to form a complex that inhibits mTOR, a serine/threonine kinase functioning downstream of the interleukin-2 (IL-2) receptor. This inhibition arrests the cell cycle at the G1/S phase in activated T lymphocytes. mTOR inhibitors are utilized clinically as calcineurin-inhibitor-sparing agents to mitigate nephrotoxicity while preserving suppression of IL-2 driven lymphocyte proliferation (Bennett et al. 2019).

Costimulation Blockers such as Interleukin-2 receptor antagonists (IL-2RAs) selectively target and block the alpha chain (CD25) of the high-affinity IL-2 receptor, which is predominantly expressed on activated T cells. The disruption of IL-2 inhibits cytokine-driven T-cell proliferation, a pivotal event in the immune response that leads to allograft rejection (Nelson et al. 2022).

Despite advancements in transplantation and immunosuppressive strategies that have markedly decreased rates of acute rejection, chronic rejection and progressive graft dysfunction remain major impediments to long-term allograft survival. Thus, while solid-organ transplantation offers life-saving benefits in patients with end-organ failure, improving short- and long-term outcomes, chronic rejection persists as a principal barrier to sustained graft survival and longevity (Demetris et al. 1997).

The long-term (chronic) rejection of transplanted organs manifests by vessel lumen occlusion and tissue fibrosis (Mundinger et al. 2013). Both these processes involve the function of macrophages infiltrating allografts within the first week after transplantation (Chen et al. 2017). During the inflammatory response to immune injury, activated macrophages and endothelial cells of the blood vessel wall produce growth factors, which, along with interferon-γ produced by T cells, induce phenotype change and proliferation of smooth muscle cells residing in the tunica media and their transmigration to the tunica intima, leading to progressive thickening of intima (formation of the neointima), and vessel occlusion (Kloc and Ghobrial 2014). Additionally, activated macrophages, by participating in all phases of tissue injury and repair, including fibrogenesis and fibrolysis, promote allograft fibrosis (Chen et al. 2023; Ogawa et al. 2021). The macrophage morphological and molecular phenotype, organelle distribution, polarity, phagocytosis, endocytosis, receptor recycling, substrate attachment, response to cellular and acellular mechanical cues, and movement depend on the organization and dynamics of the actin cytoskeleton, which is regulated by the RhoA/ROCK pathway and its reciprocal partner Rac1 (Barger et al. 2022; Kloc et al. 2019, 2021; Pollard and Cooper 2009; Schaks et al. 2019; Ulferts et al. 2024). Our gene expression analysis of chronically rejecting rat hearts showed tremendous (>20 times) upregulation of RhoA and Rac1 expression (Lisik et al. 2010). This result, together with massive accumulation of macrophages in chronically rejecting hearts, prompted us to study the involvement of RhoA and macrophages in chronic rejection. Below, we describe how actin regulates various processes in macrophages, and how interference with actin and its regulator, the RhoA/ROCK pathway, may be employed for antichronic rejection therapies in solid organ transplantation.

6.2 Cellular Actin

Vertebrates have three groups of actin isoforms: alpha, beta, and gamma. The α-actin is exclusive for muscle fibers, while β- and γ-actin coexist in all cell types. Cellular actin has two forms: globular (unpolymerized) actin monomers (G-actin) and actin filaments (F-actin) built of polymerized monomers (Arora et al. 2023; Kashina 2020; Khaitlina 2001). Actin is constantly polymerized/depolymerized within the cell, and the balance of these two processes ensures cellular homeostasis and functions. Actin is in the cell cytoplasm and nucleus. Actin filaments are located below the cellular membrane where they form a layer of cortical actin, they are part of focal adhesions that connect the cell to the substrate, and they span the cytoplasm and perinuclear area surrounding the nuclear membrane (Masters et al. 2016; Polino et al. 2023; Tribollet et al. 2022). Cell nucleus and cellular organelles, such as mitochondria, Golgi complex, and endoplasmic reticulum, are not randomly distributed within the cytoplasm but are positioned in specific places. Such nonrandom positioning, which relates to organelle functionality, depends on the attachment to and anchoring by actin filaments (Copeland et al. 2016; Coscia et al. 2023; Janota et al. 2022; Moore et al. 2021; Pain et al. 2023). Actin filaments are also involved in phagocytosis, endocytosis, exocytosis, receptor recycling from and to the cellular membrane, and mitochondria fission (Abouelezz and Almeida-Souza 2022; Hanley 2014; Horsthemke et al. 2018; Kay 2021; Miklavc and Frick 2020; Yang and

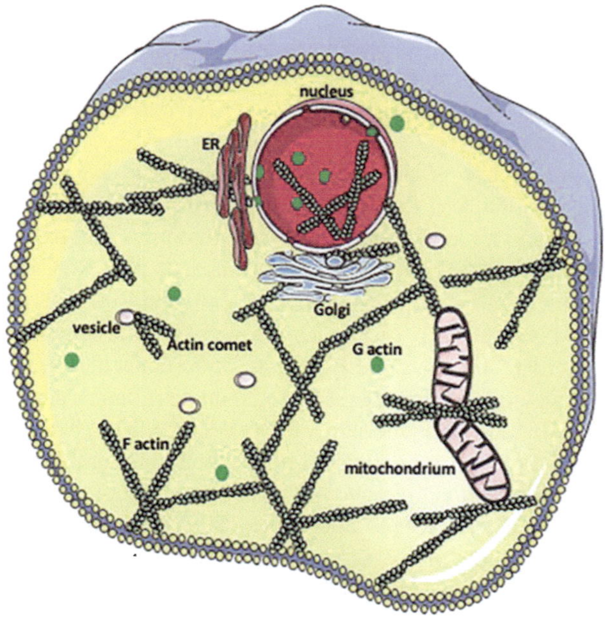

Fig. 6.1 Actin distribution within the cell. Actin filaments (F actin, green) are distributed under the cell membrane, within the cytoplasm, and around and within the cell nucleus transmitting mechanical signals and anchoring the cell nucleus and cellular organelles such as the endoplasmic reticulum (ER), Golgi complex, and mitochondria. Actin filaments participate in mitochondrial fission, and chromatin organization and compaction. They propel the movement of intracellular vesicles (through the formation of so-called actin comets). Cell cytoplasm and nucleus contain a pool of unpolymerized G actin (green spheres). Nuclear G actin regulates transcription by directly interacting with RNA polymerases

Svitkina 2019). The actin filaments in all cellular locations are interconnected, which allows any signal at the cell membrane to be transmitted to all organelles and the nucleus and vice versa. The cell nucleus contains G actin and actin filaments, which participate in nucleocytoplasmic transport, chromatin condensation, and regulate transcription (Fig. 6.1); (Fernandez et al. 2024; Kyheröinen and Vartiainen 2020; Ma et al. 2024; Serebryannyy and de Lanerolle 2020; Vartiainen 2008).

6.3 Actin Regulation by RhoA/ROCK Pathway

Although actin dynamics is coregulated by many different signaling pathways, the main controller of actin polymerization/depolymerization is the RhoA/ROCK pathway. The RhoA (Ras homolog family member A) is a small GTPase from the Rho family of GTPases that belong to the Ras superfamily of proteins. Activation of RhoA is controlled by 85 different guanine nucleotide exchange factors GEFs, which catalyze the exchange of GDP (guanosine diphosphate) to GTP (guanosine triphosphate). The deactivators of RhoA are 3 different guanine nucleotide dissociation inhibitors GDIs that inhibit the release of GDP, and 66 different GTPase-activating proteins GAPs that increase the rate of GTP hydrolysis (Mosaddeghzadeh and Ahmadian 2021).

The first downstream effectors of RhoA are Rho-associated coiled-coil-containing kinases (ROCKs) that belong to the AGC (AMP-dependent, cGMP-dependent, and protein kinase C) family of serine/threonine kinases. Mammals have two ROCK kinases: ROCK1 (ROKβ, Rho-kinase β, p160ROCK) and ROCK2 (ROKα, Rho kinase) (Julian and Olson 2014; Riento and Ridley 2003).

ROCKs phosphorylate many different substrates, which regulate actin organization, filament stability, and contractility (Fig. 6.2); (Guan et al. 2023). For example, ROCK-phosphorylated LIM (Lin-11/Isl1/Mec3) domain kinase phosphorylates and disactivates ADF (actin-depolymerizing factor)/cofilin. This stabilizes actin filaments and, thus, indirectly limits the pool of free actin monomers. Over time, such disruption of actin filament/actin monomer balance and turnover reduces cell migration (Maekawa et al. 1999; Riento and Ridley 2003). ROCKs phosphorylate the myosin light chain (MLC), increasing the ATPase activity of myosin II and inactivating MLC phosphatase, thus expanding the level of phosphorylated MLC and

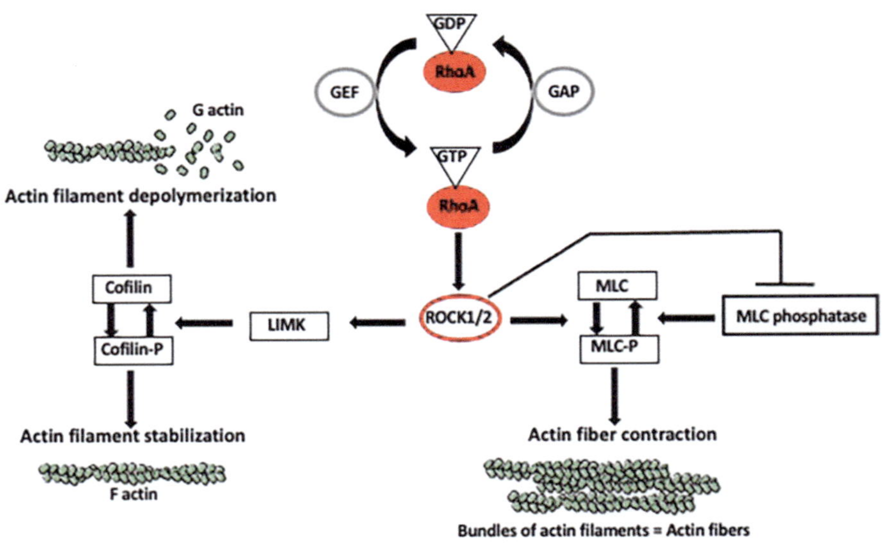

Fig. 6.2 RhoA/ROCK pathway. Guanine nucleotide exchange factors (GEFs) activate RhoA by catalyzing the exchange of GDP for GTP. GTPase-activating proteins (GAPs) deactivate RhoA by promoting hydrolysis of GTP to GDP. The first downstream effectors of RhoA are ROCK1 and ROCK2 kinases that, through various molecular targets, regulate actin polymerization and organization. One of these targets is LIMK, which, through phosphorylation of cofilin, regulates actin polymerization/depolymerization and filament stabilization. Another target is myosin light chain MLC that facilitates actin filament bundling and contraction. By affecting actin polymerization and dynamics, the RhoA and ROCKs control all actin-related functions in the cells

formation of actin stress fibers and focal contacts with the substrate (Riento and Ridley 2003). Although the amino acid sequences and kinase domains of ROCK1 and ROCK2 are 65% and 92–94% identical, respectively, and their functions often overlap, sometimes, depending on the cell- or tissue- context, their target molecules, and functions differ (Hartmann et al. 2015; Sunamura et al. 2018). ROCK 1 and 2 have different affinity for cell membrane lipids. ROCK2, in contrast to ROCK1, has a strong affinity for phosphatidylinositol (3,4,5)-trisphosphate and phosphatidylinositol (4,5)-bisphosphate (Hartmann et al. 2015). During adhesion and invasion of Y79 retinoblastoma cells, ROCK1, unlike ROCK2, phosphorylates serine LIM-kinase to inactivate cofilin, which severs actin filaments or releases actin monomers from the minus end, accelerating filament disassembly (Bernard 2007; Goode et al. 2023; Wang et al. 2014). Studies on embryonic fibroblasts from the ROCK1 or ROCK2 knockout mouse showed that ROCK1 destabilizes the actin cytoskeleton through myosin light chain phosphatase (MLCP). At the same time, ROCK2 stabilizes the actin cytoskeleton via cofilin phosphorylation (Shi et al. 2013). In an experimental setting, the RhoA/ROCK activity can be blocked pharmacologically by ROCK or GEF inhibitors or genetically by cell-specific deletion of RhoA or ROCKs. Commercially available ROCK inhibitors, such as Fasudil (HA-1077), Azaindole, Y27632, SAR407899, Fingolimod (Gilenya, FTY720), Siponimod (Mayzent), and Rezurock (Belumosudil, SLX-2119, KD-025) have an affinity for both ROCKs, but some have a much higher affinity for ROCK1 or ROCK2. For example, Rezurock IC50 values are 3 μM for ROCK1 and 100 nM for ROCK2.

In our studies on the effects of RhoA/ROCK inhibition on macrophages and chronic rejection, we have used rodent transplantation models combined with pharmacologic (RhoA/ROCK inhibitors) and genetic (macrophage-specific RhoA deletion) approaches. Below, we summarize these results.

6.4 Cellular and Molecular Effects of RhoA/ROCK Pathway Interference on Macrophages

Mouse peritoneal and bone marrow macrophages treated with various RhoA/ROCK inhibitors showed profound changes in shape, polarity, and distribution of the nucleus and organelles, which all depend on the actin cytoskeleton. The control, untreated macrophage measures around 50 μm in length, has a well-defined front and tail, and a nucleus positioned in the middle/front area. The Golgi complex is compact and located at the nucleus side, ER and mitochondria have perinuclear distribution, and focal adhesions are located away from the tail tip. The RhoA/ROCK inhibition by ROCK or GEF inhibitors (Y-27632, Rhosin, (Rho GEF-binding domain blocker), and Y16 (Rho GEF DH-PH domain blocker) causes over tenfold elongation of macrophages (Chen et al. 2017; Liu et al. 2016a, b, c). They acquire the so-called hummingbird phenotype, sometimes elongating above 800 μm. Such extreme elongation is caused by the stabilization of focal adhesions in a tail tip,

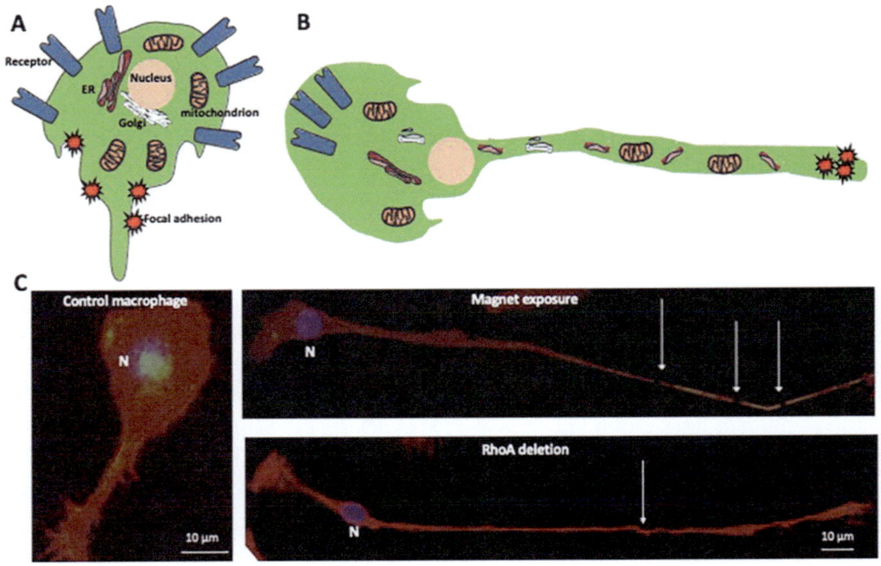

Fig. 6.3 The effects of RhoA/ROCK interference on actin-related processes in macrophages. (**a**) In control untreated macrophages, the distribution of the nucleus and all organelles, receptors, and focal adhesions is nonrandom and determined by the actin cytoskeleton. (**b**) RhoA/ROCK interference using ROCK inhibitors, RhoA deletion, or magnetic gradient force affects actin organization and polymerization, causing extreme elongation (hummingbird phenotype) and redistribution of the nucleus, organelles, and focal adhesions. It also negatively affects the recycling of the receptors to the cell membrane. (**c**) Images of control and elongated macrophages stained for actin (red, rhodamine phalloidin) and nuclei (blue, Dapi). Arrows point to the brakes caused by overstretching of macrophage tail. ER Endoplasmic reticulum, N Nucleus

which prevents macrophage detachment from the substrate. Thus, the macrophage tries to move forward, but because the tail remains attached, the macrophage stretches until, after exceeding the limit of tensile strength, breaks (Fig. 6.3); (Kloc et al. 2021; Liu et al. 2016a, b, c). Interestingly, the macrophage elongation is limited to M0 and M2 macrophages, which are initially slightly elongated and have high levels of RhoA mRNA but does not occur in M1 macrophages, which are round and have three times lower expression of RhoA mRNA (Liu et al. 2016a). This indicates that an initial elongation is a prerequisite for further stretching and hummingbird phenotype. Actin remodeling in hummingbird phenotype is accompanied by the disruption of proper localization of macrophage nucleus and organelles. The nucleus often moves to the macrophage center or tail, and the Golgi complex fragments and ER and mitochondria relocate away from the nucleus to the tail (Fig. 6.3); (Kloc et al. 2021; Chen et al. 2017; Liu et al. 2016a, b, c). Similar remodeling of the actin cytoskeleton, hummingbird phenotype, and redistribution of the nucleus and organelles also occurs in macrophages subjected to RhoA deletion or the mechanical force created by the magnetic field gradient (Fig. 6.3) (Kloc et al. 2021; Liu et al. 2017a, b; Wosik et al. 2018, 2019).

Our studies showed that remodeling of the actin cytoskeleton by RhoA/ROCK interference also affects macrophage functions and RNA and protein expression.

Treatment with Y-27632 inhibitor decreased macrophage motility, phagocytosis, and increased matrix degradation in the in vitro assay (Liu et al. 2016c). Treatment with Fingolimod not only downregulated RhoA but also upregulated the p21-activated kinase 1 (PAK-1), and amplified podosomes in mouse peritoneal macrophages (Chen et al. 2018a, b). The formation of podosomes depends on RhoA, Ras-related C3 botulinum toxin substrate 1 (Rac1), and p21-activated kinase 1 (PAK-1) (Pignatelli et al. 2012). Podosomes are the specialized extracellular matrix (ECM)-degrading organelles located on the bottom surface of macrophages. Podosomes are membrane protrusions with the actin filament core surrounded by focal adhesion proteins. Podosomes deliver matrix metalloproteinases (MMPs) for ECM degradation. Interestingly, the podosomal membrane type-1 matrix metalloproteinase (MT1-MMP), beyond the proteolytic activity, imprints spatial memory of podosome localization, speeding up their reassembly (El Azzouzi et al. 2016; Gucciardo et al. 2016).

Our recent studies of the effect of three different RhoA/ROCK inhibitors on macrophage transcriptome showed that Y-27632 and Fingolimod downregulated actin-related cell migration pathways. Rezurock downregulated fibrotic genes, such as PTX3 (pentraxin 3), CCR2 (C-C chemokine receptor type 2), and CCL2 (C-C motif chemokine ligand 2), while Fingolimod also specifically downregulated the expression of fibrosis-related NOTCH1 (neurogenic locus notch homolog protein 1) mRNA (Subuddhi et al. 2023).

We showed that Rezurock significantly modulated the phosphorylation of proto-oncogene tyrosine-protein kinase Src (c-Src) through upregulation of phosphorylation of Tyr527 and simultaneous downregulation of Tyr416, suggesting Src inactivation (unpublished data). Src is not only a master regulator in cell migration and adhesion but also mediates macrophage-mediated inflammatory responses (Byeon et al. 2012), and the upregulation of p-Tyr416Src occurs in renal interstitial fibroblasts and renal tubular cells of the fibrotic kidney (Yan et al. 2016). We also observed significant downregulation of Pak1 and Rac1 expression (unpublished data). Pak1 is an effector for Rac1 and influences processes like actin polymerization, cell migration, and cytoskeletal reorganization (Yao et al. 2020). We further observed that Rezurock treatment led to significant downregulation of proliferation rates, as evidenced by decreased Ki67 expression (proliferation marker) and dysregulation of cell cycle progression. We noticed an increase in macrophages in the G1 phase and a decrease in macrophages in the S phase of the cell cycle (unpublished data).

Using a genetic approach for the elimination of RhoA function, we showed that macrophage-specific deletion of RhoA inhibited macrophage infiltration of the graft by downregulating the expression and recycling of the G-protein-coupled fractalkine receptor CX3CR1 (chemokine (C-X3-C motif) receptor 1) that activates the RhoA pathway and regulates monocyte/macrophage migration into the vascular endothelium (Liu et al. 2017a).

6.5 RhoA/ROCK Pathway Inhibition or Macrophage-Specific Deletion of RhoA Inhibits Chronic Rejection in Rodent Transplantation Models

These above-described consequences of RhoA/ROCK genetic and pharmacologic interference on macrophages prompted us to study their effects on chronic rejection of transplanted rodent hearts.

In a genetic approach, the heart allografts from BALB/c (H-2d) mouse donors were heterotopically transplanted into Lyz2$^{Cre+/-}$ RhoA$^{flox/flox}$ recipients, which received the immunosuppressive cytotoxic T lymphocyte-associated antigen-4 immunoglobulin fusion protein (CTLA4-Ig) (0.25 mg i.p. on day 2 and day 4 post-transplantation) that blocks the CD28:CD80/86 costimulation, dampening T cell responses and preventing acute rejection (Mayer et al. 2013). Control groups were RhoA$^{flox/flox}$ (no Cre) and heterozygous Lyz2$^{Cre+/-}$RhoA$^{+/flox}$ recipients treated with CTLA4-Ig. At different time points, grafts were procured and analyzed by histology and immunostaining, and peritoneal and bone-marrow-derived monocytes/macrophages were analyzed by Western blotting and flow cytometry (Liu et al. 2017a). These studies showed that macrophage-specific deletion of RhoA decreased macrophage infiltration, neointimal hyperplasia, and chronic rejection. The RhoA deleted macrophages acquired hummingbird phenotype with a disabled CX3CR1 receptor recycling and expression. Because CX3CR1 activates the RhoA pathway and controls monocyte/macrophage migration into the vascular endothelium, its downregulation prevents macrophage infiltration and over-proliferation of smooth muscle cells in the arterial walls, and in effect prevents neointimal hyperplasia (Liu et al. 2017a).

Because the genetic approach to preventing chronic rejection is inapplicable to human transplantation, we also tested several commercially available inhibitors in chronic rejection rodent models.

First, we tested the effect of Y-27632 inhibitor on chronic rejection of cardiac transplants between donor Wistar Furth (WF) and ACI recipient rats that received the Y-27632 inhibitor (2 mg/kg, gavage feed) in addition to the subtherapeutic dose of CsA. The allografts were harvested 100 days post-transplantation for histology and immunohistochemistry. Y-27632 drastically reduced vascular occlusion and decreased immune cell infiltration (Zhang et al. 2012).

Next, we screened four other inhibitors of RhoA/ROCK in the BALB/c [H-2d]/C57BL/6 [H-2b] mouse cardiac allograft model. We found that Fasudil and Azaindole changed actin cytoskeleton and protein expression in macrophages, decreased macrophage infiltration of the allografts, inhibited vessel occlusion, and tissue fibrosis, and prevented chronic rejection. SAR-407899 decreased tissue fibrosis, and Rezurock (SLX-2119, Belumosudil) was superior in preventing tissue fibrosis but less effective in inhibiting vessel occlusion (Chen et al. 2018a, b).

In separate studies, we showed that Fingolimod (Gilenya, FTY720), combined with the inhibitors of early T cell response [CTLA4-Ig (abatacept) in mice and Everolimus in rats], prevented macrophage entry into the grafts, inhibited vessel occlusion and fibrosis, and blocked chronic rejection of cardiac transplants (Chen et al. 2021).

6.6 Clinical Application of RhoA/ROCK Inhibitors for Inhibition of Chronic Rejection

These encouraging results from animal transplantation models indicated that RhoA/ROCK pathway inhibitors may be a potential therapy for chronic rejection in human transplantation. Although most of the tested inhibitors by us are unapproved for human use, fortunately, Fingolimod, Siponimod (Mayzent), and Rezurock are FDA-approved, although for the treatments unrelated to chronic rejection of transplanted organs: Fingolimod and Siponimod for the treatment of relapsing-remitting multiple sclerosis (MS) while Rezurock for cGVHD (Bravo et al. 2022; Cutler et al. 2021; Tsai and Han 2016). The therapeutic effect of Fingolimod and Siponimod in MS relies on their ability to modulate sphingosine-1 phosphate (S1P) signaling. They bind to the S1P receptors and block the CCL2 chemokine pathway, preventing immune cells from exiting the lymphatic system and reaching the central nervous system (Ingwersen et al. 2012). Our studies discovered a crosstalk between S1P and RhoA/CX3CR1 pathways and showed that Fingolimod and Siponimod, besides affecting the S1P pathway, also block the RhoA pathway by downregulating the expression of RhoA protein and decreasing the expression of S1P1 and CX3CR1 receptors (Chen et al. 2018a, b; Uosef et al. 2020). Rezurock is especially effective in inhibiting fibrotic pathways and regulates many more fibrosis-related molecules than Fingolimod (Subuddhi et al. 2023). Based on these results, we proposed that Fingolimod, Siponimod, and Rezurock could be repurposed from their original therapeutic use to antichronic rejection therapy in transplanted patients. Currently, at the Houston Methodist Hospital, there is an ongoing clinical trial with Fingolimod in kidney transplant recipients, and the clinical trial with Rezurock is pending.

Acknowledgments We acknowledge JC Walter Jr.'s support to Houston Methodist Transplantation Center. Some drawings used to make figures were from the Servier Medical ART: SMART, smart.servier.com.

References

Abouelezz A, Almeida-Souza L (2022) The mammalian endocytic cytoskeleton. Eur J Cell Biol 101(2):151222. https://doi.org/10.1016/j.ejcb.2022.151222. Epub 2022 Apr 3

Allison AC, Eugui EM (2000) Mycophenolate mofetil and its mechanisms of action. Immunopharmacology 47(2–3):85–118. https://doi.org/10.1016/s0162-3109(00)00188-0. PMID: 10878285

Arora AS, Huang HL, Singh R, Narui Y, Suchenko A, Hatano T, Heissler SM, Balasubramanian MK, Chinthalapudi K (2023) Structural insights into actin isoforms. Elife 12:e82015. https://doi.org/10.7554/eLife.82015. PMID: 36790143; PMCID: PMC10072879

Auphan N, DiDonato JA, Rosette C, Helmberg A, Karin M (1995) Immunosuppression by glucocorticoids: inhibition of NF-kappa B activity through induction of I kappa B synthesis. Science 270(5234):286–290. https://doi.org/10.1126/science.270.5234.286

Barger SR, Vorselen D, Gauthier NC, Theriot JA, Krendel M (2022) F-actin organization and target constriction during primary macrophage phagocytosis is balanced by competing activity of myosin-I and myosin-II. Mol Biol Cell 33(14):br24. https://doi.org/10.1091/mbc.E22-06-0210. Epub 2022 Sep 21. PMID: 36129777; PMCID: PMC9727803

Bennett D, Fossi A, Marchetti L, Lanzarone N, Sisi S, Refini RM, Sestini P, Luzzi L, Paladini P, Rottoli P (2019) Postoperative acute kidney injury in lung transplant recipients. Interact Cardiovasc Thorac Surg 28(6):929–935. https://doi.org/10.1093/icvts/ivy355

Bernard O (2007) Lim kinases, regulators of actin dynamics. Int J Biochem Cell Biol 39(6):1071–1076. https://doi.org/10.1016/j.biocel.2006.11.011. Epub 2006 Nov 28

Black CK, Termanini KM, Aguirre O, Hawksworth JS, Sosin M (2018) Solid organ transplantation in the 21st century. Ann Transl Med 6(20):409. https://doi.org/10.21037/atm.2018.09.68

Bravo GÁ, Cedeño RR, Casadevall MP, Ramió-Torrentà L (2022) Sphingosine-1-phosphate (S1P) and S1P signaling pathway modulators, from current insights to future perspectives. Cells 11(13):2058. https://doi.org/10.3390/cells11132058. PMID: 35805142; PMCID: PMC9265592

Broen JCA, van Laar JM (2020) Mycophenolate mofetil, azathioprine and tacrolimus: mechanisms in rheumatology. Nat Rev Rheumatol 16(3):167–178. https://doi.org/10.1038/s41584-020-0374-8. Epub 2020 Feb 13

Byeon SE, Yi YS, Oh J, Yoo BC, Hong S, Cho JY (2012) The role of Src kinase in macrophage-mediated inflammatory responses. Mediat Inflamm 2012:512926. https://doi.org/10.1155/2012/512926. Epub 2012 Nov 11. PMID: 23209344; PMCID: PMC3504478

Chang DH, Youn JC, Dilibero D, Patel JK, Kobashigawa JA (2020) Heart transplant immunosuppression strategies at cedars-Sinai Medical Center. Int J Heart Fail 3(1):15–30. Published 2020 Sep 29. https://doi.org/10.36628/ijhf.2020.0034

Chen W, Zhao Y, Li XC, Kubiak JZ, Ghobrial RM, Kloc M (2017) Rho-specific guanine nucleotide exchange factors (Rho-GEFs) inhibition affects macrophage phenotype and disrupts Golgi complex. Int J Biochem Cell Biol 93:12–24. https://doi.org/10.1016/j.biocel.2017.10.009. Epub 2017 Oct 20

Chen W, Chen S, Chen W, Li XC, Ghobrial RM, Kloc M (2018a) Screening RhoA/ROCK inhibitors for the ability to prevent chronic rejection of mouse cardiac allografts. Transpl Immunol 50:15–25. https://doi.org/10.1016/j.trim.2018.06.002. Epub 2018 Jun 6

Chen W, Ghobrial RM, Li XC, Kloc M (2018b) Inhibition of RhoA and mTORC2/Rictor by Fingolimod (FTY720) induces p21-activated kinase 1, PAK-1 and amplifies podosomes in mouse peritoneal macrophages. Immunobiology 223(11):634–647. https://doi.org/10.1016/j.imbio.2018.07.009. Epub 2018 Jul 7

Chen W, Chen W, Chen S, Uosef A, Ghobrial RM, Kloc M (2021) Fingolimod (FTY720) prevents chronic rejection of rodent cardiac allografts through inhibition of the RhoA pathway. Transpl Immunol 65:101347. https://doi.org/10.1016/j.trim.2020.101347. Epub 2020 Oct 24

Chen M, Menon MC, Wang W, Fu J, Yi Z, Sun Z, Liu J, Li Z, Mou L, Banu K, Lee SW, Dai Y, Anandakrishnan N, Azeloglu EU, Lee K, Zhang W, Das B, He JC, Wei C (2023) HCK induces macrophage activation to promote renal inflammation and fibrosis via suppression of autophagy. Nat Commun 14(1):4297. https://doi.org/10.1038/s41467-023-40086-3. PMID: 37463911; PMCID: PMC10354075

Copeland SJ, Thurston SF, Copeland JW (2016) Actin- and microtubule-dependent regulation of Golgi morphology by FHDC1. Mol Biol Cell 27(2):260–276. https://doi.org/10.1091/mbc.E15-02-0070. Epub 2015 Nov 12. PMID: 26564798; PMCID: PMC4713130

Coscia SM, Thompson CP, Tang Q, Baltrusaitis EE, Rhodenhiser JA, Quintero-Carmona OA, Ostap EM, Lakadamyali M, Holzbaur ELF (2023) Myo19 tethers mitochondria to endoplasmic reticulum-associated actin to promote mitochondrial fission. J Cell Sci 136(5):jcs260612. https://doi.org/10.1242/jcs.260612. Epub 2023 Mar 2. PMID: 36744380; PMCID: PMC10022680

Cutler C, Lee SJ, Arai S, Rotta M, Zoghi B, Lazaryan A, Ramakrishnan A, DeFilipp Z, Salhotra A, Chai-Ho W, Mehta R, Wang T, Arora M, Pusic I, Saad A, Shah NN, Abhyankar S, Bachier C, Galvin J, Im A, Langston A, Liesveld J, Juckett M, Logan A, Schachter L, Alavi A, Howard D, Waksal HW, Ryan J, Eiznhamer D, Aggarwal SK, Ieyoub J, Schueller O, Green L, Yang Z, Krenz H, Jagasia M, Blazar BR, Pavletic S (2021) Belumosudil for chronic graft-versus-host disease after 2 or more prior lines of therapy: the ROCKstar study. Blood 138(22):2278–2289. https://doi.org/10.1182/blood.2021012021. Erratum in: Blood. 2022 Mar 17;139(11):1772. https://doi.org/10.1182/blood.2022015598. PMID: 34265047; PMCID: PMC8641099

Demetris AJ, Murase N, Lee RG et al (1997) Chronic rejection. A general overview of histopathology and pathophysiology with emphasis on liver, heart and intestinal allografts. Ann Transplant 2(2):27–44

Di Maira T, Little EC, Berenguer M (2020) Immunosuppression in liver transplant. Best Pract Res Clin Gastroenterol 46-47:101681. https://doi.org/10.1016/j.bpg.2020.101681

El Azzouzi K, Wiesner C, Linder S (2016) Metalloproteinase MT1-MMP islets act as memory devices for podosome reemergence. J Cell Biol 213(1):109–125. https://doi.org/10.1083/jcb.201510043. PMID: 27069022; PMCID: PMC4828691

Fernandez MK, Sinha M, Zidan M, Renz M (2024) Nuclear actin filaments–a historical perspective. Nucleus 15(1):2320656. https://doi.org/10.1080/19491034.2024.2320656. Epub 2024 Feb 21. PMID: 38384139; PMCID: PMC10885181

Goode BL, Eskin J, Shekhar S (2023) Mechanisms of actin disassembly and turnover. J Cell Biol 222(12):e202309021. https://doi.org/10.1083/jcb.202309021. Epub 2023 Nov 10. PMID: 37948068; PMCID: PMC10638096

Guan G, Cannon RD, Coates DE, Mei L (2023) Effect of the rho-kinase/ROCK signaling pathway on cytoskeleton components. Genes (Basel) 14(2):272. https://doi.org/10.3390/genes14020272. PMID: 36833199; PMCID: PMC9957420

Gucciardo E, Mobashir M, Lehti K (2016) Proactive for invasion: reuse of matrix metalloproteinase for structural memory. J Cell Biol 213(1):11–13. https://doi.org/10.1083/jcb.201603066. PMID: 27069020; PMCID: PMC4828696

Hanley JG (2014) Actin-dependent mechanisms in AMPA receptor trafficking. Front Cell Neurosci 8:381. https://doi.org/10.3389/fncel.2014.00381. PMID: 25429259; PMCID: PMC4228833

Hardinger KL, Brennan DC (2013) Novel immunosuppressive agents in kidney transplantation. World J Transplant 3(4):68–77. https://doi.org/10.5500/wjt.v3.i4.68

Hartmann S, Ridley AJ, Lutz S (2015) The function of rho-associated kinases ROCK1 and ROCK2 in the pathogenesis of cardiovascular disease. Front Pharmacol 6:276. https://doi.org/10.3389/fphar.2015.00276. PMID: 26635606; PMCID: PMC4653301

Horsthemke M, Wilden J, Bachg AC, Hanley PJ (2018) Time-lapse 3D imaging of phagocytosis by mouse macrophages. J Vis Exp 140:57566. https://doi.org/10.3791/57566. PMID: 30394377; PMCID: PMC6235568

Ingwersen J, Aktas O, Kuery P, Kieseier B, Boyko A, Hartung HP (2012) Fingolimod in multiple sclerosis: mechanisms of action and clinical efficacy. Clin Immunol 142(1):15–24. https://doi.org/10.1016/j.clim.2011.05.005. Epub 2011 May 26

Janota CS, Pinto A, Pezzarossa A, Machado P, Costa J, Campinho P, Franco CA, Gomes ER (2022) Shielding of actin by the endoplasmic reticulum impacts nuclear positioning. Nat Commun 13(1):2763. https://doi.org/10.1038/s41467-022-30388-3. PMID: 35589708; PMCID: PMC9120458

Julian L, Olson MF (2014) Rho-associated coiled-coil containing kinases (ROCK): structure, regulation, and functions. Small GTPases 5:e29846. https://doi.org/10.4161/sgtp.29846. Epub 2014 Jul 10. PMID: 25010901; PMCID: PMC4114931

Kashina AS (2020) Regulation of actin isoforms in cellular and developmental processes. Semin Cell Dev Biol 102:113–121. https://doi.org/10.1016/j.semcdb.2019.12.003. Epub 2020 Jan 27. PMID: 32001148; PMCID: PMC7214208

Kay RR (2021) Macropinocytosis: biology and mechanisms. Cells Dev 168:203713. https://doi.org/10.1016/j.cdev.2021.203713. Epub 2021 Jun 24

Khaitlina SY (2001) Functional specificity of actin isoforms. Int Rev Cytol 202:35–98. https://doi.org/10.1016/s0074-7696(01)02003-4

Kim IC, Youn JC, Kobashigawa JA (2018) The past, present and future of heart transplantation. Korean Circ J 48(7):565–590. https://doi.org/10.4070/kcj.2018.0189

Kloc M, Ghobrial RM (2014) Chronic allograft rejection: a significant hurdle to transplant success. Burns Trauma 2(1):3–10. https://doi.org/10.4103/2321-3868.121646. PMID: 27574640; PMCID: PMC4994504

Kloc M, Uosef A, Wosik J, Kubiak JZ, Ghobrial RM (2019) RhoA pathway and actin regulation of the Golgi/centriole complex. Results Probl Cell Differ 67:81–93. https://doi.org/10.1007/978-3-030-23173-6_5

Kloc M, Uosef A, Villagran M, Zdanowski R, Kubiak JZ, Wosik J, Ghobrial RM (2021) RhoA- and actin-dependent functions of macrophages from the rodent cardiac transplantation model perspective -timing is the essence. Biology (Basel) 10(2):70. https://doi.org/10.3390/biology10020070. PMID: 33498417; PMCID: PMC7909416

Kyheröinen S, Vartiainen MK (2020) Nuclear actin dynamics in gene expression and genome organization. Semin Cell Dev Biol 102:105–112. https://doi.org/10.1016/j.semcdb.2019.10.012. Epub 2019 Nov 15

Lisik W, Gong Y, Tejpal N, Skelton TS, Bremer EG, Kloc M, Ghobrial RM (2010) Intragraft gene expression profile associated with the induction of tolerance by allochimeric MHC I in the rat heart transplantation model. Genesis 48(1):8–19. https://doi.org/10.1002/dvg.20574

Liu Y, Chen W, Minze LJ, Kubiak JZ, Li XC, Ghobrial RM, Kloc M (2016a) Dissonant response of M0/M2 and M1 bone-marrow-derived macrophages to RhoA pathway interference. Cell Tissue Res 366(3):707–720. https://doi.org/10.1007/s00441-016-2491-x. Epub 2016 Sep 15

Liu Y, Minze LJ, Mumma L, Li XC, Ghobrial RM, Kloc M (2016b) Mouse macrophage polarity and ROCK1 activity depend on RhoA and non-apoptotic caspase 3. Exp Cell Res 341(2):225–236. https://doi.org/10.1016/j.yexcr.2016.02.004. Epub 2016 Feb 12

Liu Y, Tejpal N, You J, Li XC, Ghobrial RM, Kloc M (2016c) ROCK inhibition impedes macrophage polarity and functions. Cell Immunol 300:54–62. https://doi.org/10.1016/j.cellimm.2015.12.005. Epub 2015 Dec 17

Liu Y, Chen W, Wu C, Minze LJ, Kubiak JZ, Li XC, Kloc M, Ghobrial RM (2017a) Macrophage/monocyte-specific deletion of Ras homolog gene family member A (RhoA) downregulates fractalkine receptor and inhibits chronic rejection of mouse cardiac allografts. J Heart Lung Transplant 36(3):340–354. https://doi.org/10.1016/j.healun.2016.08.011. Epub 2016 Aug 20. PMID: 27692539; PMCID: PMC5316307

Liu Y, Kubiak JZ, Li XC, Ghobrial RM, Kloc M (2017b) Macrophages and RhoA pathway in transplanted organs. Results Probl Cell Differ 62:365–376. https://doi.org/10.1007/978-3-319-54090-0_15

Ma T, Liu X, Su H, Shi Q, He Y, Wu F, Gao C, Li K, Liang Z, Zhang D, Zhang X, Hu K, Li S, Wang L, Wang M, Yue S, Hong W, Chen X, Zhang J, Zheng L, Deng X, Wang P, Fan Y (2024) Coupling of perinuclear actin cap and nuclear mechanics in regulating flow-induced yap spatiotemporal nucleocytoplasmic transport. Adv Sci (Weinh) 11(11):e2305867. https://doi.org/10.1002/advs.202305867. Epub 2023 Dec 31. PMID: 38161226; PMCID: PMC10953556 Copy

Maekawa M, Ishizaki T, Boku S, Watanabe N, Fujita A, Iwamatsu A, Obinata T, Ohashi K, Mizuno K, Narumiya S (1999) Signaling from rho to the actin cytoskeleton through protein kinases ROCK and LIM-kinase. Science 285(5429):895–898. https://doi.org/10.1126/science.285.5429.895

Masters TA, Sheetz MP, Gauthier NC (2016) F-actin waves, actin cortex disassembly and focal exocytosis driven by actin-phosphoinositide positive feedback. Cytoskeleton (Hoboken) 73(4):180–196. https://doi.org/10.1002/cm.21287. Epub 2016 Apr 7

Mayer E, Hölzl M, Ahmadi S, Dillinger B, Pilat N, Fuchs D, Wekerle T, Heitger A (2013) CTLA4-Ig immunosuppressive activity at the level of dendritic cell/T cell crosstalk. Int

Immunopharmacol 15(3):638–645. https://doi.org/10.1016/j.intimp.2013.02.007. Epub 2013 Feb 20. PMID: 23434857; PMCID: PMC3629566
Miklavc P, Frick M (2020) Actin and myosin in non-neuronal exocytosis. Cells 9(6):1455. https://doi.org/10.3390/cells9061455. PMID: 32545391; PMCID: PMC7348895
Moini M, Schilsky ML, Tichy EM (2015) Review on immunosuppression in liver transplantation. World J Hepatol 7(10):1355–1368. https://doi.org/10.4254/wjh.v7.i10.1355
Moore AS, Coscia SM, Simpson CL, Ortega FE, Wait EC, Heddleston JM, Nirschl JJ, Obara CJ, Guedes-Dias P, Boecker CA, Chew TL, Theriot JA, Lippincott-Schwartz J, Holzbaur ELF (2021) Actin cables and comet tails organize mitochondrial networks in mitosis. Nature 591(7851):659–664. https://doi.org/10.1038/s41586-021-03309-5. Epub 2021 Mar 3. PMID: 33658713; PMCID: PMC7990722
Mosaddeghzadeh N, Ahmadian MR (2021) The RHO family GTPases: mechanisms of regulation and signaling. Cells 10(7):1831. https://doi.org/10.3390/cells10071831. PMID: 34359999; PMCID: PMC8305018
Mundinger GS, Munivenkatappa R, Drachenberg CB, Ha JS, Vaca EE, Shipley ST, Papadimitriou JC, Bartlett ST, Rodriguez ED, Barth RN (2013) Histopathology of chronic rejection in a nonhuman primate model of vascularized composite allotransplantation. Transplantation 95(10):1204–1210. https://doi.org/10.1097/TP.0b013e31828d1528. PMID: 23694932
Nelson A, Duhon B, Mehta B, Gabardi S (2022) Consensus recommendations for maintenance immunosuppression in solid organ transplantation: focused guidance for pharmacists. Pharmacotherapy 42(5):401–429. https://doi.org/10.1002/phar.2716
Ogawa T, Shichino S, Ueha S, Matsushima K (2021) Macrophages in lung fibrosis. Int Immunol 33(12):665–671. https://doi.org/10.1093/intimm/dxab040. PMID: 34270737; PMCID: PMC8633606
Pain C, Tolmie F, Wojcik S, Wang P, Kriechbaumer V (2023) intER-ACTINg: the structure and dynamics of ER and actin are interlinked. J Microsc 291(1):105–118. https://doi.org/10.1111/jmi.13139. Epub 2022 Aug 30
Pecoraro Y, Carillo C, Diso D et al (2017) Efficacy of extracorporeal photopheresis in patients with bronchiolitis obliterans syndrome after lung transplantation. Transplant Proc 49(4):695–698. https://doi.org/10.1016/j.transproceed.2017.02.035
Pignatelli J, LaLonde SE, LaLonde DP, Clarke D, Turner CE (2012) Actopaxin (α-parvin) phosphorylation is required for matrix degradation and cancer cell invasion. J Biol Chem 287(44):37309–37320. https://doi.org/10.1074/jbc.M112.385229. Epub 2012 Sep 6. PMID: 22955285; PMCID: PMC3481328
Polino AJ, Ng XW, Rooks R, Piston DW (2023) Disrupting actin filaments enhances glucose-stimulated insulin secretion independent of the cortical actin cytoskeleton. J Biol Chem 299(11):105334. https://doi.org/10.1016/j.jbc.2023.105334. Epub 2023 Oct 10. PMID: 37827287; PMCID: PMC10641669
Pollard TD, Cooper JA (2009) Actin, a central player in cell shape and movement. Science 326(5957):1208–1212. https://doi.org/10.1126/science.1175862. PMID: 19965462; PMCID: PMC3677050
Riento K, Ridley AJ (2003) Rocks: multifunctional kinases in cell behaviour. Nat Rev Mol Cell Biol 4(6):446–456. https://doi.org/10.1038/nrm1128
Safarini OA, Keshavamurthy C, Patel P (2025) Calcineurin inhibitors. 2023. In: StatPearls [Internet]. StatPearls Publishing, Treasure Island
Schaks M, Giannone G, Rottner K (2019) Actin dynamics in cell migration. Essays Biochem 63(5):483–495. https://doi.org/10.1042/EBC20190015. PMID: 31551324; PMCID: PMC6823167
Serebryannyy L, de Lanerolle P (2020) Nuclear actin: the new normal. Mutat Res 821:111714. https://doi.org/10.1016/j.mrfmmm.2020.111714. Epub 2020 Jul 21
Shi J, Wu X, Surma M, Vemula S, Zhang L, Yang Y, Kapur R, Wei L (2013) Distinct roles for ROCK1 and ROCK2 in the regulation of cell detachment. Cell Death Dis 4(2):e483. https://doi.org/10.1038/cddis.2013.10. PMID: 23392171; PMCID: PMC3734810

Small B, Au J, Brink H, Shah I, Strah H (2022) Induction and maintenance immunosuppression in lung transplantation. Indian J Thorac Cardiovasc Surg 38(Suppl 2):300–317. https://doi.org/10.1007/s12055-021-01225-x

Subuddhi A, Uosef A, Zou D, Ubelaker HV, Ghobrial RM, Kloc M (2023) Comparative transcriptome profile of mouse macrophages treated with the RhoA/Rock pathway inhibitors Y27632, Fingolimod (Gilenya), and Rezurock (Belumosudil, SLx-2119). Int Immunopharmacol 118:110017. https://doi.org/10.1016/j.intimp.2023.110017. Epub 2023 Mar 15

Sunamura S, Satoh K, Kurosawa R, Ohtsuki T, Kikuchi N, Elias-Al-Mamun M, Shimizu T, Ikeda S, Suzuki K, Satoh T, Omura J, Nogi M, Numano K, Siddique MAH, Miyata S, Miura M, Shimokawa H (2018) Different roles of myocardial ROCK1 and ROCK2 in cardiac dysfunction and postcapillary pulmonary hypertension in mice. Proc Natl Acad Sci U S A 115(30):E7129–E7138. https://doi.org/10.1073/pnas.1721298115. Epub 2018 Jul 9. PMID: 29987023; PMCID: PMC6064988

Tribollet V, Cerutti C, Géloën A, Berger E, De Mets R, Balland M, Courchet J, Vanacker JM, Forcet C (2022) ERRα coordinates actin and focal adhesion dynamics. Cancer Gene Ther 29(10):1429–1438. https://doi.org/10.1038/s41417-022-00461-6. Epub 2022 Apr 4

Tsai HC, Han MH (2016) Sphingosine-1-phosphate (S1P) and S1P signaling pathway: therapeutic targets in autoimmunity and inflammation. Drugs 76(11):1067–1079. https://doi.org/10.1007/s40265-016-0603-2

Ulferts S, Lopes M, Miyamoto K, Grosse R (2024) Nuclear actin dynamics and functions at a glance. J Cell Sci 137(6):jcs261630. https://doi.org/10.1242/jcs.261630. Epub 2024 Apr 2

Uosef A, Vaughn N, Chu X, Elshawwaf M, Abdelshafy AAA, Elsaid KMK, Ghobrial RM, Kloc M (2020) Siponimod (Mayzent) downregulates RhoA and cell surface expression of the S1P1 and CX3CR1 receptors in mouse RAW 264.7 macrophages. Arch Immunol Ther Exp (Warsz) 68(3):19. https://doi.org/10.1007/s00005-020-00584-4

Vartiainen MK (2008) Nuclear actin dynamics—from form to function. FEBS Lett 582(14):2033–2040. https://doi.org/10.1016/j.febslet.2008.04.010. Epub 2008 Apr 16

Wang J, Liu X-H, Yang Z-J, Xie B, Zhong Y-S (2014) The effect of ROCK-1 activity change on the adhesive and invasive ability of Y79 retinoblastoma cells. BMC Cancer 14:89. https://doi.org/10.1186/1471-2407-14-89

Wojciechowski D, Wiseman A (2021) Long-term immunosuppression management: opportunities and uncertainties. Clin J Am Soc Nephrol 16(8):1264–1271. https://doi.org/10.2215/CJN.15040920

Wosik J, Chen W, Qin K, Ghobrial RM, Kubiak JZ, Kloc M (2018) Magnetic field changes macrophage phenotype. Biophys J 114(8):2001–2013. https://doi.org/10.1016/j.bpj.2018.03.002. PMID: 29694876; PMCID: PMC5937143

Wosik J, Suarez-Villagran M, Miller JH Jr, Ghobrial RM, Kloc M (2019) Macrophage phenotype bioengineered by magnetic, genetic, or pharmacologic interference. Immunol Res 67(1):1–11. https://doi.org/10.1007/s12026-019-9066-3

Yan Y, Ma L, Zhou X, Ponnusamy M, Tang J, Zhuang MA, Tolbert E, Bayliss G, Bai J, Zhuang S (2016) Src inhibition blocks renal interstitial fibroblast activation and ameliorates renal fibrosis. Kidney Int 89(1):68–81. https://doi.org/10.1038/ki.2015.293. Epub 2016 Jan 4. PMID: 26444028; PMCID: PMC4824676

Yang C, Svitkina TM (2019) Ultrastructure and dynamics of the actin-myosin II cytoskeleton during mitochondrial fission. Nat Cell Biol 21(5):603–613. https://doi.org/10.1038/s41556-019-0313-6. Epub 2019 Apr 15. PMID: 30988424; PMCID: PMC6499663

Yao D, Li C, Rajoka MSR, He Z, Huang J, Wang J, Zhang J (2020) P21-Activated kinase 1: emerging biological functions and potential therapeutic targets in cancer. Theranostics 10(21):9741–9766. https://doi.org/10.7150/thno.46913. PMID: 32863957; PMCID: PMC7449905

Zhang L, Kloc M, Tejpal N, You J, Cordero-Reyes A, Youker K, Ghobrial R (2012) Rock1 inhibitor abrogates chronic rejection in rat cardiac model system. Open J Organ Transpl Surg 2:46–51. https://doi.org/10.4236/ojots.2012.24012

Chapter 7
Donor-Derived Cell-Free DNA in Acute and Chronic Rejection of Solid Allograft Transplantation

Javier Muro-Perez, Manuel Muro-Perez, Jose Antonio Galian, Noelia Pérez-López, Isabel Legaz, and Manuel Muro

Abstract Today, there have been spectacular advances in surgical techniques, organ preservation for transplantation, optimal and efficient donor and recipient selection, more efficient diagnosis of transplant complications, and major progress in pharmacological immunosuppression procedures. In this regard, survival rates after transplantation of various organs have been gaining ground, particularly in the case of lung transplants, whose average survival rate is lower than other types of transplants. In this regard, it is important to detect acute and subclinical clinical rejection, as well as chronic allograft rejection. This is especially important in heart and lung transplants. In the latter type of transplant, and due to the chronic dysfunction of the lung allograft, it is key to detect rejection early and promptly, as it can affect up to half of the transplant patient population. Therefore, effective diagnostic tools are needed to visualize the level of allograft damage using genomic methods such as those that measure donor-derived cell-free DNA (dd-cfDNA). The plasma concentration of dd-cfDNA increases after graft injury or infection. Our team has experience quantifying this parameter in allograft injury progression, and our experience and comparison with the published literature will be presented in the following sections, discussing validated and non-validated results.

The role of these authors (Javier Muro-Perez, Manuel Muro-Perez, Isabel Legaz, and Manuel Muro) is the same and their order is arbitrary.

J. Muro-Perez · M. Muro-Perez · J. A. Galian · N. Pérez-López · M. Muro (✉)
Immunology Service, Clinical University Hospital "Virgen de la Arrixaca", Murcian Institute of Biomedical Research – IMIB, El Palmar, Murcia, Spain
e-mail: manuel.muro@carm.es

I. Legaz
Legal and Forensic Science Department, University of Murcia, Murcia, Spain

M. Kloc, R. M. Ghobrial (eds.), *Cellular and Molecular Mechanisms of Chronic Rejection in Organ Transplantation*, Results and Problems in Cell Differentiation 77, https://doi.org/10.1007/978-3-032-07686-1_7

7.1 Introduction

Even today, in the various types of transplants, a biopsy of the grafted organ is often the gold standard for proper monitoring of the transplanted graft.

In this sense, a biopsy of a grafted organ consists of taking a small sample of the transplanted tissue and sending it to a pathologist for microscopic examination to ensure proper viability of the allograft. The biopsy is performed to detect any evidence or signs of potential rejection or other problems that may arise in the transplanted organ.

However, as any detection method that depends on the subjective evaluation of the human eye is invasive and can cause discomfort to the patient, it is not absolute, risk-free, or ideal. In this sense, it has been reported that approximately 1% of biopsies can lead to complications, with a risk of macroscopic hematuria above 3.5% (Schwarz et al. 2005), although these are usually mild and localized (pain and bleeding). However, there are more serious complications, such as the risk of organ perforation, which could even lead to severe bleeding, sepsis, and death.

In this sense, we must strive for safety profiles that minimize the risk of post-transplant complications (Rush and Gibson 2019) with specific procedures.

Fortunately, the detection of subclinical rejection has avoided the need for routine protocol biopsies, and new noninvasive biomarkers have been developed that allow us to obtain more information on the allograft status and, if necessary, detect rejection in order to proceed in a timely manner, increase or modify the immunosuppression administered, and implement anti-rejection therapies.

These new noninvasive biomarkers that allow the detection of subclinical rejection include urinary and serum markers, such as formulas or biochemical ratios (urinary albumin–creatinine ratio), the detection of donor-specific antibodies (DSA), the presence of specific cell populations that vary during rejection, or more powerful imaging techniques (magnetic resonance imaging and ultrasound) to detect changes in the graft that could indicate rejection (Legaz et al. 2020; El Band et al. 2021; Boix et al. 2018, 2021, 2023; Alfaro et al. 2021a). Similarly, at the genomic level, we have a significant array of promising biomarkers: gene expression (serum and in mTOR biopsies), differential microRNA expression, mitochondrial DNA analysis, or circulating cell-free DNA analysis that can aid in the diagnosis of subclinical rejection (Loupy et al. 2017; Farcas et al. 2024; Legaz et al. 2021, 2024a; Alfaro et al. 2021a, 2022, 2023; Jimenez-Coll et al. 2023a) and allow for early intervention.

Historically, in kidney transplants, serum creatinine, glomerular filtration rate, and urinary markers were the only ones that served the purpose of measuring kidney function and assessing graft status due to their low cost, simple procedure, and relative reliability. However, their sensitivity and specificity for detecting allograft damage are extremely limited.

Today, the determination and quantification of donor-derived cell-free DNA (dd-cfDNA), which measures graft damage, offers a real possibility for noninvasive recognition of allogeneic damage in transplant recipients.

7.2 How Is dd-cfDNA Evaluated in Solid Organ Transplantation?

The quantification of the different degraded DNA fragments released from cells or tissues into the blood or other fluids is known as cfDNA measurement. A summary of donor-derived circulating free DNA extraction and analysis is shown in Fig. 7.1.

Historically, the first detection was reported as early as 1948 in patients with systemic lupus erythematosus, but many years passed before the role of their different concentrations or release in the disease could be established, primarily in tumor pathology (Ulrich and Guibert 2017). Thus, it is important not only to quantify the concentrations released by tumor cells into the blood but also to identify genetic and epigenetic variations, which can be biomarkers for various pathologies (Jeon et al. 2016). There are several types of cfDNA reported in the literature: ccf-mtDNA (circulating cell-free mitochondrial DNA), ctDNA (circulating tumor DNA), cffDNA

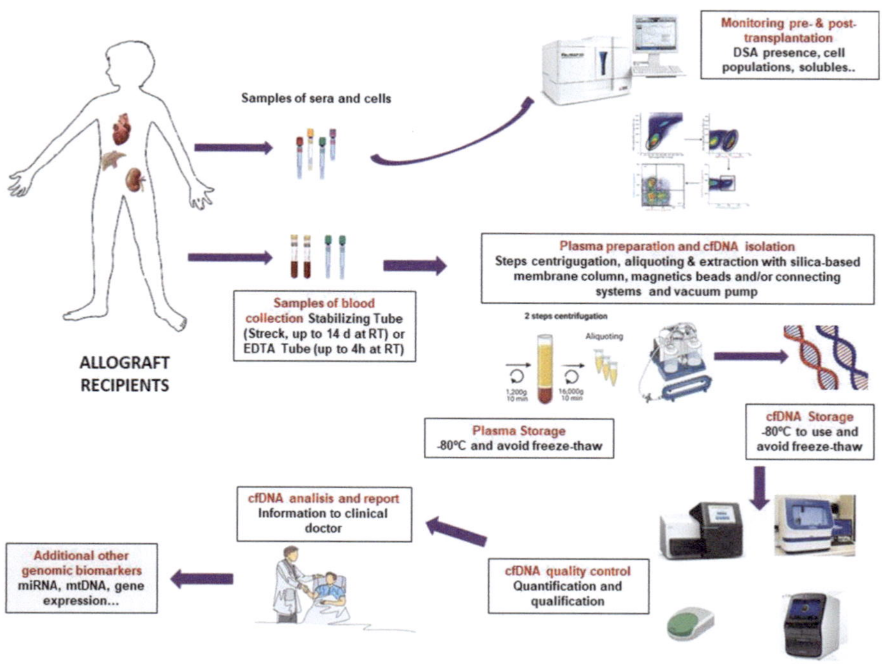

Fig. 7.1 Diagram of donor-derived circulating free DNA determination. A summary of dd-cfDNA extraction and analysis is shown. The following steps are indicated: Blood samples collection in stabilizing tubes, plasma preparation, and cfDNA isolation (with plasma and cfDNa storage), cfDNA quality control, and cfDNA analysis and report. This report is evaluated by the doctor and could indicate and modify immunosuppressant treatment. Other additional biomarkers such as genomic (miRNA, mtDNA, and gene expression), cellular (regulatory and functional populations), or serological (DSA monitoring, cytokines, chemokines, or soluble proteins) biomarkers can be added and indicated for allograft recipient outcome

(cell-free fetal DNA), and dd-cfDNA (donor-derived cell-free DNA) or mdd-cfDNA (donor-derived methylated cell-free DNA). Furthermore, dd-cfDNA released from a transplanted graft can be quantified and detected in blood, urine, or other biological fluids (Jimenez-Coll et al. 2023a).

In this sense, at the time of implantation, the concentration of dd-cfDNA increases above 5% (possibly due to the traumatic process suffered by the organ during its removal from the donor and its surgical implantation in the recipient), but these values decrease rapidly and become almost zero 1 week post-transplant (Botella et al. 2024).

In the case of complications such as graft rejection, the release and concentration of dd-cfDNA can increase up to five times the baseline values, making it important in the early detection of subclinical rejection (our cfDNA).

The revolution in efficient laboratory diagnosis began in 2011 with methods for detecting dd-cfDNA by digital polymerase chain reaction (PCR), based on differences in single-nucleotide polymorphisms (SNP) between donor and recipient (Snyder et al. 2011), to minimize the likelihood that two unrelated individuals share them.

However, after much research in the field, commercially available dd-cfDNA detection kits show unquestionable limitations related to twin transplantation or its evolution in retransplant patients where there may be more than two different genomes (Jimenez-Coll et al. 2023a).

Currently, there are several commercial dd-cfDNA detection kits available for use in transplants, Allosure by CareDx, Prospera by Natera or Trac by Viracor Eurofins, the first of which is the most widely used (Huang and Jordan 2022), where 1% dd-cfDNA is established as a cut-off point to discriminate the presence or absence of graft rejection, with a specificity of 85%, a sensitivity of 59%, a positive predictive value (PPV) of 61%, and a negative predictive value (NPV) with accuracy of 84%. Another thing is its true value in detecting differences between humoral rejection and cellular rejection (Botella et al. 2024).

However, although these types of commercial kits for dd-cfDNA determination are attractive for the early detection of subclinical rejection, they are prohibitively expensive for most healthcare systems and may present significant cost-effectiveness barriers. Perhaps other technologies such as digital PCR could minimize this cost.

Aside from the economic factors, dd-cfDNA quantification also has a relatively low PPV, making it difficult to assess high dd-cfDNA levels without clinical information from other parameters. However, clearly, having a higher NPV can help avoid unnecessary biopsies (Botella et al. 2024).

Another potential confounding factor is that other pathologies, besides rejection, can show increased dd-cfDNA levels. Leukopenia and leukocytosis can also alter the dd-cfDNA fraction and produce false-positive or false-negative results, as can BK virus viremia (Oellerich et al. 2022) or obesity. However, the enormous potential of dd-cfDNA is driving commercial companies to improve their products, while research centers, hospitals, and universities are trying to refine the technology and scientific knowledge of this biomarker, with its limitations and advantages depending on the methods used.

In PCR-based technologies, qPCR is limited by the lack of multiplexing and only detects known mutations, while digital PCR and PCR coupled with spectrometry allow for limited multiplexing and the detection of known mutations (Jimenez-Coll et al. 2023a).

In targeted NGS (next-generation sequencing) technologies, the methods or techniques [Tam-Seq, eTam-Seq, Safe-SeqS, duplex sequencing, TEC-Seq, single primer extension (SPE), UMI-duplex SPE, CAPP-Seq, iDES eCAPP-Seq, or Ig-HTS] have limitations, such as multiplex PCR amplicon methods (fragment size-dependent), which are less comprehensive than whole-genome sequencing or whole-exome sequencing. If we use nontargeted methods, the limitations would be low sensitivity (greater depth comes at a high cost), the need for bioinformatics experts, high cost, variable sensitivity, and the large amount of data generated (Jimenez-Coll et al. 2023a).

7.3 Role of dd-cfDNA Determination in Kidney

In kidney transplantation, biopsy remains the gold standard for monitoring transplanted kidneys, from early detection to routine protocol screening (which is increasingly less commonly performed), as well as for assessing graft progression (Botella et al. 2024). However, as it is a truly invasive method, it presents problems regarding safety and acceptability for kidney recipients. In this regard, approximately 1% of biopsies produce complications with an increased risk of macroscopic hematuria (Schwarz et al. 2005). Furthermore, with current immunosuppression, detection of subclinical rejection is rare, leading to the need to invest in noninvasive methods to measure kidney graft function.

In kidney transplantation, we use serum creatinine and urinary indicators of kidney function (urinary albumin–creatinine ratio) to measure organ function because they are inexpensive, reliable, and relatively easy to interpret. However, it must be acknowledged that the sensitivity and specificity for measuring kidney graft damage are limited.

However, there are also other non-invasive procedures in kidney and heart transplantation for the monitoring of patients with rejection, such as detection of HLA and non-HLA antibodies, gene and miRNA expression models, cellular immunity parameters, urinary chemokine levels (CXCL9 and CXCL10), or proteomic and peptide signatures of rejection in urine and blood (Legaz et al. 2020; Alfaro et al. 2021a, 2022, 2023; Boix et al. 2018, 2021, 2023), although none of them reach complete validation, complete scientific demonstration, and financial backing (some are quite expensive) to use them as routine in clinical practice. Therefore, the search for and evaluation of new biomarkers in kidney transplantation are essential, and dd-cfDNA may be that marker with great potential for the early detection of kidney graft damage (Botella et al. 2024). Thus, early identification of increased dd-cfDNA levels could make biopsy unnecessary and allow for modification of immunosuppressive therapy.

In this regard, there are many single-center and multicenter studies evaluating this method in kidney transplantation, such as Allosure with CareDx, Prospera with Natera, and Trac with Viracor-Eurofins (Martuszewski et al. 2021; Jimenez-Coll et al. 2023a; Botella et al. 2024), which present the discriminatory levels for defining or not defining kidney rejection (dd-cfDNA >1%), as well as the sensitivity, specificity, NPV, and PPV of dd-cfDNA quantification.

In a relatively recent study, our group presented the results of 30 kidney transplant patients in whom dd-cfDNA levels were assessed for 2 years (Botella et al. 2024) and found that 60% of our patients had positive or elevated values, coinciding with the diagnosis by biopsy and the detection of DSA antibodies like other studies already mentioned previously, but in our study, we described dd-cfDNA as a better biomarker to predict cellular rejection than humoral rejection. Another very recent study (Aubert et al. 2024) shows that elevated dd-cfDNA levels correlated with the occurrence and severity of all types of kidney rejection, including subclinical rejection, even in recipients with stable allograft function. A high NPV of dd-cfDNA was used to avoid biopsies. However, our results demonstrate that a low NPV, due to the potential impact of other post-transplant pathologies or biological complications (obesity, recurrent autoimmune diseases, and infections) on dd-cfDNA levels, should be taken into account when interpreting the results (Botella et al. 2024), and cases of dd-cfDNA may be considered "false positives." On the other hand, the aforementioned study by Aubert et al. (2024) also demonstrated an association with subclinical rejection in recipients with stable allograft function and may be a powerful diagnostic tool in the near future.

In summary, it appears that in kidney transplantation, dd-cfDNA levels could be correlated with biopsy results and seem to be more effective than biochemical markers in detecting clinical and subclinical rejection with dd-cfDNA values >0.5%, as well as the association of dd-cfDNA with dnDSA and underlying alloimmune-mediated injury in kidney allograft recipients, especially in the combination of dd-cfDNA fraction and quantity (Jimenez-Coll et al. 2023a; Botella et al. 2024). Along the same lines, other authors (Aubert et al. 2024) have shown that the addition of dd-cfDNA, combined with conventional kidney transplant markers, increases the discriminatory capacity of rejection diagnosis. Therefore, in conclusion, plasma levels of dd-cfDNA could be considered an alternative biomarker for monitoring graft rejection in the early stages of kidney transplantation, up to several months before its clinical presentation.

7.4 Role of dd-cfDNA Determination in Liver Transplantation

In liver transplantation, as in kidney transplantation, liver biopsy also remains the gold standard for assessing various allograft complications, despite the potential risks associated with its performance.

As in kidney transplantation, the emergence and validation of new biomarkers for earlier detection and prediction of liver graft damage, without the potential risks of biopsy, must play a leading role without having a major clinical impact.

In this sense, technology for detecting dd-cfDNA is a promising tool for sensitively detecting transplanted liver graft dysfunction (Jimenez-Coll et al. 2023a).

Several published articles also show that dd-cfDNA was correlated with the aspartate aminotransferase to alanine aminotransferase ratio (Knight et al. 2019; Fernandez-Galan et al. 2022), demonstrating good diagnostic performance for acute rejection. Likewise, another recent prospective cohort study suggests that dd-cfDNA by SNP and NGS outperformed conventional liver function tests in predicting acute rejection (Levitsky et al. 2022).

Also, a very recent study in a Japanese population by Kanamori et al. (2024) characterized the dynamics of dd-cfDNA levels after transplantation and evaluated their use in monitoring graft integrity and diagnosing acute rejection. Furthermore, dd-cfDNA levels were aligned with liver allograft fibrosis and C4d scores to assess the potential of using dd-cfDNA as a general indicator of graft integrity. The diagnostic accuracy of SNP-directed dd-cfDNA in this study demonstrated an AUC of 0.975 for acute rejection. Quantification of dd-cfDNA showed a sensitivity and specificity of 100% and 90.9%, respectively, in diagnosing acute rejection, providing an NPV of 100%, which suggests avoiding invasive biopsy when dd-cfDNA is below the cut-off value.

Theoretically, the destruction of donor liver tissue releases two copies of dd-cfDNA into the blood, derived from damage to cellular sources, such as endothelial cells or other hepatic stromal cells.

An additional point regarding cfDNA is that the short half-life of cfDNA (minutes to 1.5 hours) continuously reflects graft damage in real time, while biochemical liver function tests have long half-lives, limiting the ability to monitor ongoing changes in the liver graft.

However, a major limitation is the elevated dd-cfDNA levels associated with active viral and bacterial infections, as well as vascular and biliary complications (Jimenez-Coll et al. 2023a).

In conclusion, in recipients with abnormal liver function tests, it may be the preferred evaluation, implying a benefit in screening for rejection and in prospective trials to minimize immunosuppression. The absence of elevated dd-cfDNA levels, along with normal liver function tests, could eliminate the need for biopsies in the future. Finally, samples with suspected chronic rejection (fibrosis with inflammation and C4d) may show persistently elevated dd-cfDNA levels (Levitsky et al. 2022).

7.5 Role of dd-cfDNA Determination in Heart Transplantation

Firstly, heart transplantation is usually the final treatment for the disease known as terminal heart failure, allowing patients undergoing this procedure to live longer, with a better quality of life after the intervention (Farcas et al. 2024). However, there are still problems in this type of transplant despite the best surgical procedures, the improvement in immunosuppression protocols, and the more suitable management of the patient outcome after heart transplantation.

In a similar manner to other types of transplants, one of the main immune problems in heart transplantation is rejection, which occurs when the heart recipient's immune system sees and recognizes the grafted heart as foreign and triggers an inflammatory response to attack it.

We must try to understand what mechanisms promote the rejection of the heart graft or its acceptance and tolerance to update and implement new developments, procedures, and protocols that increase the safety and prediction that the grafted heart will evolve more favorably and, if it occurs, to be able to manage and treat it more effectively. We must also assume that it is a more complicated transplant than kidney or liver transplantation, since in the case of graft failure after heart transplantation, the death of the patient is more likely, which can occur quickly if it is not re-transplanted or connected to a mechanical circulatory support device. Thus, as a result, the immunological rejection of this type of transplant must be stricter in its effective diagnosis, follow-up, and immunosuppressant treatment.

Particularly, the types of rejection that can occur in the heart graft are acute cell rejection (ACR), antibody-mediated rejection (AMR), or cardiac graft vasculopathy (CAV). Being able to identify with appropriate biomarkers any of these processes harmful to the heart graft, such as genomic markers of the dd-cfDNA type, will help to control the patient otcome. We will now develop what is known about these tools and procedures in heart transplantation.

Regarding the diagnosis of rejection, endomyocardial biopsy (EMB) appears to be useful, but new approaches and less invasive methods are needed to assess the immune status of the graft.

Among these noninvasive methods, developments such as microRNA (miRNA) expression analysis, Luminex determination of donor-specific antibodies (DSA), dd-cfDNA levels, the releasing of certain soluble immune molecules, or gene expression profiling exist and are already operational and are in validation studies to assess the state of graft tolerance or rejection (Loupy et al. 2017; Palmieri et al. 2021).

In this sense, it is logical to think that with the destruction of allograft cardiomyocytes, associated with a rejection event, small DNA fragments are released that end up in the recipient's cardiac circulation.

Breaking down the various publications in heart transplantation, a first study (De Vlaminck et al. 2014) that analyzed 565 plasma samples to quantify the performance of dd-cfDNA in detecting graft rejection yielded results that revealed that the sensitivity and specificity of the dd-cfDNA biomarker were comparable to those of endomyocardial biopsy. Another subsequent report shows an increase in circulating dd-cfDNA levels immediately post-transplant, but this decreases to 0.13% at 28 days (Agbor-Enoh et al. 2021). These disparate data report that dd-cfDNA levels show stabilization at baseline levels 1 week post-transplant (Knuttgen et al. 2022; Zangwill et al. 2020).

However, it is sometimes difficult to reliably assign clinical data in heart transplantation with specific biomarkers, but it is possible to obtain NPV. In this sense, dd-cfDNA levels appear to correlate with the degree of acute rejection with a high NPV, increasing when a rejection event occurs and decreasing when anti-rejection treatment has been successful (Agbor-Enoh et al. 2021; Qian et al. 2022).

But there are other factors that can be confusing in considering dd-cfDNA levels as totally determining to define rejection and should be considered with caution. These factors, which can lead to finding certain elevated dd-cfDNA levels without the presence of organ rejection, include myocardial ischemia, myocardial trauma, pregnancy, and multiorgan transplantation (Holzhauser et al. 2023).

In addition to the data referenced above, other groups report that in approximately more than 20% of cases, serum dd-cfDNA levels may be elevated, yet recipients do not present symptoms or clinical signs of rejection (Afzal et al. 2022).

Finally, the method of quantifying plasma dd-cfDNA levels is difficult because it cannot determine whether the cardiac rejection event is cellular or antibody-mediated. Therefore, while the results may be promising in heart transplantation, in the current situation, further studies and additional validated results will be necessary to confirm and reliably implement this diagnostic procedure as a definitive method for monitoring tolerance or rejection in recipients undergoing heart transplantation (Crespo-Leiro et al. 2015).

In summary, other new and less invasive diagnostic methods, such as DSA monitoring, dd-cfDBA, miRNA, or gene expression profiling, have proven useful, reducing the number of endomyocardial biopsies required.

7.6 Role of dd-cfDNA Determination in Lung Transplantation

Finally, lung transplantation is different from other transplants, such as kidney transplants, because it is a vital therapeutic option, and its long-term outcomes are determined by chronic lung allograft dysfunction (CLAD), which is the leading cause of post-transplant mortality. This is an irreversible deterioration in lung function, defined as a decrease in forced expiratory volume in 1 s (FEV1) of ≥20% compared to baseline (Beeckmans et al. 2024). Classically, there are two main phenotypes of CLAD: bronchiolitis obliterans syndrome (BOS) and restrictive allograft syndrome (RAS).

The first type shows a decrease in FEV1 ≥20% from baseline with an obstructive pulmonary function test (PFT), in the absence of persistent radiological opacities or decreased total lung capacity (TLC). The second type of CLAD, RAS, is characterized by a decrease in FEV1 ≥20% from baseline, accompanied by a restrictive pattern on PFTs (a decrease in TLC ≥10% from baseline) and persistent opacities on chest X-ray or CT due to interstitial or pleural fibrosis (Beeckmans et al. 2024).

The early stages of allograft dysfunction may manifest as changes in small airway function or lung structure, as evidenced by other diagnostic modalities, despite stable FEV1 levels (known as preclinical CLAD). For further information, there is a 2001 ISHLT consensus document on the diagnosis of BOS (Estenne et al. 2002).

Standard allograft monitoring typically includes regular clinical evaluations, pulmonary function tests, imaging studies, such as chest X-rays or CT scans, and periodic bronchoscopy with transbronchial biopsies, and bronchoalveolar lavage, which is essential for diagnosing rejection, infection, and other post-transplant complications (Beeckmans et al. 2024).

However, like other types of transplantation, invasive biopsy carries risks such as bleeding and pneumothorax, and is prone to sampling errors. Lung function tests may also be confounded by factors such as obesity, diaphragmatic dysfunction, and patient exertion (Beeckmans et al. 2024), which can make it difficult to accurately estimate allograft function.

Therefore, sensitive and noninvasive biomarkers are needed to assess graft lung function, such as dd-cfDNA. Values for assessing allograft damage can range, as in other types of transplants previously discussed, between 0.5% and 1% (Botella et al. 2024).

Regarding how dd-cfDNA levels can affect the different degrees of CLAD and its progression, very little is known, although some studies report increases in dd-cfDNA in CLAD and its relationship to FEV1 deterioration (Ju et al. 2023), a fact corroborated by other studies (Rosenheck et al. 2022; Glanville et al. 2019; Nikolova et al. 2024; Charya et al. 2022). However, these trials failed to distinguish indeed between bronchiolitis obliterans syndrome and allograft restriction syndrome.

It is important to consider that the percentage of dd-cfDNA can be altered by several parameters, depending on the timing of sampling, the calculation methods, or the development of an infectious process (Novo et al. 2024; Nikolova et al. 2024).

A recently published study exploring the role of plasma dd-cfDNA levels as a biomarker of CLAD over time (at 3 consecutive time points, without infection or acute rejection, during stable, preclinical CLAD, and established CLAD) (Beeckmans et al. 2024) shows elevated dd-cfDNA levels in 47% of stable CLAD, 66% of preclinical CLAD, and 71% of CLAD samples, indicating ongoing lung allograft injury, albeit with high intra- and inter-patient variability, suggesting that further studies are needed to take this biomarker as fully established in the diagnosis of lung damage.

7.7 Future Directions in dd-cfDNA Analysis in Transplantation

The use of dd-cfDNA methods is increasingly important as a diagnostic tool and as a biomarker of graft organ function, primarily due to their sensitivity and specificity, which allow for early and reliable detection of potential allograft damage. They are easy to use and interpret, do not require excessive sample manipulation, and can be performed with a single sample (Mac et al. 2019).

In cases of organ damage or malfunction, dd-cfDNA levels increase rapidly, allowing for faster and more effective treatment of subclinical rejection than with biochemical measures. This reduces the number of unnecessary biopsies and allows physicians to modify immunosuppressive therapy promptly, preventing the progression of damage, thus reducing costs and improving patient outcomes. However, there are still factors of doubt and how different parameters of the donor and the recipient can influence the quantification of dd-cfDNA and alter results, as well as the lack of consensus on which clinical data should be taken into account and agree on whether to standardize the absolute value or the percentage of dd-cfDNA with respect to improving patient management (Martuszewski et al. 2021; Jimenez-Coll et al. 2023a).

Other limitations, especially in the specific case of the United States, include the need to use commercial kits that require sending the recipient's blood samples to external laboratories, generating results that often differ little from clinical findings. CFDNA is somewhat unstable, and during transport, if specific tubes are not used, the result will be altered, something that generally does not occur in European transplant centers (Botella et al. 2024).

One of the most critical aspects of this technique, especially if performed using NGS, with respect to healthcare systems is its high cost, which greatly hinders its implementation in post-transplant routines.

Therefore, its future implementation is subject to addressing a series of problems and limitations related to economics, logistics, variability, improved validation, precise demonstration of validity, and the achievement of consensus. Furthermore, the potential for longitudinal follow-up for personalized immunosuppression must also be addressed (Bu et al. 2022).

In conclusion, further research is needed to ensure that dd-cfDNA quantification can become a safe, validated, and reliable tool for predicting the early onset of graft damage, allowing for the modification of immunosuppression and the administration of appropriate treatment to optimize outcomes in transplant recipients. In conclusion, dd-cfDNA quantification is a very promising biomarker for organ transplant rejection, but further studies are needed for its full implementation in post-transplant follow-up of graft recipients.

Acknowledgments We are grateful for the support from "Fundación Mutua Madrileña" Fund. Nª Project: AP183152023

Conflict of Interests None of the authors has a conflict of interests

References

Afzal A, Alam A, van Zyl JS, Zafar H, Felius J, Hall SA, Carey SA (2022) Observed elevated donor-derived cell free DNA in orthotopic heart transplant recipients without clinical evidence of rejection. Clin Transpl 36:e14549

Agbor-Enoh S, Shah P, Tunc I, Hsu S, Russell S, Feller E, Shah K, Rodrigo ME, Najjar SS, Kong H et al (2021) Cell-free DNA to detect heart allograft acute rejection. Circulation 143:1184–1197

Alfaro R, Legaz I, González-Martínez G, Jimenez-Coll V, Martínez-Banaclocha H, Galián JA, Botella C, de la Peña-Moral J, Moya-Quiles MR, Campillo JA, Minguela A, Llorente S, Muro M (2021a) Monitoring of B cell in kidney transplantation: development of a novel clusters analysis and role of transitional B cells in transplant outcome. Diagnostics 11:641. https://doi.org/10.3390/diagnostics11040641

Alfaro R, Lorente S, Jimenez-Coll V, Martínez-Banaclocha H, Galián JA, Botella C, Moya-Quiles MR, Muro-Pérez M, de la Peña-Moral J, Minguela A, Legaz I, Muro M (2022) Evaluating the link between BAFF system gene expression and acute rejection development in kidney transplantation. J Clin Med 11:3956. https://doi.org/10.3390/jcm11143956

Alfaro R, Rodríguez-Aguilar L, Llorente S, Jimenez-Coll V, Martínez-Banaclocha H, Galián JA, Botella C, Moya-Quiles MR, Muro-Perez M, Minguela A, Legaz I, Muro M (2023) Early cytomegalovirus reactivation in renal recipients is associated with high levels of B cell maturation antigen transcript expression prior to transplantation. Int J Mol Sci 24:10491. https://doi.org/10.3390/ijms241310491

Aubert O, Ursule-Dufait C, Brousse R, Gueguen J, Racapé M, Raynaud M, Van Loon E, Pagliazzi A, Huang E, Jordan SC et al (2024) Cell-free DNA for the detection of kidney allograft rejection. Nat Med 30:2320–2327

Beeckmans H, Pagliazzi A, Kerckhof P, Hofkens R, Debackere F, Zajacova A, Bos S, Vanaudenaerde BM, de Loor H, Naesens M, Vos R (2024) Donor-derived cell-free DNA in chronic lung allograft dysfunction phenotypes: a pilot study. Front Transplant 3:1513101. https://doi.org/10.3389/frtra.2024.1513101

Boix F, Trujillo C, Muro M (2018) Cell-mediated immunity (CMI) as the instrument to assess the response against the allograft: present and future. Curr Protein Pept Sci 19:1092–1106. https://doi.org/10.2174/1389203719666180730164542

Boix F, Legaz I, Minhas A, Alfaro R, Jiménez-Coll V, Mrowiec A, Martínez-Banaclocha H, Galián JA, Botella C, Moya-Quiles MR, Sanchez-Bueno F, Robles R, de la Peña-Moral J, Ramirez P, Pons JA, Minguela A, Muro M (2021) Identification of peripheral $CD154^+$ T cells and HLA-DRB1 as biomarkers of acute cellular rejection in adult liver transplant recipients. Clin Exp Immunol 203:315–328. https://doi.org/10.1111/cei.13533

Boix F, Jimenez-Coll V, Legaz I, Alfaro R, Moya-Quiles MR, Peña-Moral J, Minguela A, Llorente S, Muro M (2023) Higher expression of activated $CD8^+$ T lymphocytes ($CD8^+CD25^+$, $CD8^+CD69^+$ and $CD8^+CD95^+$) mediate early post-transplant acute tubular injury in kidney recipients. Front Biosci 28:119. https://doi.org/10.31083/j.fbl2806119

Botella C, Galián JA, Jiménez-Coll V, Fernández-González M, Morales F, Martínez-Gómez G, González-López R, Alegría MJ, Moya MR, Martinez-Banaclocha H, Minguela A, Legaz I, Llorente S, Muro M (2024) Early monitoring of donor-derived cell-free DNA in kidney allograft recipients followed-up for two years: experience of one center. Life 14:1491. https://doi.org/10.3390/life14111491

Bu L, Gupta G, Pai A, Anand S, Stites E, Moinuddin I, Bowers V, Jain P, Axelrod DA, Weir MR et al (2022) Clinical outcomes from the Assessing Donor-derived cell-free DNA Monitoring Insights of kidney Allografts with Longitudinal surveillance (ADMIRAL) study. Kidney Int 101:793–803

Charya A, Ponor IL, Jang M, Kong H, Shah P, Mathew J et al (2022) Restrictive allograft syndrome patients have higher cell-free DNA assessed allograft injury prior to diagnosis. J Heart Lung Transplant 41 Supplement:S20–S21. https://doi.org/10.1016/j.healun.2022.01.043

Crespo-Leiro MG, Stypmann J, Schulz U, Zuckermann A, Mohacsi P, Bara C, Ross H, Parameshwar J, Zakliczynski M, Fiocchi R et al (2015) Performance of gene-expression profiling test score variability to predict future clinical events in heart transplant recipients. BMC Cardiovasc Disord 15:120

De Vlaminck I, Valantine HA, Snyder TM, Strehl C, Cohen G, Luikart H, Neff NF, Okamoto J, Bernstein D, Weisshaar D et al (2014) Circulating cell-free DNA enables noninvasive diagnosis of heart transplant rejection. Sci Transl Med 6:241ra77

El Band JEK, Llorente S, Martinez-Garcia P, Alfaro R, Jimenez-Coll V, Boix F, Galián JA, Martinez-Banaclocha H, Botella C, Moya-Quiles MR, Minguela A, Legaz I, Muro M (2021) Evaluation of antibodies directed against two GPCRs, anti-AT1R and anti-ETAR, on kidney transplant outcome. Curr Protein Pept Sci 22:745–757. https://doi.org/10.2174/1389203722666210706163149

Estenne M, Maurer JR, Boehler A, Egan JJ, Frost A, Hertz M et al (2002) Bronchiolitis obliterans syndrome 2001: an update of the diagnostic criteria. J Heart Lung Transplant 21:297–310. https://doi.org/10.1016/S1053-2498(02)00398-4

Farcas AO, Stoica MC, Maier IM, Maier AC, Sin AI (2024) Heart transplant rejection: from the endomyocardial biopsy to gene expression profiling. Biomedicine 12:1926. https://doi.org/10.3390/biomedicines12081926

Fernandez-Galan E, Badenas C, Fondevila C, Jimenez W, Navasa M, Puig-Butille JA et al (2022) Monitoring of donor-derived cell-free DNA by short tandem repeats: concentration of total cell-free DNA and fragment size for acute rejection risk assessment in liver transplantation. Liver Transpl 28:257–268. https://doi.org/10.1002/lt.26272

Glanville AR, Verleden GM, Todd JL, Benden C, Calabrese F, Gottlieb J et al (2019) Chronic lung allograft dysfunction: definition and update of restrictive allograft syndrome-a consensus report from the pulmonary council of the ISHLT. J Heart Lung Transplant 38:483–492. https://doi.org/10.1016/j.healun.2019.03.008

Holzhauser L, DeFilippis EM, Nikolova A, Byku M, Contreras JP, De Marco T, Hall S, Khush KK, Vest AR (2023) The end of endomyocardial biopsy?: a practical guide for noninvasive heart transplant rejection surveillance. JACC Heart Fail 11:263–276. https://www.sciencedirect.com/science/article/pii/S2213177922006436.

Huang E, Jordan SC (2022) Donor-derived cell-free DNA in kidney transplantation: evolving concepts and potential limitations. Kidney Int 101:676–677

Jeon SH, Lee HJ, Bae K, Yoon KA, Lee ES, Cho Y (2016) Efficient capture and isolation of tumor-related circulating cell-free DNA from cancer patients using electroactive conducting polymer nanowire platforms. Theranostics 6:828

Jimenez-Coll V, El Kaaoui El Band J, Llorente S, González-López R, Fernández-González M, Martínez-Banaclocha H, Galián JA, Botella C, Moya-Quiles MR, Minguela A, Legaz I, Muro M (2023a) All that glitters in cfDNA analysis is not gold or its utility is completely established due to graft damage: a critical review in the field of transplantation. Diagnostics 13:1982. https://doi.org/10.3390/diagnostics13121982

Ju C, Xu X, Zhang J, Chen A, Lian Q, Liu F et al (2023) Application of plasma donor-derived cell free DNA for lung allograft rejection diagnosis in lung transplant recipients. BMC Pulm Med 23:37. https://doi.org/10.1186/s12890-022-02229-y

Kanamori H, Yamada Y, Ito Y, Shirosaki K, Yamagishi S, Maeda Y et al (2024) Noninvasive graft monitoring using donor-derived cell-free DNA in Japanese liver transplantation. Hepatol Res 54:300–314. https://doi.org/10.1111/hepr.13978

Knight SR, Thorne A, Lo Faro ML (2019) Donor-specific cell-free DNA as a biomarker in solid organ transplantation. A systematic review. Transplantation 103:273–283. https://doi.org/10.1097/tp.0000000000002482

Knuttgen F, Beck J, Dittrich M, Oellerich M, Zittermann A, Schulz U, Fuchs U, Knabbe C, Schutz E, Gummert J et al (2022) Graft-derived cell-free DNA as a noninvasive biomarker of cardiac allograft rejection: a cohort study on clinical validity and confounding factors. Transplantation 106:615–622

Legaz I, Boix F, López M, Alfaro R, Galián JA, Llorente S, Campillo JA, Botella C, Ramírez P, Sánchez-Bueno F, Pons JA, Moya-Quiles MR, Minguela A, Muro M (2020) Influence of preformed antibodies in liver transplantation. J Clin Med 9:708. https://doi.org/10.3390/jcm9030708

Legaz I, Bernardo MV, Alfaro R, Martínez-Banaclocha H, Galián JA, Jimenez-Coll V, Boix F, Mrowiec A, Salmeron D, Botella C, Parrado A, Moya-Quiles MR, Minguela A, Llorente S, de la Peña-Moral J, Muro M (2021) PCR array technology in biopsy samples identifies up-regulated mTOR pathway genes as potential rejection biomarkers after kidney transplantation. Front Med (Lausanne) 8:547849. https://doi.org/10.3389/fmed.2021.547849

Legaz I, Morales R, Bolarín JM, Collados-Ros A, Pons JA, Muro M (2023) Is the development of ascites in alcoholic liver patients influenced by specific KIR/HLA gene profiles? Biomedicine 11:2405. https://doi.org/10.3390/biomedicines11092405

Legaz I, Jimenez-Coll V, González-López R, Fernández-González M, Alegría-Marcos MJ, Galián JA, Botella C, Moya-Quiles R, Muro-Pérez M, Minguela A, Llorente S, Muro M (2024a) MicroRNAs as potential graft rejection or tolerance biomarkers and their dilemma in clinical routines behaving like devilish, angelic, or frightening elements. Biomedicine 12:116. https://doi.org/10.3390/biomedicines12010116

Levitsky J, Kandpal M, Guo K, Kleiboeker S, Sinha R, Abecassis M (2022) Donor-derived cell-free DNA levels predict graft injury in liver transplant recipients. Am J Transplant 22:532–540. https://doi.org/10.1111/ajt.16835

Loupy A, Van Huyen JPD, Hidalgo L, Reeve J, Racapé M, Aubert O, Venner JM, Falmuski K, Bories MC, Beuscart T et al (2017) Gene expression profiling for the identification and classification of antibody-mediated heart rejection. Circulation 135:917–935

Mac QD, Mathews DV, Kahla JA, Stoffers CM, Delmas OM, Holt BA, Adams AB, Kwong GA (2019) Non-invasive early detection of acute transplant rejection via nanosensors of granzyme B activity. Nat Biomed Eng 3:281–291

Martuszewski A, Paluszkiewicz P, Król M, Banasik M, Kepinska M (2021) Donor-derived cell-free DNA in kidney transplantation as a potential rejection biomarker: a systematic literature review. J Clin Med 10:193

Nikolova A, Agbor-Enoh S, Bos S, Crespo-Leiro M, Ensminger S, Jimenez-Blanco M et al (2024) European Society for Organ Transplantation (ESOT) consensus statement on the use of non-invasive biomarkers for cardiothoracic transplant rejection surveillance. Transpl Int 37:12445. https://doi.org/10.3389/ti.2024.12445

Novo M, Nordén R, Westin J, Dellgren G, Böhmer J, Ricksten A et al (2024) Donor fractions of cell-free DNA are elevated during CLAD but not during infectious complications after lung transplantation. Transpl Int 37:12772. https://doi.org/10.3389/ti.2024.12772

Oellerich M, Budde K, Osmanodja B, Bornemann-Kolatzki K, Beck J, Schütz E, Walson PD (2022) Donor-derived cell-free DNA as a diagnostic tool in transplantation. Front Genet 13:1031894

Palmieri V, Mansueto G, Coscioni E, Maiello C, Benincasa G, Napoli C (2021) Novel biomarkers useful in surveillance of graft rejection after heart transplantation. Transpl Immunol 67:101406

Qian X, Shah P, Agbor-Enoh S (2022) Noninvasive biomarkers in heart transplant: 2020–2021 year in review. Curr Opin Organ Transplant 27:7–14

Rosenheck JP, Ross DJ, Botros M, Wong A, Sternberg J, Chen YA et al (2022) Clinical validation of a plasma donor-derived cell-free DNA assay to detect allograft rejection and injury in lung transplant. Transplant Direct 8:e1317. https://doi.org/10.1097/TXD.0000000000001317

Rush DN, Gibson IW (2019) Subclinical inflammation in renal transplantation. Transplantation 103:E139–E145

Schwarz A, Gwinner W, Hiss M, Radermacher J, Mengel M, Haller H (2005) Safety and adequacy of renal transplant protocol biopsies. Am J Transplant 5:1992–1996

Snyder TM, Khush KK, Valantine HA, Quake SR (2011) Universal noninvasive detection of solid organ transplant rejection. Proc Natl Acad Sci USA 108:6229–6234

Ulrich BC, Guibert N (2017) Towards a comprehensive framework for cell-free DNA analysis: lessons from TRACERx. Ann Transl Med 5:446–451

Zangwill SD, Kindel SJ, Ragalie WS, North PE, Pollow A, Hidestrand M, Tomita-Mitchell A, Stamm KD, Mitchell ME (2020) Early changes in cell-free DNA levels in newly transplanted heart transplant patients. Pediatr Transplant 24:e13622

Part III
Shared and Organ-Specific Features of Chronic Rejection and Therapeutic Approaches

Chapter 8
Vascular Remodeling in Chronic Rejection: Insights into Transplant Vasculopathy Across Solid Organ Allografts

Samantha L. Nelson, Sariah Hyacinth, Basmala Aldamak, and Benjamin J. Kopecky

Abstract Solid organ transplantation is an established treatment method for individuals with end-stage organ dysfunction. However, despite the common perception that transplantation offers a lasting cure, transplant recipients have limited long-term treatment options for chronic rejection. Transplant vasculopathy is recognized as one of the defining pathological indicators of chronic rejection, characterized by diffuse intimal wall thickening compromising vessel function, and, in severe cases, leads to allograft failure. While the increasing number of solid organ transplants reflects the modern advancements in treating end-stage organ dysfunction, it also highlights the continuing challenge of understanding and developing effective therapies for managing chronic rejection.

8.1 Introduction

Solid organ transplantation has allowed individuals with end-stage organ dysfunction to receive life-saving treatment. Advancements in surgical technique have accompanied the growing demand for organ transplantation since its origins in clinical practice in the mid-twentieth century (Black et al. 2018). Improvements, such as effective immunosuppression, clinical trials, research, drug development, and greater donor awareness, have all contributed to improving patient survival (Neuberger et al. 2017). However, a solid organ transplant does not yet provide a

S. L. Nelson · S. Hyacinth · B. J. Kopecky (✉)
Department of Medicine, Division of Cardiology, University of Colorado Anschutz Medical Campus, Aurora, CO, USA
e-mail: Benjamin.Kopecky@cuanschutz.edu

B. Aldamak
Department of Chemical and Biological Engineering, Iowa State University, Ames, IA, USA

M. Kloc, R. M. Ghobrial (eds.), *Cellular and Molecular Mechanisms of Chronic Rejection in Organ Transplantation*, Results and Problems in Cell Differentiation 77, https://doi.org/10.1007/978-3-032-07686-1_8

definitive treatment solution for end-stage organ dysfunction; the long-term outcomes of the solid organ allograft are limited due to chronic rejection (Cozzi et al. 2017). Chronic rejection is the leading cause of graft rejection, but is yet to be fully understood due to its complex immunological and nonimmunological components. Immunosuppressive therapies are designed as lifelong, multidrug regimens to suppress the recipient's immune response against the transplanted tissue and prevent or treat rejection, but have limitations. These therapies are broadly immunosuppressive rather than targeted, which greatly reduce acute immune-mediated rejection, but are largely ineffective against chronic rejection (Ebrahimi et al. 2014). A distinctive feature of chronic rejection is vascular remodeling of vessels within the allograft, more commonly known as transplant vasculopathy (TV). As chronic rejection remains the leading cause of solid organ graft failure, developing effective therapies requires a thorough understanding of its underlying mechanisms, including a clear definition of its common pathophysiology.

8.2 Transplant Vasculopathy Is a Persistent Challenge in Solid Organ Transplantation

Chronic rejection remains the leading cause of long-term graft failure across all solid organ transplants. TV is the most common form of chronic rejection (Franco-Acevedo et al. 2023). Rather than being organ-specific, TV presents as a pervasive and graft-limiting transplant complication across multiple organ systems, including the heart, lungs, kidney, and liver. Histological changes resembling TV were first documented by pioneering surgeon Alexis Carrel in 1910 at the University of Chicago. In an effort to improve his vascular surgery techniques, Carrel documented rejection-related vascular disease in those allografts (Young 1999). Now, over a century later, transplant recipients are routinely counseled about the risk of TV, and it is well recognized in the transplant community as a major limitation to allograft survival.

Blood vessels are composed of an outer, middle, and inner layer. The tunica adventitia is the outermost layer and is comprised of connective tissue for vessel structure and contains stem/progenitor cells that respond to arterial injury (Majesky et al. 2011). The tunica media is the middle layer, thick with smooth muscle cells and elastic fibers that allow the vessels to expand and contract. Lastly, the tunica intima is the innermost layer that is composed of both endothelial cells which serves as a barrier between the blood and the other layers of the vessel walls. TV is widely considered the principal pathological manifestation of chronic rejection because of its identifiable thickening of the intima wall within blood vessels in the allograft (Geraghty et al. 1996). This thickening results from neointimal hyperplasia, a process involving the migration and proliferation of various immune and vascular cells into the intimal layer, which leads to luminal narrowing, progressive vascular dysfunction, graft ischemia, and, ultimately, failure (Franco-Acevedo et al. 2023;

Geraghty et al. 1996; Kloc and Ghobrial 2014). Despite immunosuppressive advances in the management of acute rejection, TV remains a long-term complication with limited progress in its prevention or treatment. Although TV affects all solid organ transplants, its pathobiology has been most extensively investigated in the context of heart transplantation. And while this knowledge remains incomplete, insights gained from vasculopathy in highly researched transplanted organs may provide an important foundation for advancing the study of and creating clinical therapies for vasculopathy in other, less well-characterized transplanted organs.

8.3 Transplant Vasculopathy Is Ubiquitous Across All Solid Organ Transplantation but Is the Most Well Characterized in Heart Transplantation

TV has been most extensively studied in the context of heart transplantation. Known as cardiac allograft vasculopathy (CAV), this form of TV is the major cause of death among individuals following heart transplantation (Stomberski and Colvin 2025). Furthermore, the incidence of CAV in heart transplant patients is significant, with approximately 50% of patients developing CAV in the first 10 years (Shetty and Chowdhury 2023). Although CAV is sometimes described as an accelerated form of coronary artery disease (CAD), the pathological features differ significantly. CAD primarily affects the intimal layer of proximal coronary arteries and typically lacks significant inflammation. In contrast, CAV commonly involves intramyocardial vessels, which are usually spared in CAD (Ramzy et al. 2005). It is thought the inflammation and remodeling of these vessels may be triggered by both immunological and nonimmunological factors of chronic rejection such as donor characteristics like age and cause of death, HLA antibodies, organ procurement methods, and ischemia-reperfusion injury (Shetty and Chowdhury 2023). Epicardial and intramyocardial coronary arteries within the transplanted heart develop progressive, diffuse thickening due to outward and inward vascular remodeling of the blood vessel (Pober et al. 2021). However, neither the outward nor inward vascular remodeling is fully understood. It is thought that these changes may reflect endothelial-mediated responses to cytokines acting directly on smooth muscle cells found within the media layer, causing breaks in the elastic lamina of the vessel wall from migration and proliferation to the intima, and fibrosis in the connective tissue of the adventitia (Aranda Jr. and Hill 2000; Pober et al. 2021; Ramzy et al. 2005).

Diagnosing CAV is challenging, as many patients remain asymptomatic until the disease has progressed. The clinical significance of CAV is determined when the compensatory outward remodeling of the vessel fails to offset the pathological inward remodeling, leading to narrowing of the vessel and adverse effects such as disrupted blood flow, severe diastolic dysfunction, and progressive heart failure (Pober et al. 2021). When symptoms do appear, they often indicate an advanced stage of CAV. Therefore, routine screening and surveillance are essential for heart transplant recipients (Aranda Jr. and Hill 2000). To enable earlier detection, many transplant centers use diagnostic tools such as echocardiograms, coronary angiograms, and intravascular ultrasounds (IVUS) (Skorić et al. 2014).

Invasive procedures such as a coronary angiography and IVUS remain the "clinical gold standard" for diagnosing CAV. Coronary angiograms evaluate the luminal space of the epicardial coronary arteries to accurately identify areas of narrowing or obstruction within the vessel. Unfortunately, angiograms have a relatively low sensitivity that make early detection difficult (Dedieu et al. 2014; Stomberski and Colvin 2025). IVUS has slowly gained popularity among transplant centers over the last few decades for evaluating arterial wall structure with ultrasound, offering enhanced sensitivity for early detection of the characteristic intimal thickening seen in CAV (Kamel et al. 2024). Unfortunately, some patients are not candidates for invasive evaluations.

To identify lower risk alternatives, positron emission tomography/computed tomography (PET/CT) with myocardial blood flow quantification has shown potential in effectively ruling out moderate-to-severe CAV in transplant recipients (Abadie et al. 2023). The PET/CT technique measures blood flow through epicardial arteries and microvasculature vessels in the transplanted heart to give both a functional view and quantitative measurement of potential vessel dysfunction (Bateman et al. 2021; Valenta et al. 2013). Regardless of whether using invasive or noninvasive imaging methods, it's important to establish a standardized framework essential for evaluating disease severity, such as the grading system implemented by the International Society for Heart and Lung Transplantation (ISHLT), which categorizes CAV based on angiographic findings and allograft function.

ISHLT currently defines the grading system for CAV based on angiographic evidence of lumen narrowing, or stenosis, and allograft function. The classification ranges from CAV0 to CAV3. Since CAV is a progressive disease, the prognosis of patients once developing CAV is generally poor, with patients who experience a higher severity of disease such as CAV2 or CAV3, are more at risk for cardiac complications, such as arrhythmias and sudden cardiac death (Skorić et al. 2014). However, there are arguments that suggest that the ISHLT grading system for CAV should be more comprehensive. Recent research by Prasad et al. (2025) suggests that further stratification of the CAV1 group by using both angiographic data and PET/CT evaluations demonstrated enhanced prognostic precision in stratifying outcomes for patients with CAV. Their study proposes subgroups CAV1a and CAV1b, with <30% and ≥30% stenosis, respectively. Despite both groups traditionally being classified under CAV1, the CAV1b patient group exhibited worse outcomes than CAV1a. This finding affirms the value of understanding stenosis thresholds in CAV grading to better predict patient outcomes. Additionally, integrating both invasive and noninvasive imaging modalities in patient care can provide functional insights, further refining risk stratification in heart transplant recipients.

Despite advancements in diagnostic imaging and risk stratification for CAV, effective, targeted therapies remain limited. Current treatment strategies primarily focus on nonspecific preventative measures, including the use of statins, mammalian target of rapamycin (mTOR) inhibitors, and aspirin. Studies from Kobashigawa et al. (2005) and Wenke et al. (2003) suggest that the use of statin therapies like pravastatin or simvastatin, respectively, may reduce CAV incidence by reducing cholesterol levels. More recent work by Stomberski and Colvin (2025) suggests that mTOR inhibitors may reduce the proliferation of fibroblasts and smooth muscle cells within the medial layer, potentially slowing CAV progression. Kim et al. (2017), in a 7-year-long longitudinal study of 120 patients, found the early use of aspirin as an anti-platelet and an anti-inflammatory agent may reduce rates of moderate-to-severe CAV. While these interventions show promise, they do not provide a definitive solution for preventing chronic rejection in heart transplantation, as they fail to fully address the comprehensive pathobiological factors underlying CAV. The only known definitive treatment for severe CAV is re-transplantation (Barghash and Pinney 2020). Looking beyond heart transplantation, to include how TV manifests in other solid organ transplants, such as in the lung, may offer insights into both shared and organ-specific mechanisms of chronic rejection.

8.4 Chronic Lung Allograft Vasculopathy Shares Mechanistic Pathways with Other Solid Organs but Displays Unique Remodeling Characteristics

Although it has not been studied as extensively as CAV, research into chronic rejection in lung transplantation is actively evolving and remains more advanced than other solid organ transplants, such as those of the kidneys and liver. Transplanted lungs are particularly susceptible to chronic lung allograft dysfunction (CLAD), a form of chronic rejection characterized by a progressive and often irreversible decline in pulmonary function, similarly leading to allograft failure. CLAD significantly impacts patient outcomes, as lung transplant recipients experience markedly lower long-term survival rates compared to recipients of other solid organ transplants (Verleden et al. 2019). CLAD is diagnosed in three progressive stages: possible, probable, and definite CLAD, each marked by a persistent decline in forced expiratory volume compared to the patient's post-transplant baseline. In the definite stage, the CLAD sub-phenotype [bronchiolitis obliterans syndrome (BOS), restrictive allograft syndrome (RAS), mixed, or undetermined] is identified based on pulmonary function, imaging (X-ray or CT), or tissue pathology findings (Gauthier et al. 2018; Yoshiyasu and Sato 2020).

Historically, BOS has been recognized as the predominant obstructive phenotype of CLAD, affecting approximately 50% of lung transplant recipients within 5 years post-transplant (Gauthier et al. 2018). It wasn't until recently that studies expanded to include RAS, a restrictive phenotype of chronic rejection, which was initially identified by Sato et al. in 2011 (Bedair and Hachem 2021; Gauthier et al. 2018). BOS primarily involves the remodeling of small airways, leading to small airway damage known as obliterative bronchiolitis (OB) (Saggar et al. 2008; Sato et al. 2011). OB is an airway obstruction resulting from both immune-mediated and structural lung injury causing peri-airway inflammation and luminal fibrosis (Saggar et al. 2008). Notably, BOS tends to spare other critical lung structures, such as the alveoli and pleura, which are essential for essential gas exchange and maintaining lung structure (Sato 2020). RAS, when compared to BOS, is a more recently recognized and broader classification, involving more widespread lung tissue damage that extends beyond the airways into the peripheral lung parenchyma and alveoli (Sato 2020).

Belperio et al. (2002) and Saggar et al. (2008) demonstrated that OB/BOS is associated with significant perivascular infiltration of mononuclear immune cells within the airway circulation, whose recruitment potentially originated from prior rejection. This immune infiltration is partly driven by the expression of chemokines, which act as attractants and facilitate the continuous recruitment of mononuclear cells to the graft. This results in chronic inflammation and initiates a cascade that promotes progressive vascular remodeling. Due to this chronic cascade, the damage originating in the pulmonary arteries will eventually extend to the airways and lung tissue. Saggar et al. (2008) and, more recently, Vanstapel et al. (2022) detailed histopathological evaluations of lung allografts with OB/BOS, Saggar et al. (2008) and, more recently, Vanstapel et al. (2022) detailed histopathological evaluations of lung allografts with OB/BOS, identifying pulmonary arteriopathy as evidenced by intimal hyperplasia and/or media hypertrophy and a presence of bronchial arterial vasculopathy. The arteriopathy was noted due to abnormalities of the intimal layer consistently revealing both eccentric and concentric fibroproliferative patterns, such as intimal thickening due to mononuclear cell infiltration, deposition of extracellular matrix (ECM) proteins, and fibrous tissue deposition within the medial layer. Pulmonary arteries often exhibited severe bronchial vasculopathy and luminal stenosis, while the medial layer simultaneously showed smooth muscle cell atrophy, most likely a consequence of progressive vessel stiffening due to the fibrotic remodeling. Vanstapel et al. (2022) suggested that fibrosis and arterial vasculopathy may appear to share many of the same disease pathways, which are triggered by similar biological mediators involved in angiogenesis, fibrosis, and vascular regulation. It was also noted that vasculopathy may occur before fibrotic changes develop.

In the context of CLAD, microscopic evidence of long-term damage to the small lung blood vessels has been identified as a strong predictor of chronic graft failure. One potential link between fibrosis and vasculopathy is endothelial-to-mesenchymal transition (EMT). During EMT, endothelial cells that line blood vessels in the graft respond to stress or injury by losing their normal characteristics and acquire more mesenchymal traits, causing them to behave more like cells involved in scarring and tissue repair. These changes can allow the transitioned cells an increased capability to migrate and invade tissues, produce ECM, and promote fibrosis.

Previous studies on the treatment of CLAD have largely focused on the BOS subtype. One of the few evidence-based therapies for managing chronic rejection after lung transplantation is the use of azithromycin (Bedair and Hachem 2021). Initial findings were reported by Gerhardt et al. in 2003, recognizing azithromycin's potential benefits in a small number of patients. However, more recent studies suggest that azithromycin may only be effective primarily in the early stages of BOS, emphasizing the need for additional treatment strategies (Jain et al. 2010; Yoshiyasu and Sato 2020).

While CLAD treatment continues to evolve, other organ transplants such as the kidney experience related yet distinct vascular rejection processes. Kidney TV shares common mechanisms of allograft rejection but differs in incidence and presents unique, organ-specific pathological features.

8.5 Kidney Transplant Vasculopathy Shares Common Allograft Rejection Mechanisms but Presents with Unique Patterns and Organ-Specific Pathology

Chronic rejection in kidney transplantation, or chronic allograft nephrology (CAN), is less understood, with only a moderate level of insight into its underlying pathophysiological mechanisms. Kidney TV, a major contributor to CAN, shares immunological and nonimmunological features while also demonstrating distinct rejection mechanisms specific to the kidney (Fletcher et al. 2009). Chronic rejection in kidney allografts leads to chronic renal allograft arteriopathy (CRA), interstitial fibrosis and tubular atrophy, and transplant glomerulopathy (Filippone et al. 2018; Fletcher et al. 2009; Grimm et al. 2001; Shimizu et al. 2023). Like CAV, chronic renal allograft arteriopathy (CRA) involves both inward vascular remodeling through concentric intimal thickening and a potential compensatory outward remodeling response. CRA is commonly associated with interstitial fibrosis and tubular atrophy (IFTA), which reflects chronic ischemic injury in the kidney allograft. Studies by Grimm et al. (2001) and Shimizu et al. (2023) reported infiltration and proliferation of recipient-derived smooth muscle and inflammatory cells within the intimal layer of renal arteries, leading to elastic lamina disruption and neointima formation. This vascular remodeling can result in stenosis of the renal arteries and their branches, reducing blood flow to the renal tubules. The resulting ischemia contributes to graft tissue damage and promotes further IFTA, ultimately leading to a decline in kidney function (Lusco et al. 2016).

Organ-specific mechanisms such as transplant glomerulopathy in kidney allografts arise as downstream consequences of chronic rejection processes, acting on the unique architecture and function of the kidney. Transplant glomerulopathy is characterized by duplication of the glomerular basement membrane resulting from chronic endothelial injury often caused by immune responses or HLA antibodies and is detectable on biopsy via electron microscopy (Filippone et al. 2018). IFTA is typically detected through traditional histological examination of kidney biopsy tissue; however, new technologies such as dynamic full-field optical coherence tomography are being investigated as rapid, on-site methods for histopathological analysis, potentially enhancing diagnostic efficiency in transplant pathology (Zuccarelli et al. 2025).

Nonimmunological factors also play a significant role in chronic kidney transplant rejection. These include donor characteristics such as age, pre-existing conditions or injuries, HLA mismatch, organ size, and ischemia-reperfusion injury (Fletcher et al. 2009). In a study by Imperiali et al. (2018), risk factors for TV/CRA in kidney transplant recipients were investigated, suggesting risk associations with older, male donors whose cause of death was vascular brain accident.

Although diagnostic advances have improved the detection of graft rejection, effective treatment for CAN remains limited. New pharmacologic desensitization strategies aim to both remove donor-specific antibodies and target their cellular sources, such as B cells (with rituximab) and plasma cells (with proteasome inhibitors). While these approaches can reduce the risk of acute rejection, they do not prevent long-term antibody reconstitution. Peri-transplant interventions like normothermic regional perfusion (NRP), instead of the standard in situ cold perfusion (ICP), to reduce cold ischemia and minimizing organ transport may improve patient and graft outcomes, but concerns about safety, efficacy, and cost require further research (Pittappilly et al. 2025). Despite testing various therapies alone and in combination, no regimen has fully addressed the challenges of chronic rejection. Management of chronic rejection in kidney transplant recipients remains largely supportive, with re-transplantation as the only definitive option (Pittappilly et al. 2025).

Like the kidney, liver transplant vasculopathy follows shared chronic rejection pathways but is less well understood, partly due to its rare occurrence.

8.6 Liver Transplant Vasculopathy Aligns with Shared Chronic Rejection Pathways but Is Complicated by Its Rare Manifestation and Limited Understanding of Antibody-Mediated Injury

In liver transplantation, chronic liver allograft rejection is primarily characterized by chronic obliterative arteriopathy (COA), which predominantly affects the medium and large first- and second-order branches of the hepatic artery at the liver hilum and the common bile duct (Angelico et al. 2021; Demetris et al. 1995). Similar to chronic rejection in other solid organ transplants, arteriopathy affecting the hepatic arterial branches and bile ducts can lead to luminal narrowing, resulting in significant ischemic damage to the graft and potentially irreversible injury requiring re-transplantation (Angelico et al. 2021; Oguma et al. 1989). Histologically, COA is characterized by progressive intimal thickening results from the accumulation of macrophages, lymphocytes, smooth muscle cells, and other inflammatory cells. Additional histopathological features in chronic liver allograft rejection include endothelial cell hypertrophy, transmural inflammation, and sclerosis of the vessel adventitia (Demetris et al. 1995). Additionally, endothelial cell hypertrophy of the outer intima of these hepatic arteries likely reflects mechanical stress or vessel wall damage (Gifre-Renom and Jones 2021; Katoh 2023). While small arterioles in the portal tracts of the transplanted liver are usually spared from intimal changes, they may demonstrate medial hypertrophy, reflecting secondary vascular remodeling most likely as a compensatory change to maintain adequate vessel function in the allograft (Demetris et al. 1995).

Unlike in other solid organ transplants, acute cellular rejection of the liver, occurring in 15–25% of liver transplantations, does not significantly impact long-term graft survival (Choudhary et al. 2017). Until recently, antibody-mediated rejection (AMR) was thought to spare liver transplants; however, it is now recognized that AMR can indeed affect liver allografts, albeit rare and with clinical manifestation that remains not fully understood (Angelico et al. 2021).

Management of liver TV follows the same routine to other treatment forms of chronic rejection, involving an escalation of immunosuppressive therapies or re-transplantation in advanced dysfunction. As with other solid organ transplants, the complex and poorly understood pathogenesis of chronic rejection generally renders the resulting damage irreversible; however, mTOR inhibitors, which have shown promise in treating CAV progression, may offer a potential rescue therapy in cases of chronic rejection following liver transplantation (Nishida et al. 2001).

8.7 Conclusion

Transplant vasculopathy remains a major graft-limiting factor across all solid organ transplantations. Recognizing both the shared and organ-specific features of this condition is essential for developing targeted therapies to prevent and manage chronic rejection. Ongoing research into effective treatments is critical—not only to extend graft longevity but also to improve overall patient survival.

Acknowledgments BJK is supported by the NHLBI (K08HL159359), the Boettcher Foundation, the Department of Medicine at the University of Colorado COLLABORATE grant, the Ludeman Family Center, and the SCORE grant.

Conflicts of Interest No authors have a conflict of interest.

References

Abadie BQ, Chan N, Sharalaya Z, Bhat P, Harb S, Jacob M, Starling RC, Tang WHW, Cremer PC, Jaber WA (2023) Negative predictive value and prognostic associations of Rb 82 PET/CT with myocardial blood flow in CAV. JACC Heart Fail 11(5):555–565. https://doi.org/10.1016/j.jchf.2022.11.012

Angelico R, Sensi B, Manzia TM, Tisone G, Grassi G, Signorello A, Milana M, Lenci I, Baiocchi L (2021) Chronic rejection after liver transplantation: opening the Pandora's box. World J Gastroenterol 27(45):7771–7783. https://doi.org/10.3748/wjg.v27.i45.7771

Aranda JM Jr, Hill J (2000) Cardiac transplant vasculopathy. Chest 118(6):1792–1800. https://doi.org/10.1378/chest.118.6.1792

Barghash MH, Pinney SP (2020) Heart retransplantation: candidacy, outcomes, and management. Curr Transplant Rep 7(1):12–17. https://doi.org/10.1007/s40472-019-00257-y

Bateman TM, Heller GV, Beanlands R, Calnon DA, Case J, deKemp R, DePuey EG, Di Carli M, Guler EC, Murthy VL, Rosenblatt J, Sher R, Slomka P, Ruddy TD (2021) Practical guide for interpreting and reporting cardiac PET measurements of myocardial blood flow: an information statement from the American Society of Nuclear Cardiology, and the Society of Nuclear Medicine and Molecular Imaging. J Nucl Med 62(11):1599–1615. https://doi.org/10.2967/jnumed.121.261989

Bedair B, Hachem RR (2021) Management of chronic rejection after lung transplantation. J Thorac Dis 13(11):6645–6653. https://doi.org/10.21037/jtd-2021-19

Belperio JA, Keane MP, Burdick MD, Lynch JP III, Xue YY, Li K, Ross DJ, Strieter RM (2002) Critical role for CXCR3 chemokine biology in the pathogenesis of bronchiolitis obliterans syndrome. J Immunol 169(2):1037–1049. https://doi.org/10.4049/jimmunol.169.2.1037

Black CK, Termanini KM, Aguirre O, Hawksworth JS, Sosin M (2018) Solid organ transplantation in the 21st century. Ann Transl Med 6(20):409. https://doi.org/10.21037/atm.2018.09.68

Choudhary NS, Saigal S, Bansal RK, Saraf N, Gautam D, Soin AS (2017) Acute and chronic rejection after liver transplantation: what a clinician needs to know. J Clin Exp Hepatol 7(4):358–366. https://doi.org/10.1016/j.jceh.2017.10.003

Cozzi E, Colpo A, De Silvestro G (2017) The mechanisms of rejection in solid organ transplantation. Transfus Apher Sci 56(4):498–505. https://doi.org/10.1016/j.transci.2017.07.005

Dedieu N, Greil G, Wong J, Fenton M, Burch M, Hussain T (2014) Diagnosis and management of coronary allograft vasculopathy in children and adolescents. World J Transplant 4(4):276–293. https://doi.org/10.5500/wjt.v4.i4.276

Demetris AJ, Murase N, Delaney CP, Woan M, Fung JJ, Starzl TE (1995) The liver allograft, chronic (ductopenic) rejection, and microchimerism: what can they teach us? Transplant Proc 27(1):67–70

Ebrahimi A, Hosseini SA, Rahim F (2014) Immunosuppressive therapy in allograft transplantation: from novel insights and strategies to tolerance and challenges. Cent Eur J Immunol 39(3):400–409. https://doi.org/10.5114/ceji.2014.45955

Filippone EJ, McCue PA, Farber JL (2018) Transplant glomerulopathy. Mod Pathol 31(2):235–252. https://doi.org/10.1038/modpathol.2017.123

Fletcher JT, Nankivell BJ, Alexander SI (2009) Chronic allograft nephropathy. Pediatr Nephrol 24(8):1465–1471. https://doi.org/10.1007/s00467-008-0869-z

Franco-Acevedo A, Comes J, Mack JJ, Valenzuela NM (2023) New insights into maladaptive vascular responses to donor specific HLA antibodies in organ transplantation. Front Transplant 2:1146040. https://doi.org/10.3389/frtra.2023.1146040

Gauthier JM, Ruiz-Pérez D, Li W, Hachem RR, Puri V, Gelman AE, Kreisel D (2018) Diagnosis, pathophysiology and experimental models of chronic lung allograft rejection. Transplantation 102(9):1459–1466. https://doi.org/10.1097/TP.0000000000002250

Geraghty JG, Stoltenberg RL, Sollinger HW, Hullett DA (1996) Vascular smooth muscle cells and neointimal hyperplasia in chronic transplant rejection. Transplantation 62(4):502–509. https://doi.org/10.1097/00007890-199608270-00001

Gifre-Renom L, Jones EAV (2021) Vessel enlargement in development and pathophysiology. Front Physiol 12:639645. https://doi.org/10.3389/fphys.2021.639645

Grimm PC, Nickerson P, Jeffery J, Savani RC, Gough J, McKenna RM, Stern E, Rush DN (2001) Neointimal and tubulointerstitial infiltration by recipient mesenchymal cells in chronic renal-allograft rejection. N Engl J Med 345(2):93–97. https://doi.org/10.1056/NEJM200107123450203

Imperiali NC, Giordani MC, Groppa SR, Mombelli C, Cortez SKV, Fabriani DO, Hanglin MMG,Ocampo ML, Christiansen SB, Diez GJR (2018) Risk factors for transplant vasculopathy in kidneytransplant. Transplantation, 102 Suppl 1, S489. https://doi.org/10.1097/01.tp.0000543303.05302.0e

Jain R, Hachem RR, Morrell MR, Trulock EP, Chakinala MM, Yusen RD, Huang HJ, Mohanakumar T, Patterson GA, Walter MJ (2010) Azithromycin is associated with increased survival in lung transplant recipients with bronchiolitis obliterans syndrome. J Heart Lung Transplant 29(5):531–537. https://doi.org/10.1016/j.healun.2009.11.008

Kamel MA, Scalia IG, Badr AT, Baba Ali N, Farina JM, Pereyra M, Abbas MT, Mahmoud AK, Scott RL, Steidley DE, Rosenthal JL, Lemond LM, Sell-Dottin KA, Hardaway BW, Barry T, Yang M, Chao CJ, Jokerst CE, Ayoub C, Arsanjani R (2024) Cardiac allograft vasculopathy: challenges and advances in invasive and non-invasive diagnostic modalities. J Cardiovasc Dev Dis 11(3):95. https://doi.org/10.3390/jcdd11030095

Katoh K (2023) Effects of mechanical stress on endothelial cells in situ and in vitro. Int J Mol Sci 24(22):16518. https://doi.org/10.3390/ijms242216518

Kim M, Bergmark BA, Zelniker TA, Mehra MR, Stewart GC, Page DS, Woodcome EL, SmallwoodJA, Gabardi S, Givertz MM (2017) Early aspirin use and the development of cardiac allograft vasculopathy.The Journal of heart and lung transplantation : the official publication of the International Society for HeartTransplantation, 36(12):1344–1349. https://doi.org/10.1016/j.healun.2017.06.015

Kloc M, Ghobrial RM (2014) Chronic allograft rejection: a significant hurdle to transplant success. Burns Trauma 2(1):3–10. https://doi.org/10.4103/2321-3868.121646

Kobashigawa JA, Moriguchi JD, Laks H, Wener L, Hage A, Hamilton MA, Cogert G, Marquez A, Vassilakis ME, Patel J, Yeatman L (2005) Ten-year follow-up of a randomized trial of pravastatin in heart transplant patients. J Heart Lung Transplant 24(11):1736–1740. https://doi.org/10.1016/j.healun.2005.02.017

Lusco MA, Fogo AB, Najafian B, Alpers CE (2016) Tubular atrophy. Am J Kidney Dis 67(6):e33–e34. https://doi.org/10.1053/j.ajkd.2016.04.007

Majesky MW, Dong XR, Hoglund V, Mahoney WM Jr, Daum G (2011) The adventitia: a dynamic interface containing resident progenitor cells. Arterioscler Thromb Vasc Biol 31(7):1530–1539. https://doi.org/10.1161/ATVBAHA.110.221549

Neuberger JM, Bechstein WO, Kuypers DRJ, Burra P, Citterio F, De Geest S, Duvoux C, Jardine AG, Kamar N, Krämer BK, Metselaar HJ, Nevens F, Pirenne J, Rodríguez-Perálvarez ML, Samuel D, Schneeberger S, Serón D, Trunečka P, Tisone G, van Gelder T (2017) Practical recommendations for long-term management of modifiable risks in kidney and liver transplant recipients: a guidance report and clinical checklist by the consensus on managing modifiable risk in transplantation (COMMIT) group. Transplantation 101(4S):S1–S56. https://doi.org/10.1097/TP.0000000000001651

Nishida S, Pinna A, Verzaro R, Levi D, Kato T, Khan F, Nery J, Weppler D, Tzakis A (2001) Sirolimus (rapamycin)-based rescue treatment following chronic rejection after liver transplantation. Transplant Proc 33(1–2):1495. https://doi.org/10.1016/S0041-1345(00)02566-5

Oguma S, Zerbe T, Banner B, Belle S, Starzl TE, Demetris AJ (1989) Chronic liver allograft rejection and obliterative arteriopathy: possible pathogenic mechanisms. Transplant Proc 21(1 Pt 2):2203–2207

Pittappilly M, Sharshir M, Paramesh A (2025) Chronic allograft nephropathy-a narrative review of its pathogenesis, diagnosis, and evolving management strategies. Biomedicine 13(4):929. https://doi.org/10.3390/biomedicines13040929

Pober JS, Chih S, Kobashigawa J, Madsen JC, Tellides G (2021) Cardiac allograft vasculopathy:current review and future research directions. Cardiovascular research, 117(13):2624–2638.https://doi.org/10.1093/cvr/cvab259

Prasad N, Harris E, Yuzefpolskaya M, DeFilippis EM, Colombo PC, Sayer G, Chernovolenko M, Fried J, Bae D, Oh KT, Raikhelkar J, Topkara VK, Castillo M, Lam EY, Latif F, Takeda K, Uriel N, Einstein AJ, Clerkin KJ (2025) Can the grading of mild cardiac allograft vasculopathy be further refined? An angiographic and physiologic assessment of heart transplant recipients

with ISHLT CAV 1. J Heart Lung Transplant 44(6):905–912. https://doi.org/10.1016/j.healun.2025.02.005
Ramzy D, Rao V, Brahm J, Miriuka S, Delgado D, Ross HJ (2005) Cardiac allograft vasculopathy: a review. Can J Surg 48(4):319–327
Sato M (2020) Bronchiolitis obliterans syndrome and restrictive allograft syndrome after lung transplantation: why are there two distinct forms of chronic lung allograft dysfunction? Ann Transl Med 8(6):418. https://doi.org/10.21037/atm.2020.02.159
Sato M, Waddell TK, Wagnetz U, Roberts HC, Hwang DM, Haroon A, Wagnetz D, Chaparro C, Singer LG, Hutcheon MA, Keshavjee S (2011) Restrictive allograft syndrome (RAS): a novel form of chronic lung allograft dysfunction. J Heart Lung Transplant 30(7):735–742. https://doi.org/10.1016/j.healun.2011.01.712
Skorić B, Čikeš M, Ljubas Maček J, Baričević Ž, Škorak I, Gašparović H, Biočina B, Miličić D (2014)Cardiac allograft vasculopathy: diagnosis, therapy, and prognosis. Croatian medical j, 55(6):562–576.https://doi.org/10.3325/cmj.2014.55.562
Shetty M, Chowdhury YS (2023) Heart transplantation allograft vasculopathy. In: StatPearls. StatPearls Publishing. https://www.ncbi.nlm.nih.gov/books/NBK555912/
Shimizu T, Omoto K, Inui M, Nozaki T, Takagi T, Ishida H (2023) Clinicopathological Analyses of ChronicRenal Allograft Arteriopathy after Kidney Transplantation. Nephron, 147 Suppl 1:22–27.https://doi.org/10.1159/000531177
Stomberski CT, Colvin MM (2025) Cardiac allograft vasculopathy: a focus on advances in diagnosis and management. Methodist Debakey Cardiovasc J 21(3):58–71. https://doi.org/10.14797/mdcvj.1580
Saggar R, Ross DJ, Saggar R, Zisman DA, Gregson A, Lynch JP III, Keane MP, Weigt SS, Ardehali A, Kubak B, Lai C, Elashoff D, Fishbein MC, Wallace WD, Belperio JA (2008) Pulmonary hypertension associated with lung transplantation obliterative bronchiolitis and vascular remodeling of the allograft. Am J Transplant 8(9):1921–1930. https://doi.org/10.1111/j.1600-6143.2008.02338.x
Valenta I, Dilsizian V, Quercioli A, Ruddy TD, Schindler TH (2013) Quantitative PET/CT measures of myocardial flow reserve and atherosclerosis for cardiac risk assessment and predicting adverse patient outcomes. Curr Cardiol Rep 15(3):344. https://doi.org/10.1007/s11886-012-0344-0
Vanstapel A, Verleden SE, Verbeken EK, Braubach P, Goos T, De Sadeleer L, Kaes J, Vanaudenaerde BM, Jonigk D, Ackermann M, Ceulemans LJ, Van Raemdonck DE, Neyrinck AP, Vos R, Verleden GM, Weynand B (2022) Beyond bronchiolitis obliterans: in-depth histopathologic characterization of bronchiolitis obliterans syndrome after lung transplantation. J Clin Med 11(1):111. https://doi.org/10.3390/jcm11010111
Verleden GM, Glanville AR, Lease ED, Fisher AJ, Calabrese F, Corris PA, Ensor CR, Gottlieb J, Hachem RR, Lama V, Martinu T, Neil DAH, Singer LG, Snell G, Vos R (2019) Chronic lung allograft dysfunction: definition, diagnostic criteria, and approaches to treatment – a consensus report from the Pulmonary Council of the ISHLT. J Heart Lung Transplant 38(5):493–503
Wenke K, Meiser B, Thiery J, Nagel D, von Scheidt W, Krobot K, Steinbeck G, Seidel D, Reichart B (2003) Simvastatin initiated early after heart transplantation: 8-year prospective experience. Circulation 107(1):93–97. https://doi.org/10.1161/01.CIR.0000043241.32523.EE
Yoshiyasu N, Sato M (2020) Chronic lung allograft dysfunction post-lung transplantation: the era of bronchiolitis obliterans syndrome and restrictive allograft syndrome. World J Transplant 10(5):104–116. https://doi.org/10.5500/wjt.v10.i5.104
Young JB (1999) Allograft vasculopathy: diagnosing the nemesis of heart transplantation. Circulation 100(5):458–460. https://doi.org/10.1161/01.CIR.100.5.458
Zuccarelli L, Bernard Q, Tarris G, Martin L, Funes de la Vega M, Jacq A, Rebibou J-M, Tinel C, Boccara C, Thouvenin O, Chassot J-M, Rabant M, Zuber J, Legendre C, Quenot J-P, Peraldi M-N, Amrouche L, Scemla A, Chavarot N, Anglicheau D, Legendre M, Maldiney T (2025) Rapid on-site histopathological analysis of kidney biopsy with dynamic full-field optical coherence tomography. Kidney Int Rep 10(5):1529–1537. https://doi.org/10.1016/j.ekir.2025.03.010

Chapter 9
The Myeloid–T Lymphocyte Axis in Chronic Lung Allograft Rejection: Emerging Insights into the Immune Circuitry Driving Fibrosis

Sajad Moshkelgosha, Sumiha Ramendra, Ke Fan Bei, and Stephen Juvet

Abstract While lung transplantation is a life-saving treatment option for patients with end-stage lung diseases, its long-term outcome is hampered by exceedingly high rates of chronic rejection, manifesting as chronic lung allograft dysfunction (CLAD). It has long been recognized that exposure of the lung to the external environment via the airways, coupled with its rich immune microenvironment, makes the lung uniquely susceptible to heightened alloimmunity. Extensive immunosuppressive drugs do not provide effective control of the process of chronic rejection, and so, in recent years, investigators have turned their attention toward the identification of underlying cellular mechanisms that are either inherently resistant to or evolve to escape these medications. In this review, we explore recent advances in identifying the cellular mediators of chronic lung allograft rejection, with a focus on the myeloid–T cell immune axis, which will require targeted therapeutic approaches in the future to prevent irreversible allograft fibrosis.

Sajad Moshkelgosha and Sumiha Karunagaran contributed equally with all other contributors.

S. Moshkelgosha
Latner Thoracic Research Laboratories, Toronto General Hospital Research Institute, University Health Network, Toronto, ON, Canada

S. Ramendra · K. F. Bei
Latner Thoracic Research Laboratories, Toronto General Hospital Research Institute, University Health Network, Toronto, ON, Canada

University of Toronto, Toronto, ON, Canada

S. Juvet (✉)
Latner Thoracic Research Laboratories, Toronto General Hospital Research Institute, University Health Network, Toronto, ON, Canada

Division of Respirology, Department of Medicine, University of Toronto, Toronto, ON, Canada
e-mail: stephen.juvet@uhn.ca

M. Kloc, R. M. Ghobrial (eds.), *Cellular and Molecular Mechanisms of Chronic Rejection in Organ Transplantation*, Results and Problems in Cell Differentiation 77, https://doi.org/10.1007/978-3-032-07686-1_9

9.1 Introduction

Chronic allograft rejection and fibrosis—manifesting clinically as chronic lung allograft dysfunction (CLAD)—continues to occur at a much higher rate in lung allografts than in other transplanted organs (Opelz et al. 2013), limiting median survival to only 6 years following lung transplantation (Perch et al. 2022). Despite the improved quality of life for recipients, there is a pressing need to improve its long-term outcome.

CLAD may manifest as bronchiolitis obliterans syndrome (BOS), in which there is an obstructive ventilatory defect accompanied by chest imaging that is either normal or compatible with hyperinflation (Burke et al. 1984; Cooper et al. 1993). A more aggressive form of CLAD, restrictive allograft syndrome (RAS), presents with a restrictive ventilatory defect and interstitial opacities on chest imaging (M. Sato et al. 2011). Mixed and undefined forms of CLAD are also recognized (Verleden et al. 2019). In both BOS and RAS, small airway obliteration (bronchiolitis obliterans) is present; RAS additionally has features of interstitial fibrosis—classically pleuroparenchymal fibroelastosis—but features of nonspecific interstitial pneumonia and organizing pneumonia may also be present (Ofek et al. 2013). Why these two clinical phenotypes of CLAD exist is unknown, although the more extensive fibrosis in RAS, involving multiple pulmonary anatomical compartments, may reflect a broader and more intense state of immune activation within the graft (Kawashima and Juvet 2020).

Ongoing post-transplant exposure of the lung allograft to inflammatory insults from the external environment via the airways is often cited as the main reason for the high incidence of CLAD. However, a central question is why the very high doses of immunosuppression used in lung transplant recipients are so frequently insufficient to prevent CLAD. Does this observation indicate that these already high doses are still simply inadequate, or does it reflect the existence of alloimmune pathways resistant to existing drugs? To address these questions, a deeper mechanistic understanding of the pathways leading to lung allograft fibrosis in the context of immunosuppression is required. High-dimensional analytical approaches, applied to lung allograft tissue, have begun to unravel the cellular profile of rejection and fibrosis after lung transplantation. In this review, we outline recent advances in our understanding of the major cellular players involved in CLAD pathogenesis. Particular focus is given to the dynamic interactions between myeloid cells and T lymphocytes, which are increasingly recognized as central coordinators of chronic inflammation and fibrosis and make up a large proportion of the graft-resident immune compartment (Fig. 9.1).

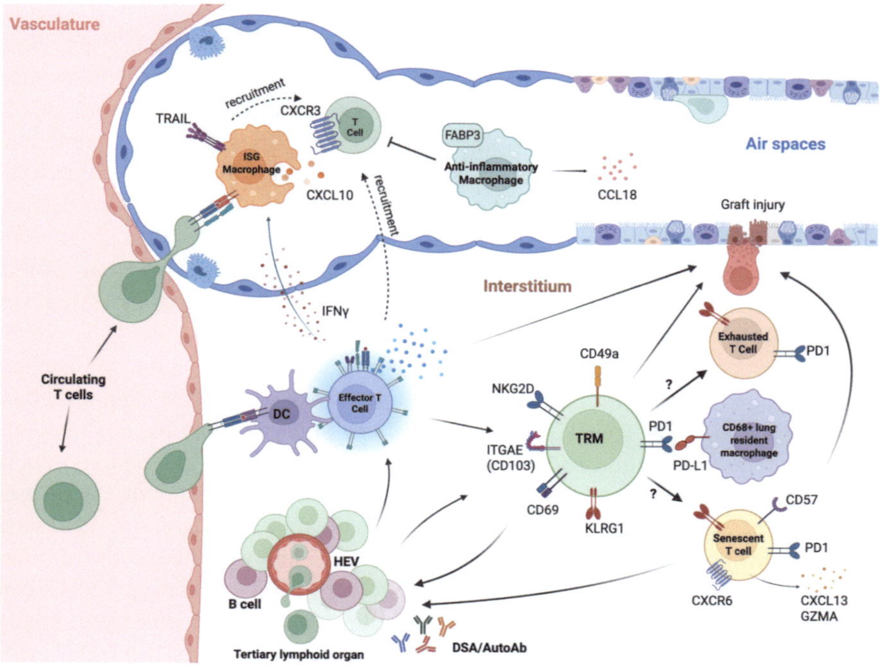

Fig. 9.1 Overview of the myeloid–T lymphocyte axis in chronic lung allograft rejection. T cells are recruited from the circulation into the graft via graft endothelium. Upon entry into the interstitium or airspaces, they interact with myeloid cells such as dendritic cells (DC) and macrophages. Recent work has revealed alveolar macrophages with pro-inflammatory interferon-stimulated gene expression (ISG; TRAIL+ CXCL10+) that likely recruit effector T cells, and those with an anti-inflammatory program (FABP3+ CCL18+) that may inhibit CD8+ T cells. Interstitial CD68+ macrophages provide activating signals to CD4+ and CD8+ tissue-resident memory (TRM) T cells, which can contribute to ongoing graft injury. Repeated alloantigen stimulation can lead TRM and other intragraft T cells to adopt either exhausted or senescent properties, allowing them to persist in the graft and likely making them resistant to conventional immunosuppression. Tertiary lymphoid organs (TLOs) are organized hubs containing T and B cells that can recruit CD62L+ T cells via high endothelial venules (HEVs). TLOs are sites of immune activation that can drive T-cell differentiation as well as B-cell class switch recombination, affinity maturation, and secretion of donor-specific antibodies (DSA) and autoantibodies (AutoAb). (Figure created in Biorender.com)

9.2 Overview of Inflammatory Mechanisms in the Pulmonary Allograft

Early lung graft injury is dominated by innate immune activation triggered by ischemia–reperfusion injury (IRI) and alloimmune-independent insults. Warm ischemia and cold ischemia damage airway epithelium and endothelium, releasing damage-associated molecular patterns (DAMPs) and chemokines that recruit macrophages and neutrophils into the graft (Shah et al. 2012). Activated donor tissue-resident macrophages and monocytes can also generate chemokine gradients that promote neutrophil infiltration, where experimental depletion of these myeloid cells markedly attenuates primary graft dysfunction (PGD) (Querrey et al. 2022; Kurihara et al. 2021; Dhaliwal et al. 2012). Accumulation of neutrophil extracellular traps (NETs) can further amplify injury in IRI, leading to PGD (Sayah et al. 2015; Caldarone et al. 2019). Cell crosstalk also further amplifies innate inflammation. Graft-resident donor nonclassical monocytes mediate the secretion of monocyte chemoattractants and play an important role in recruiting recipient spleen-derived classical monocytes to the graft. Recipient-derived monocytes can then also recruit neutrophils into the allograft in an IL-1β-dependent manner (Hsiao et al. 2018). Consistent with this, the blockade of IL-1β or depletion of monocytes reduces neutrophil-mediated vascular injury in murine lung IRI (Hsiao et al. 2018). Together, these findings highlight the central role of donor and recipient myeloid cell interactions in shaping the early innate inflammatory environment of the lung allograft, laying the groundwork for subsequent alloimmune injury and chronic dysfunction.

Following innate immune activation, the adaptive immune system mounts donor-specific responses through both direct allorecognition, where recipient T cells recognize intact donor MHC molecules, and indirect allorecognition, involving presentation of processed alloantigens by recipient antigen-presenting cells (Martinu et al. 2009; Benzimra et al. 2017). As many as 5–15% of circulating T cells can respond to donor alloantigen, depending on the degree of human leukocyte antigen (HLA) mismatch and recipient immune status (Greenland et al. 2016). $CD8^+$ cytotoxic T cells recognize donor HLA class I antigens and respond to the donor cells directly, while $CD4^+$ helper T-cell subsets orchestrate diverse immune responses including providing help to B cells. Th1 cells secrete IL-2, IL-12, and IFN-γ to drive activation of macrophages and cytotoxic T cells; in lung allografts, Th17 cells are also thought to contribute via secretion of IL-17, which promotes further innate cell recruitment (Harper et al. 2015).

Interleukin-17A, produced by Th17 cells and some $CD8^+$ T cells, is strongly implicated in pulmonary graft injury. IL-17 drives neutrophil recruitment by inducing chemokines (e.g., IL-8/CXCL8), and IL-17 blockade attenuates obliterative bronchiolitis in lung allograft models (Yamada et al. 2019; Kimura and Kishimoto 2010; Watanabe et al. 2023). IL-6 acts as a master regulator by promoting Th17 differentiation and IL-17 production (while inhibiting regulatory T cells) and sustains acute phase and fibroproliferative responses (Kimura and Kishimoto 2010). The Th1 cytokines IFN-γ and TNF-α can amplify macrophage activation; indeed, studies show that IFN-γ and other pro-inflammatory cytokines surge during rejection and drive tissue injury (Hodge et al. 2007; Mu et al. 2014).

9.3 Myeloid Cells in Lung Allograft Rejection

The coordinated actions of macrophages and dendritic cells (DCs) are central to initiating adaptive immune responses in the lung allograft (Fig. 9.1). These antigen-presenting cells (APCs) efficiently process and present donor-derived antigens to T cells, thereby bridging innate and adaptive immunity. However, their role extends beyond antigen presentation. In the context of lung transplantation, macrophages and DCs are highly responsive to "danger" signals generated by tissue injury. Together, the collaboration of donor and recipient APCs ensures that lung transplant antigens are efficiently presented to the adaptive immune system.

Beyond specific antigen presentation, macrophages and DCs in the transplanted lung respond vigorously to "danger" signals associated with transplantation injury. Ischemia–reperfusion injury and surgical stress lead to the release of damage-associated molecular patterns (DAMPs), such as high-mobility group box 1 (HMGB1) and extracellular ATP from the graft tissue. These endogenous ligands engage pattern recognition receptors (e.g., Toll-like receptors and NOD-like receptors) on macrophages and DCs much as pathogen-associated molecular patterns (PAMPs) from microbes would (Janeway and Medzhitov 2002). The result is the activation of innate immunity even in the absence of infection. For instance, signaling through TLR4 on donor graft cells and host myeloid cells has been shown to exacerbate transplant injury—TLR4-deficient mouse grafts endure less reperfusion damage (Shen et al. 2007). In the lung, resident alveolar macrophages are primed to sense inhaled microbial products like endotoxin, as well as sterile particulates, and they rapidly initiate inflammatory cascades. Upon reperfusion of a transplanted lung, these macrophages detect DAMPs and secrete cytokines (TNFα, IL-1β, and IL-6) and chemokines that recruit neutrophils and monocytes. Such innate activation can closely mimic an acute antimicrobial response, except that here it is triggered by tissue injury and alloantigen exposure. The phenomenon of heterologous immunity refers to cross-reactive immune memory where previous exposure to a pathogen can predispose to an anti-allograft response. Virus-specific memory T cells, for instance, may recognize allogeneic MHC/peptide complexes that happen to resemble viral epitopes and become activated in a transplant recipient (Selin et al. 1999). Such memory T cells are not naïve—they resist conventional immunosuppression and can respond rapidly upon transplant, often bypassing the need for co-stimulation. Myeloid cells play an indirect but crucial role in heterologous immunity by presenting the alloantigens that trigger these memory T cells and by creating an inflammatory milieu that lowers the threshold for T-cell activation. In lung transplantation, this issue is clinically relevant: many patients have been exposed to common viruses (like EBV, CMV, or influenza) and harbor memory T cells that could cross-react with donor antigens. Dendritic cells can uptake and present viral antigens alongside alloantigens, potentially activating virus-primed T cells that then attack the graft (Ford 2016; Morris et al. 2020).

Myeloid cells themselves can also be influenced by heterologous or bystander activation. Trained immunity is a concept describing how innate immune cells can retain a form of memory of past stimuli leading to either enhanced or suppressed responses upon a secondary challenge (Netea et al. 2016). A macrophage primed by fungal β-glucan or endotoxin might later secrete an exaggerated burst of cytokines when it encounters interferons or graft-derived molecules, thereby amplifying tissue inflammation and T-cell recruitment. Additionally, emerging research demonstrates that macrophages and monocytes can develop specific memory for alloantigens. In a striking mouse study, Dai et al. showed that murine monocytes/macrophages exposed to allogeneic MHC class I during one transplant acquired a memory that accelerated rejection of a second transplant with the same allo-MHC, independent of T cells (Oberbarnscheidt et al. 2014; Dai et al. 2020). This innate memory was mediated by paired immunoglobulin-like receptors (PIR-A/B) on myeloid cells recognizing non-self MHC molecules, and blocking these receptors improved graft survival (Dai et al. 2020). Such findings highlight that innate cells can be primed by environmental or graft-related stimuli in ways that influence future allograft immunity.

One of the most illuminating insights in recent years is that "macrophages" and "dendritic cells" are not monolithic cell types, but consist of multiple functionally distinct subpopulations. High-dimensional profiling techniques—especially multi-parameter flow cytometry and single-cell RNA sequencing—have uncovered a remarkable heterogeneity among myeloid cells infiltrating solid organ transplants. In the lung, the traditional dichotomy of alveolar macrophages versus interstitial macrophages is now understood to encompass further subsets with unique origins and activation states. Recent single-cell RNA-seq analyses by our group have identified at least two phenotypically distinct AM subsets that emerge during lung allograft dysfunction. We profiled BAL cells and explanted lungs from transplant recipients with chronic lung allograft dysfunction (CLAD) and discovered one subset of AMs enriched in interferon-stimulated gene expression and another characterized by high metallothionein expression (Moshkelgosha et al. 2022). Notably, these specialized macrophage subsets were largely absent in stable transplant BAL fluid, but dominated in patients with acute or chronic rejection, underscoring their relevance to pathology. In addition, some graft-infiltrating macrophages express high levels of MHC class II and co-stimulatory molecules, resembling DC-like "antigen-presenting macrophages," while others exhibit scavenger receptor and wound-healing profiles. In a healthy lung, resident AMs are typically immunoregulatory, producing IL-10 and keeping inflammation in check. But after transplant, the influx of bone marrow–derived cells can populate the lung with macrophages that are more prone to produce IL-1β, TNFα, and promote fibrosis (Wang et al. 2024). Nonclassical monocytes (CD14low/CD16high in humans) patrol the vasculature and are generally anti-inflammatory; yet, one study in lung transplantation suggests that donor-derived nonclassical monocytes can assume an inflammatory role during reperfusion injury (Kurihara et al. 2021). This indicates that even within a given subset defined by classical immunology, functional behavior can be altered by the transplant environment. DC subsets are likewise diverse in the lung. Conventional DCs can be broadly divided into type 1 (cDC1, typically CD103+ or CD8+ in mice) and type 2 (cDC2, CD11b+), which specialize in cross-presenting antigen to CD8+ T cells and priming CD4+ T cells, respectively. The lung contains both, and after transplant, these DCs may differentially regulate immunity: cDC1s are efficient in stimulating cytotoxic T cells (important in acute cellular rejection), whereas cDC2 can produce IL-6 and IL-23, and contribute to Th17 responses; both may induce Treg cells under certain conditions. In animal models, cDCs have been shown to stimulate or attenuate lung allograft rejection depending on the context (Benson et al. 2012; Watanabe et al. 2023). Plasmacytoid DCs (pDCs), another subset, circulate in low numbers, but can infiltrate the graft; they secrete large amounts of type I interferon during viral infection but can also exert tolerogenic effects (some tolerance induction protocols in animal models have even utilized donor or host pDCs to promote graft acceptance (Huang et al. 2024)). In summary, the application of high-dimensional profiling has revealed a complex landscape of myeloid cell subpopulations in lung transplants. Each subset—distinguished by lineage (donor vs. recipient, fetal vs. adult origin) and by transcriptional or surface marker profiles—may play a specialized role in either propagating rejection or, potentially, in repairing tissue and promoting tolerance. Recognizing this heterogeneity is crucial, as it opens the door to targeted therapies rather than treating "macrophages" as a single uniform target.

Myeloid cells additionally play an important role in the development of antibody-mediated rejection. In the lung, as in other organs, acute AMR is characterized by complement deposition and microvascular injury. Infiltrating monocytes/macrophages are frequently observed in peritubular or alveolar capillaries along with neutrophils during AMR (Wei et al. 2020). These macrophages can express Fcγ receptors that bind to donor-specific antibodies attached to graft endothelial cells. Ligation of FcγR on macrophages triggers their activation and release of proteases and reactive oxygen species that damage the endothelium. It also induces the secretion of pro-inflammatory cytokines that amplify local immune responses. Valenzuela et al. (2013) demonstrated that engagement of FcγRs by HLA antibodies on endothelial cells causes monocytes to adhere and transmigrate, partly via upregulation of P-selectin on endothelium (Valenzuela et al. 2013). Blocking P-selectin or downstream complement reduced monocyte recruitment and vascular inflammation. In a lung transplant context, this means that macrophages can act as executors of the damage initiated by antibodies.

While the focus is often on how myeloid cells promote adaptive immune responses that destroy the graft, it is important to note that they are also involved in regulating and tuning those responses. Dendritic cells, in particular, can induce not only immunity but also tolerance, depending on their maturation state and the signals they deliver to T cells. "Immature" or semi-mature DCs that present antigen in the absence of strong co-stimulation can lead to T-cell anergy or even drive the development of regulatory T cells (Tregs). This phenomenon has been leveraged in experimental transplant models: for example, infusion of donor-derived tolerogenic DCs (generated ex vivo to be IL-10 producing and low on co-stimulatory molecules) prolonged organ allograft survival in rodents and nonhuman primates (Ezzelarab et al. 2013). In one study, treating mouse cardiac allograft recipients with host-type "immature" DCs pulsed with donor antigen led to antigen-specific hyporesponsiveness and prolonged graft survival. These tolerogenic DCs appear to promote adaptive regulatory mechanisms—such as expanding Foxp3+ Tregs or inducing T-cell deletion—rather than effector immunity. Similarly, certain macrophage subsets can suppress adaptive responses. A notable example is regulatory macrophages (Mregs), a phenotype of macrophages induced by conditions like exposure to immune complexes or apoptotic cells. Mregs secrete high levels of IL-10 and TGF-β and express surface molecules (e.g., PD-L1, CD206) that dampen

T-cell responses. These Mregs can drive the expansion of Tregs in vivo; for instance, Riquelme et al. (2018) showed that human regulatory macrophages induce the formation of adaptive Tregs that express TIGIT and effectively suppress alloreactive T cells (Riquelme et al. 2018). Thus, the way macrophages and DCs “educate” adaptive immunity can be either to incite rejection or to foster tolerance, depending on their activation state and the context. In the rejecting lung, unfortunately, the balance is usually tipped toward pro-inflammatory activation. The cytokine milieu (high IFN-γ, IL-1, and alarmins) drives DCs and macrophages into an immunogenic profile. They express high IL-12 and B7 co-stimulatory molecules that favor Th1 and CTL responses, and low IL-10, thereby promoting rejection. Nevertheless, insights from tolerant transplant models provide a blueprint for the opposite scenario. For example, myeloid-derived suppressor cells (MDSCs)—which are immature monocyte/DC-like cells—accumulate in some models of long-term graft acceptance and can directly inhibit effector T cells (Dugast et al. 2008). These cells often require the presence of regulatory T cells or signals like GM-CSF and IL-10 to expand. In one study, transplant tolerance achieved via co-stimulatory blockade was dependent on the migration of suppressive myeloid cells; blocking the complement receptor C5aR1 disrupted MDSC trafficking and broke tolerance, underscoring a myeloid link in maintaining unresponsiveness (Pengam et al. 2019). Although lung transplants in humans rarely achieve operational tolerance, it is conceivable that modulating lung myeloid cells could tilt the immune response toward adaptation rather than destruction. For instance, approaches to promote an IL-10+ alveolar macrophage phenotype or to deliver tolerogenic donor DCs to lung transplant recipients are areas of active research.

9.4 T Cells in Lung Allograft Rejection

In contrast to conventional circulating T cells, TRMs are distinct subpopulations that persist in their respective tissues allowing for localized maintenance of homeostasis and a more efficient local immune response (B. V. Kumar et al. 2017). TRMs arise from effector T cells that have been activated by local antigens—such as those that arise during a respiratory viral infection. A small population of TRMs adapts to the tissue microenvironment and becomes long-lived. Unlike effector T cells, TRMs do not circulate and instead reside in the tissue to provide efficient localized protection from re-infection (Dijkgraaf et al. 2021). TRMs in the lung exhibit distinct dynamics compared to other barrier tissues, displaying both potent effector capacity and a unique capacity for egress, possibly reflecting adaptations to the lung’s delicate architecture or exposure to airborne stimuli (Carbone 2023). However, unlike in viral infections where lung TRMs gradually wane, in the context of alloimmunity and CLAD, these cells appear to persist, which is likely driven by sustained antigen exposure and chronic inflammation. Despite systemic immunosuppression, recipient-derived TRMs can accumulate within the graft, often bearing a cytotoxic phenotype and demonstrating clonal expansion within the tissue itself (Snyder et al.

2022). This suggests that even in the face of high-dose glucocorticoids, these TRMs may resist inhibition, contributing to sustained intragraft inflammation and tissue injury (Fig. 9.1).

Recently, Moghbeli et al. found that NKG2D is expressed on a subset of CD8+ T cells in CLAD using single-cell RNA sequencing of CLAD lung tissue and lung-draining lymph nodes (Moghbeli et al. 2025). These NKG2D+ CD8+ T cells expressed high levels of granzyme transcripts, along with NKG7 and perforin, suggesting a cytotoxic phenotype (Moghbeli et al. 2025). Spatial analysis of NKG2D+ CD8+ T-cell localization using mouse model of CLAD revealed that they localized in proximity to Scgb1a1+ epithelial cells and were associated with CLAD-like pathology (Moghbeli et al. 2025). Blocking NKG2D with a monoclonal antibody reversed pathophysiology associated with CLAD in treated mice (Moghbeli et al. 2025); however, this may have been due to a global decrease in cytotoxic CD8+ T cells as NKG2D is expressed upon CD8+ T-cell activation in mice (Prajapati et al. 2018). TCR sequencing revealed a high degree of clonal overlap between lung tissue and draining lymph node CD8+ TRM cells characterized by expression of CD69, ITGAE, SELL, and S1P1, indicating a potential alloantigen-driven expansion of CD8+ T cells that differentiated into TRMs. Khatri et al. analyzed four CLAD lung samples obtained during re-transplantation and compared these to three donor lungs using single-cell transcriptome profiling. CD8+ T cells were enriched in CLAD lungs, although CD4+ T cells were not similarly increased.[38] Within CD8+ T cells, PTPRC (encoding CD45RA/RO), granzymes, and interferon genes were increased in CLAD lung CD8+ T cells compared to donor lung CD8+ T cells (Khatri et al. 2023). CD8+ effector T cells were predicted to interact with basal cells in CLAD airways through spatial analysis (Khatri et al. 2023). Similarly, we used imaging mass cytometry to analyze four BOS, four RAS, and four control lungs with a 35-parameter panel to detect cell populations in CLAD. Both memory CD4+ and memory CD8+ T cells expressing CD45, CD45RO, PD1, and CD196 were observed in CLAD lungs compared to control lungs, along with a CD4+CD57+PD1+ population described above (Renaud-Picard et al. 2025; Moshkelgosha et al. 2025). Also, a higher frequency of regulatory T cells was present in RAS lungs compared to both BOS and control lungs, suggesting differences in the immune environment between BOS and RAS phenotypes of CLAD (Renaud-Picard et al. 2025). Together, these data reveal the complexity of T-cell differentiation within the lung allograft and illustrate the utility of high-dimensional transcriptomic and protein-based analytical techniques in unraveling the immune landscape of CLAD. The precise role of TRM cells in CLAD pathogenesis remains to be determined, but the apparent resistance of these cells to augmented immunosuppression—discussed in greater detail below—suggests that they may serve to drive ongoing tissue damage and remodeling.

Activation of T cells through antigen presentation initiates the differentiation of naïve T cells into effector T cells, which are responsible for targeting and eliminating antigens. However, in situations where antigen cannot be cleared—such as in cancer or organ transplantation—this stimulation becomes chronic. Prolonged exposure to antigen leads to T-cell exhaustion or senescence (Fig. 9.1). While these two states share many features, they are distinct in their phenotype and function. Although both cell states exhibit impaired proliferation, exhausted T cells display decreased effector function in a sequential manner with prolonged inflammation (Wherry 2011) while senescent T cells are dysfunctional, with loss of telomere length and DNA damage (Akbar and Henson 2011; Crespo et al. 2013). Proteins associated with T-cell exhaustion include PD-1 (Hui et al. 2017; Fourcade et al. 2010; Sakuishi et al. 2010), LAG-3 (Richter et al. 2010), TIM-3 (Fourcade et al. 2010; Sakuishi et al. 2010; Jones et al. 2008), and CTLA-4 (Akbar and Henson 2011), whereas senescent T cells are defined by short telomere length and telomerase activity. Some putative protein markers of exhaustion and senescence exhibit an imperfect relationship to these states. For example, while KLRG1 has been used to define senescent T cells (Akbar and Henson 2011), it can also be expressed by effector memory T cells (Lucas et al. 2024; Herndler-Brandstetter et al. 2018; Renkema et al. 2017). Similarly, while TIM-3 has been associated with exhaustion, it is also expressed by some nonexhausted effector memory T cells (Avery et al. 2018; Sabins et al. 2017). The development of T-cell exhaustion and senescence is well described in chronic viral infections and cancer; in lung and other solid organ transplantation, persistent alloantigen exposure may also drive these altered T-cell states, and the role of these states in lung transplant outcome is just beginning to be examined. Similar to other cases of chronic antigen stimulation, PD1 is highly expressed in lung allografts and has been associated with CLAD development. In an observational study comparing BOS and RAS, Righi et al. analyzed the expression of PD1, PDL1, CTLA4, and TOX with immunohistochemistry in lung grafts obtained at the time of re-transplantation for CLAD (Righi et al. 2021). The authors found that intragraft T cells expressed more PD1, PDL1, and CTLA4 in RAS lungs compared to BOS lungs. Furthermore, FOXP3-expressing T cells—which may or may not have had regulatory properties—were higher in RAS compared to BOS lungs. CD8+ T cells expressing the exhaustion marker TOX were also elevated in RAS lungs (Righi et al. 2021). The same group also went on to perform a retrospective, observational pilot study of transbronchial biopsies and BAL (Righi et al. 2023). While they observed that BAL eosinophils were associated with a greater frequency of lymphocytes expressing PD1 (Righi et al. 2023), neither T-cell lineage markers nor putative exhaustion and senescence molecules were analyzed, making interpretation of these findings difficult. As PD1 expression is a feature of both exhausted T cells and activated T cells (Patsoukis et al. 2020), the precise role of PD1-expressing lymphocytes in lung transplantation remains unknown. However, in a heterotopic tracheal allograft model of lung transplantation, treatment with PD-L1 Fc recombinant protein—which triggers inhibitory signaling via PD1—decreased IL-6 and IFNγ expression and increased IL-10 expression compared to treatment with an anti-PD-L1 antibody (Kaiho et al. 2023). Treatment with PDL1 Fc recombinant protein also decreased the CD4/CD8 T-cell ratio and did not cause CD4+CD44+ memory T cells to express PD1 or TIM3 (Kaiho et al. 2023). Furthermore, PD1 expression on CD8 T cells seems to restrain their interaction with dendritic cells and limits their differentiation into effector T cells (Takahashi et al. 2018). Unlike in cancer settings, PD1 expression on CD8 T cells may be protective of lung transplant and limit allograft injury (Takahashi et al. 2018). Nevertheless, data show that lung-resident PD-L1+ macrophages can provide activating signals to PD1+ TRM (Snyder et al. 2021), in keeping with the notion that inhibitory PDL1-PD1 signaling may be overcome by stronger activating signals in the same microenvironment.

Another marker of T-cell senescence is CD57. Brugière et al. analyzed lung transplant recipients from the COLT cohort study of 11 French lung transplant centers and found that early post-transplant, a low proportion of CD4+CD57+ILT2+ T cells were associated with reduced incidence of CLAD at 3 years where a greater than 2.64-fold increase of CD57+ILT2+ cells within the CD4 compartment was associated with a higher incidence of CLAD when compared to CD57+ILT2+ cells lower than 2.64-fold (Brugière et al. 2022; Pison et al. 2023). When tested for their functional properties, CD4+CD57+ILT2+ T cells showed higher degranulation capabilities with cell-surface expression of CD107a compared to CD4+CD57+ILT2− cells. The ILT2+ cells could be selectively inhibited by HLA-G, a nonclassical HLA molecule associated with maternal-fetal tolerance and lung transplant outcome (Brugière et al. 2022; Di Cristofaro et al. 2015; Lazarte et al. 2019), via interaction through ILT2 (LILRB2) (Brugière et al. 2022; Pison et al. 2023). We recently reported that CD4+CD57+ T cells expressing PD1 emerge in high frequencies in the BAL in patients at risk for CLAD (Moshkelgosha et al. 2025). Using imaging mass cytometry, CD4+CD57+PD1+ T cells were found to be adjacent to trafficking B cells (CD20+CCR7+) and epithelial cells in CLAD lung tissue. CD27+ and CD27− subpopulations of CD4+CD57+PD1+ T cells were further described using CITE-seq with putative peripheral helper (CD27−) and cytotoxic effector (CD27+) functions (Moshkelgosha et al. 2025). The CD27− population was characterized by expression of CXCL13, TOX, IL21, CD40L, ICOS, TIGIT, and TNFRSF18 with greater clonal expansion than the CD27+ population, which was characterized by granzymes and NKG7 transcripts (Moshkelgosha et al. 2025). In vitro, the cells failed to proliferate and were resistant to re-stimulation-induced cell death, yet secreted IL-21 and CXCL8 (Moshkelgosha et al. 2025), consistent with a senescent phenotype. Thus, evidence suggests that CLAD lungs contain T cells that could be classified as exhausted or senescent. Deeper transcriptional and functional characterization of these cells—including the role of exhaustion and senescence in TRM cells—will be required to understand their role in CLAD development.

9.5 Infection and Alloimmunity Crosstalk in Lung Allografts

The lung allograft is uniquely vulnerable to immune-mediated injury due to its constant exposure to environmental factors, including pathogens and pollutants, which perpetually stimulate the immune system (Righi et al. 2024). This persistent immune challenge, compounded by the inherent alloimmune response against the foreign tissue, initiates a multifaceted inflammatory cascade that profoundly influences both immediate postoperative recovery and long-term graft viability (Parulekar and Kao 2019). Bacterial pathogens are the most common infectious culprits early after lung transplantation and have well-defined roles in precipitating rejection. Gram-negative organisms, such as *Pseudomonas aeruginosa*, *Staphylococcus aureus*, and Enterobacteriaceae, dominate post-transplant pneumonias (Paglicci et al. 2021; McCort et al. 2021). These bacteria provoke vigorous innate responses. For example, *P. aeruginosa* infection induces strong granulocyte-colony-stimulating factor (G-CSF)-driven neutrophil influx into the graft. Moreover, allograft neutrophils responding to *P. aeruginosa* upregulate the co-stimulatory B7 molecules (CD80/CD86), greatly enhancing recipient T-cell activation against donor antigens (Kulkarni et al. 2020). Recently, *P. aeruginosa* was shown to selectively deplete Tregs via an exotoxin delivered via a type 3 secretion system, resulting in loss of tolerance and antibody-mediated rejection in a co-stimulation blockade-induced lung allograft tolerance model (Liao et al. 2025). In essence, the infected graft becomes a hub of antigen presentation and T-cell co-stimulation that amplifies alloimmune attack.

Viral infections likewise have multifaceted effects on lung transplant rejection. Cytomegalovirus (CMV) is a classic example, as it is the second most common post-transplant infection after bacterial pneumonia (Zamora 2004). CMV establishes lifelong latent infection, and its reactivation in the immunosuppressed recipient causes marked expansions of CMV-specific T cells. These CMV-specific $CD8^+$ and $CD4^+$ T cells can occupy a large fraction of the immune repertoire and are critical for controlling the virus, but their chronic activation also drives systemic immune aging known as immunosenescence. Over time, the massive clonal expansion of CMV-reactive cells narrows overall T-cell diversity, while many of these cells express exhaustion markers (PD-1 and KLRG1) that alter immune regulation (Meesing et al. 2019; van den Berg et al. 2019; Müller and Di Benedetto 2024; Courivaud et al. 2012). In practice, CMV reactivation causes graft inflammation and cellular injury that substantially raise the risk of both acute and chronic rejection. The mechanisms of CMV-driven rejection are incompletely understood but may involve heterologous immunity (cross-reactivity of anti-CMV and alloreactive TCRs, bystander activation, and innate immune activation) via DAMP release. Beyond CMV, community-acquired respiratory viruses (influenza, RSV, adenovirus, parainfluenza, and metapneumovirus) and herpesviruses (HSV and VZV) are common in lung recipients and can contribute to rejection risk (Trachuk et al. 2020; Gottlieb et al. 2009; D. Kumar et al. 2005). These viruses often lead to pneumonia in transplant recipients and trigger robust inflammatory cytokine responses (IL-1, TNF-α, IL-6, and IL-8) that chemoattract alloreactive leukocytes into the graft (Abedi et al. 2018). The resulting inflammatory milieu is highly conducive to immune-mediated injury: infected cells secrete danger signals that heighten T-cell and neutrophil infiltration, effectively "priming" the lung for rejection. Remarkably, respiratory viral infections also provoke the release of circulating exosomes carrying lung self-antigens, viral proteins, and proteasome fragments; in animal studies, these exosomes break tolerance and trigger immune responses against self-Ags, leading to CLAD-like pathology (Gunasekaran et al. 2020). Thus, both CMV and community-acquired respiratory viruses undermine graft survival by inducing inflammatory damage, skewing T-cell immunity, and creating self-antigen-driven autoimmunity within the graft (de Zwart et al. 2022). *Aspergillus* and *Candida* species predominate among these fungal pathogens. Invasive aspergillosis is firmly linked to chronic graft loss, and even mere airway colonization with *Aspergillus* (without tissue invasion) has been correlated with CLAD and increased mortality (Valentine et al. 2009). The mechanism appears to involve innate immune responses to fungal products: for example, chitin in the fungal cell wall triggers epithelial wounding and potent cytokine signaling in the lung. Gene expression studies show that *Aspergillus* colonization upregulates defense and inflammatory pathways in airway epithelium, suggesting that chronic fungal presence sets off low-grade inflammation and tissue remodeling (Samanta et al. 2021; Weigt et al. 2018). This persistent inflammatory state may compromise graft tolerance through dysregulated myeloid–T cell interactions.

9.6 Tertiary Lymphoid Organs as Hubs for Myeloid–T Cell Interactions

Tertiary lymphoid organs (TLOs), also known as tertiary lymphoid structures, are ectopic lymphoid aggregates that arise in sites of chronic inflammation, including lung allografts undergoing chronic rejection (Fig. 9.1). Unlike secondary lymphoid organs (SLOs), which are developmentally programmed and anatomically fixed, TLOs form de novo in peripheral tissues and serve as local hubs for immune activation and cell–cell interaction. In the setting of chronic lung allograft rejection, TLOs facilitate direct interactions between T cells, B cells, and myeloid cells. This localized immune architecture enables antigen presentation, cytokine signaling, and even class switching, thereby contributing to both T cell- and B cell-mediated alloimmune responses. Indeed, cultured rat and human lung allograft tissue has been shown to produce both donor-specific antibodies and autoantibodies (Miyamoto et al. 2017, 2025). The intimate proximity of $CD8^+$ T cells and myeloid cells within TLOs may also potentiate antibody-dependent cell-mediated cytotoxicity (ADCC), a pathway increasingly recognized in chronic lung allograft dysfunction (CLAD).

Unlike their generally favorable role in infections and cancer, where TLOs often correlate with improved immune surveillance and clinical outcomes, TLOs in transplantation are often associated with poor prognosis and graft loss. Sato et al. highlighted this contrast, noting that their presence in solid organ allografts is associated with chronic inflammation and fibrosis (Y. Sato et al. 2023). This divergence underscores the unique immunopathological context of allografts, where sustained antigen exposure and alloimmune activation reprogram TLO function from immunoprotection to immunopathology. Experimental models have shown that SLOs are not strictly required for graft rejection when TLOs are present within the graft itself. Both Gelman et al. and Wagnetz et al. independently demonstrated that lung and tracheal allograft rejection can proceed in the absence of SLOs if TLOs have formed within the allograft, affirming the autonomous immunological capacity of these ectopic structures (Gottlieb et al. 2009; Wagnetz et al. 2012). Moreover, TLOs are not merely passive markers of inflammation—they actively contribute to tissue remodeling and fibrosis. Studies linked TLO formation with fibrotic progression in CLAD, suggesting that fibroblasts and lymphoid cells may interact reciprocally to promote collagen deposition and airway remodeling (Watanabe et al. 2019; Smirnova et al. 2022). Yet, TLOs are not universally pathogenic. In animal models of lung transplant in which tolerance is induced with dual co-stimulation blockade, TLOs can harbor regulatory T cells (Tregs) that mitigate immune injury (Li et al. 2019). These Tregs can dampen B-cell activity and suppress antibody-mediated rejection, pointing to a dichotomous role for TLOs in both immune activation and regulation, depending on the context. In sum, TLOs are central players in the immunobiology of lung allografts, enabling localized adaptive immunity that can either sustain rejection or, under certain conditions, promote immune regulation. Understanding the balance of effector and regulatory pathways within these structures may hold the key to modulating chronic rejection and promoting long-term graft survival.

9.7 Conclusions and Future Directions

The lung's complex internal immune milieu—whose primary role is to provide defense against inhaled pathogens—appears to act as a stage from which alloimmune responses can be amplified (Fig. 9.1). As in other barrier tissues, TRM cells are abundant, and the TRM niche appears to fill up with alloreactive cells over time post-transplant. Evidence accrued to date shows that CD8+ TRM cells contribute to injury and are refractory to very high doses of corticosteroids. Equally, early observations implicate that specific transcriptional states in alveolar macrophages may have a key bearing on the impact of acute inflammatory insults on the development of CLAD. Another important feature of chronically rejected lung allograft is the propensity of intragraft lymphocytes to organize into TLOs. Clearly, although TLOs may be Treg-rich and associated with good outcomes in specifically tolerized animal models, there are no compelling data linking TLOs to tolerance in human lung allograft recipients. Rather, they may be associated with intragraft allo- and autoantibody production (Miyamoto et al. 2017, 2025) and the B cells therein could provide co-stimulatory and/or cytokine support to alloreactive T cells.

We have therefore only begun to understand the complexity of the immune landscape in chronic lung allograft rejection. Further work is now needed to address several key questions. First, what new diagnostic and prognostic tools can be applied in lung transplant recipients before CLAD onset to identify those at risk for CLAD? Secondly, are there specific ways to target TRM, B cells, and myeloid cells, to prevent CLAD? Third, can the processes leading to the infiltration and activation of these harmful cell populations be interrupted in a way that prevents damage to allograft structural cells, and possibly redirects them to promote allograft tolerance while preserving host defense? In our view, further exploration of these possibilities will lead to advances in precision medicine that can be applied in clinical lung transplantation, with the ultimate goal of preventing CLAD.

References

Abedi A, Hall R, Levine DJ (2018) Respiratory viruses and other relevant viral infections in the lung transplant recipient. In: Raghu G, Carbone RG (eds) Lung transplantation: evolving knowledge and new horizons. Springer International Publishing, Cham, pp 241–262. https://doi.org/10.1007/978-3-319-91184-7_15

Akbar AN, Henson SM (2011) Are senescence and exhaustion intertwined or unrelated processes that compromise immunity? Nat Rev Immunol 11(4):289–295. https://doi.org/10.1038/nri2959

Avery L, Filderman J, Szymczak-Workman AL, Kane LP (2018) Tim-3 co-stimulation promotes short-lived effector T cells, restricts memory precursors, and is dispensable for T cell exhaustion. Proc Natl Acad Sci 115(10):2455–2460. https://doi.org/10.1073/pnas.1712107115

Benson HL, Suzuki H, Lott J, Fisher AJ, Walline C, Heidler KM, Brutkiewicz R, Blum JS, Wilkes DS (2012) Donor lung derived myeloid and plasmacytoid dendritic cells differentially regulate T cell proliferation and cytokine production. Respir Res 13(1):25. https://doi.org/10.1186/1465-9921-13-25

Benzimra M, Calligaro GL, Glanville AR (2017) Acute rejection. J Thorac Dis 9(12):5440–5457. https://doi.org/10.21037/jtd.2017.11.83

Brugière O, Mouren D, Trichereau J, Vallée A, Kuzniak I, Hirschi S, Renaud-Picard B et al (2022) Chronic lung allograft dysfunction is associated with an early increase of circulating cytotoxic CD4+CD57+ILT2+ T cells, selectively inhibited by the immune check-point HLA-G. J Heart Lung Transplant 41(5):626–640. https://doi.org/10.1016/j.healun.2022.01.013

Burke CM, Theodore J, Dawkins KD, Yousem SA, Blank N, Billingham ME, Van Kessel A, Jamieson SW, Oyer PE, Baldwin JC (1984) Post-transplant obliterative bronchiolitis and other late lung sequelae in human heart-lung transplantation. Chest 86(6):824–829. https://doi.org/10.1378/chest.86.6.824

Caldarone L, Mariscal A, Sage A, Khan M, Juvet S, Martinu T, Zamel R et al (2019) Neutrophil extracellular traps in *ex vivo* lung perfusion perfusate predict the clinical outcome of lung transplant recipients. Eur Respir J 53(4):1801736. https://doi.org/10.1183/13993003.01736-2018

Carbone FR (2023) Unique properties of tissue-resident memory T cells in the lungs: implications for COVID-19 and other respiratory diseases. Nat Rev Immunol 23(5):329–335. https://doi.org/10.1038/s41577-022-00815-z

Cooper JD, Billingham M, Egan T, Hertz MI, Higenbottam T, Lynch J, Mauer J, Paradis I, Patterson GA, Smith C (1993) A working formulation for the standardization of nomenclature and for clinical staging of chronic dysfunction in lung allografts. International Society for Heart and Lung Transplantation. J Heart Lung Transplant 12(5):713–716

Courivaud C, Bamoulid J, Gaugler B, Roubiou C, Arregui C, Chalopin J-M, Borg C et al (2012) Cytomegalovirus exposure, immune exhaustion and cancer occurrence in renal transplant recipients. Transpl Int 25(9):948–955. https://doi.org/10.1111/j.1432-2277.2012.01521.x

Crespo J, Sun H, Welling TH, Tian Z, Zou W (2013) T cell Anergy, exhaustion, senescence, and stemness in the tumor microenvironment. Curr Opin Immunol 25(2):214–221. https://doi.org/10.1016/j.coi.2012.12.003

Dai H, Lan P, Zhao D, Abou-Daya K, Liu W, Chen W, Friday AJ et al (2020) PIRs mediate innate myeloid cell memory to nonself MHC molecules. Science (New York, NY) 368(6495):1122–1127. https://doi.org/10.1126/science.aax4040

de Zwart A, Riezebos-Brilman A, Lunter G, Vonk J, Glanville AR, Gottlieb J, Permpalung N, Kerstjens H, Alffenaar J-W, Verschuuren E (2022) Respiratory syncytial virus, human metapneumovirus, and parainfluenza virus infections in lung transplant recipients: a systematic review of outcomes and treatment strategies. Clin Infect Dis 74(12):2252–2260. https://doi.org/10.1093/cid/ciab969

Dhaliwal K, Scholefield E, Ferenbach D, Gibbons M, Duffin R, Dorward DA, Morris AC et al (2012) Monocytes control second-phase neutrophil emigration in established lipopolysaccharide-induced murine lung injury. Am J Respir Crit Care Med 186(6):514–524. https://doi.org/10.1164/rccm.201112-2132OC

Di Cristofaro J, Reynaud-Gaubert M, Carlini F, Roubertoux P, Loundou A, Basire A, Frassati C, Thomas P, Gomez C, Picard C (2015) HLA-G*01:04~UTR3 recipient correlates with lower survival and higher frequency of chronic rejection after lung transplantation. Am J Transplant 15(9):2413–2420. https://doi.org/10.1111/ajt.13305

Dijkgraaf FE, Kok L, Schumacher TNM (2021) Formation of tissue-resident CD8 + T-cell memory. Cold Spring Harb Perspect Biol 13(8):a038117. https://doi.org/10.1101/cshperspect.a038117

Dugast A-S, Haudebourg T, Coulon F, Heslan M, Haspot F, Poirier N, de Silly RV et al (2008) Myeloid-derived suppressor cells accumulate in kidney allograft tolerance and specifically suppress effector T cell expansion. J Immunol 180(12):7898–7906. https://doi.org/10.4049/jimmunol.180.12.7898

Ezzelarab MB, Zahorchak AF, Lu L, Morelli AE, Chalasani G, Demetris AJ, Lakkis FG et al (2013) Regulatory dendritic cell infusion prolongs kidney allograft survival in nonhuman primates. Am J Transplant 13(8):1989–2005. https://doi.org/10.1111/ajt.12310

Ford ML (2016) Virally-induced heterologous immunity in renal transplant recipients: important or inconsequential? Am J Transplant 16(5):1348–1349. https://doi.org/10.1111/ajt.13657

Fourcade J, Sun Z, Benallaoua M, Guillaume P, Luescher IF, Sander C, Kirkwood JM, Kuchroo V, Zarour HM (2010) Upregulation of Tim-3 and PD-1 expression is associated with tumor antigen–specific CD8+ T cell dysfunction in melanoma patients. J Exp Med 207(10):2175–2186. https://doi.org/10.1084/jem.20100637

Gottlieb J, Schulz TF, Welte T, Fuehner T, Dierich M, Simon AR, Engelmann I (2009) Community-acquired respiratory viral infections in lung transplant recipients: a single season cohort study. Transplantation 87(10):1530–1537. https://doi.org/10.1097/TP.0b013e3181a4857d

Greenland JR, Wong CM, Ahuja R, Wang AS, Uchida C, Golden JA, Hays SR et al (2016) Donor-reactive regulatory T cell frequency increases during acute cellular rejection of lung allografts. Transplantation 100(10):2090–2098. https://doi.org/10.1097/TP.0000000000001191

Gunasekaran M, Bansal S, Ravichandran R, Sharma M, Perincheri S, Rodriguez F, Hachem R et al (2020) Respiratory viral infection in lung transplantation induces exosomes that trigger chronic rejection. J Heart Lung Transplant 39(4):379–388. https://doi.org/10.1016/j.healun.2019.12.009

Harper SJF, Ali JM, Wlodek E, Negus MC, Harper IG, Manu Chhabra M, Qureshi S et al (2015) CD8 T-cell recognition of acquired alloantigen promotes acute allograft rejection. Proc Natl Acad Sci USA 112(41):12788–12793. https://doi.org/10.1073/pnas.1513533112

Herndler-Brandstetter D, Ishigame H, Shinnakasu R, Plajer V, Stecher C, Zhao J, Lietzenmayer M et al (2018) KLRG1+ effector CD8+ T cells lose KLRG1, differentiate into all memory T cell lineages, and convey enhanced protective immunity. Immunity 48(4):716–729.e8. https://doi.org/10.1016/j.immuni.2018.03.015

Hodge G, Hodge S, Chambers D, Reynolds PN, Holmes M (2007) Acute lung transplant rejection is associated with localized increase in T-cell IFNγ and TNFα proinflammatory cytokines in the airways. Transplantation 84(11):1452–1458. https://doi.org/10.1097/01.tp.0000290679.94163.e1

Hsiao H-M, Fernandez R, Tanaka S, Li W, Spahn JH, Chiu S, Akbarpour M et al (2018) Spleen-derived classical monocytes mediate lung ischemia-reperfusion injury through IL-1β. J Clin Invest 128(7):2833–2847. https://doi.org/10.1172/JCI98436

Huang C-H, Chen WY, Chen R-F, Ramachandran S, Liu K-F, Kuo Y-R (2024) Cell therapies and its derivatives as immunomodulators in vascularized composite allotransplantation. Asian J Surg 47(10):4251–4259. https://doi.org/10.1016/j.asjsur.2024.04.094

Hui E, Cheung J, Zhu J, Xiaolei S, Taylor MJ, Wallweber HA, Sasmal DK et al (2017) T cell costimulatory receptor CD28 is a primary target for PD-1–mediated inhibition. Science 355(6332):1428–1433. https://doi.org/10.1126/science.aaf1292

Janeway CA, Medzhitov R (2002) Innate immune recognition. Annu Rev Immunol 20:197–216. https://doi.org/10.1146/annurev.immunol.20.083001.084359

Jones RB, Ndhlovu LC, Barbour JD, Sheth PM, Jha AR, Long BR, Wong JC et al (2008) Tim-3 expression defines a novel population of dysfunctional T cells with highly elevated frequencies in progressive HIV-1 infection. J Exp Med 205(12):2763–2779. https://doi.org/10.1084/jem.20081398

Kaiho T, Suzuki H, Hata A, Matsumoto H, Tanaka K, Sakairi Y, Motohashi S, Yoshino I (2023) Targeting PD-1/PD-L1 inhibits rejection in a heterotopic tracheal allograft model of lung transplantation. Front Pharmacol 14(November):1298085. https://doi.org/10.3389/fphar.2023.1298085

Kawashima M, Juvet SC (2020) The role of innate immunity in the Long-term outcome of lung transplantation. Ann Transl Med 8(6):412. https://doi.org/10.21037/atm.2020.03.20

Khatri A, Todd JL, Kelly FL, Nagler A, Ji Z, Jain V, Gregory SG, Weinhold KJ, Palmer SM (2023) JAK-STAT activation contributes to cytotoxic T cell–mediated basal cell death in human chronic lung allograft dysfunction. JCI Insight 8(6):e167082. https://doi.org/10.1172/jci.insight.167082

Kimura A, Kishimoto T (2010) IL-6: regulator of Treg/Th17 balance. Eur J Immunol 40(7):1830–1835. https://doi.org/10.1002/eji.201040391

Kulkarni HS, Ramphal K, Ma L, Brown M, Oyster M, Speckhart KN, Takahashi T et al (2020) Local complement activation is associated with primary graft dysfunction after lung transplantation. JCI Insight 5(17):e138358. https://doi.org/10.1172/jci.insight.138358

Kumar D, Erdman D, Keshavjee S, Peret T, Tellier R, Hadjiliadis D, Johnson G, Ayers M, Siegal D, Humar A (2005) Clinical impact of community-acquired respiratory viruses on bronchiolitis obliterans after lung transplant. Am J Transplant 5(8):2031–2036. https://doi.org/10.1111/j.1600-6143.2005.00971.x

Kumar BV, Ma W, Miron M, Granot T, Guyer RS, Carpenter DJ, Senda T et al (2017) Human tissue-resident memory T cells are defined by core transcriptional and functional signatures in lymphoid and mucosal sites. Cell Rep 20(12):2921–2934. https://doi.org/10.1016/j.celrep.2017.08.078

Kurihara C, Lecuona E, Qiang W, Yang W, Núñez-Santana FL, Akbarpour M, Liu X et al (2021) Crosstalk between nonclassical monocytes and alveolar macrophages mediates transplant ischemia-reperfusion injury through classical monocyte recruitment. JCI Insight 6(6):147282. https://doi.org/10.1172/jci.insight.147282

Lazarte J, Ma J, Martinu T, Levy L, Klement W, White M, Pelling J et al (2019) Donor human leukocyte antigen-G single nucleotide polymorphisms are associated with post-lung transplant mortality. Eur Respir J 54(2):1802126. https://doi.org/10.1183/13993003.02126-2018

Li W, Gauthier JM, Higashikubo R, Hsiao H-M, Tanaka S, Vuong L, Ritter JH et al (2019) Bronchus-associated lymphoid tissue-resident Foxp3+ T lymphocytes prevent antibody-mediated lung rejection. J Clin Invest 129(2):556–568. https://doi.org/10.1172/JCI122083

Liao F, Zhou D, Cano M, Liu Z, Scozzi D, Tague LK, Byers DE et al (2025) Pseudomonas aeruginosa infection induces intragraft lymphocytotoxicity that triggers lung transplant antibody-mediated rejection. Sci Transl Med 17(784):eadp1349. https://doi.org/10.1126/scitranslmed.adp1349

Lucas ED, Huggins MA, Peng C, O'Connor C, Gress AR, Thefaine CE, Dehm EM, Kubota Y, Jameson SC, Hamilton SE (2024) Circulating KLRG1+ Long-lived effector memory T cells retain the flexibility to become tissue resident. Sci Immunol 9(96):eadj8356. https://doi.org/10.1126/sciimmunol.adj8356

Martinu T, Chen D-F, Palmer SM (2009) Acute rejection and humoral sensitization in lung transplant recipients. Proc Am Thorac Soc 6(1):54–65. https://doi.org/10.1513/pats.200808-080GO

McCort M, MacKenzie E, Pursell K, Pitrak D (2021) Bacterial infections in lung transplantation. J Thorac Dis 13(11):6654–6672. https://doi.org/10.21037/jtd-2021-12

Meesing A, Abraham RS, Razonable RR (2019) Clinical correlation of cytomegalovirus infection with CMV-specific CD8+ T-cell immune competence score and lymphocyte subsets in solid organ transplant recipients. Transplantation 103(4):832–838. https://doi.org/10.1097/TP.0000000000002396

Miyamoto E, Motoyama H, Sato M, Aoyama A, Menju T, Shikuma K, Sowa T et al (2017) Association of Local Intrapulmonary Production of antibodies specific to donor major his-

tocompatibility complex class I with the progression of chronic rejection of lung allografts. Transplantation 101(5):e156–e165. https://doi.org/10.1097/TP.0000000000001665
Miyamoto E, Vosoughi D, Wang J, Al-Refaee J, Berra G, Daigneault T, Duong A et al (2025) Local intragraft humoral immune responses in chronic lung allograft dysfunction. J Heart Lung Transplant 44(1):105–117. https://doi.org/10.1016/j.healun.2024.07.019
Moghbeli K, Lipp MA, Bueno M, Craig A, Rojas M, Abbas M, Lakkis ZI et al (2025) NKG2D blockade impairs tissue-resident memory T cell accumulation and reduces chronic lung allograft dysfunction. JCI Insight 10(4):e184048. https://doi.org/10.1172/jci.insight.184048
Morris AB, Pinelli DF, Liu D, Wagener M, Ford ML (2020) Memory T cell–mediated rejection is mitigated by FcγRIIB expression on CD8+ T cells. Am J Transplant 20(8):2206–2215. https://doi.org/10.1111/ajt.15837
Moshkelgosha S, Duong A, Wilson G, Andrews T, Berra G, Renaud-Picard B, Liu M et al (2022) Interferon-stimulated and metallothionein-expressing macrophages are associated with acute and chronic allograft dysfunction after lung transplantation. J Heart Lung Transplant 41(11):1556–1569. https://doi.org/10.1016/j.healun.2022.05.005
Moshkelgosha S, Levy L, Safavi S, Karunagaran S, Wilson G, Renaud-Picard B, Madu G et al (2025) Emergence of a senescent and inflammatory pulmonary CD4(+) T cell population prior to lung allograft failure. Sci Adv 11(12):eadp9052. https://doi.org/10.1126/sciadv.adp9052
Mu H-j, Xie P, Chen J-y, Gao F, Zou J, Zhang J, Zhang B (2014) Association of TNF-α, TGF-β1, IL-10, IL-6, and IFN-γ gene polymorphism with acute rejection and infection in lung transplant recipients. Clin Transpl 28(9):1016–1024. https://doi.org/10.1111/ctr.12411
Müller L, Di Benedetto S (2024) Immunosenescence and cytomegalovirus: exploring their connection in the context of aging, health, and disease. Int J Mol Sci 25(2):753. https://doi.org/10.3390/ijms25020753
Netea MG, Joosten LAB, Latz E, Mills KHG, Natoli G, Stunnenberg HG, O'Neill LAJ, Xavier RJ (2016) Trained immunity: a program of innate immune memory in health and disease. Science 352(6284):aaf1098. https://doi.org/10.1126/science.aaf1098
Oberbarnscheidt MH, Zeng Q, Li Q, Dai H, Williams AL, Shlomchik WD, Rothstein DM, Lakkis FG (2014) Non-self recognition by monocytes initiates allograft rejection. J Clin Invest 124(8):3579–3589. https://doi.org/10.1172/JCI74370
Ofek E, Sato M, Saito T, Wagnetz U, Roberts HC, Chaparro C, Waddell TK et al (2013) Restrictive allograft syndrome post lung transplantation is characterized by pleuroparenchymal fibroelastosis. Mod Pathol 26(3):350–356. https://doi.org/10.1038/modpathol.2012.171
Opelz G, Döhler B, Ruhenstroth A, Cinca S, Unterrainer C, Stricker L, Scherer S et al (2013) The collaborative transplant study registry. Transplant Rev (Orlando) 27(2):43–45. https://doi.org/10.1016/j.trre.2013.01.004
Paglicci L, Borgo V, Lanzarone N, Fabbiani M, Cassol C, Cusi MG, Valassina M et al (2021) Incidence and risk factors for respiratory tract bacterial colonization and infection in lung transplant recipients. Eur J Clin Microbiol Infect Dis 40(6):1271–1282. https://doi.org/10.1007/s10096-021-04153-1
Parulekar AD, Kao CC (2019) Detection, classification, and management of rejection after lung transplantation. J Thorac Dis 11(Suppl 14):S1732–S1739. https://doi.org/10.21037/jtd.2019.03.83
Patsoukis N, Wang Q, Strauss L, Boussiotis VA (2020) Revisiting the PD-1 pathway. Sci Adv 6(38):eabd2712. https://doi.org/10.1126/sciadv.abd2712
Pengam S, Durand J, Usal C, Gauttier V, Dilek N, Martinet B, Daguin V et al (2019) SIRPα/CD47 axis controls the maintenance of transplant tolerance sustained by myeloid-derived suppressor cells. Am J Transplant 19(12):3263–3275. https://doi.org/10.1111/ajt.15497
Perch M, Hayes DJ, Cherikh WS, Zuckermann A, Harhay MO, Hsich E, Potena L et al (2022) The International Thoracic Organ Transplant Registry of the International Society for Heart and Lung Transplantation: thirty-ninth adult lung transplantation report-2022; focus on lung transplant recipients with chronic obstructive pulmonary disease. J Heart Lung Transplant 41(10):1335–1347. https://doi.org/10.1016/j.healun.2022.08.007

Pison C, Tissot A, Bernasconi E, Royer P-J, Roux A, Koutsokera A, Coiffard B et al (2023) Systems prediction of chronic lung allograft dysfunction: results and perspectives from the cohort of lung transplantation and systems prediction of chronic lung allograft dysfunction cohorts. Front Med 10(March):1126697. https://doi.org/10.3389/fmed.2023.1126697

Prajapati K, Perez C, Rojas LBP, Burke B, Guevara-Patino JA (2018) Functions of NKG2D in CD8+ T cells: an opportunity for immunotherapy. Cell Mol Immunol 15(5):470–479. https://doi.org/10.1038/cmi.2017.161

Querrey M, Chiu S, Lecuona E, Qiang W, Sun H, Anderson M, Kelly M et al (2022) CD11b suppresses TLR activation of nonclassical monocytes to reduce primary graft dysfunction after lung transplantation. J Clin Invest 132(14):e157262. https://doi.org/10.1172/JCI157262

Renaud-Picard B, Moshkelgosha S, Berra G, Cheung M, Hwang D, Hedley D, Juvet S, Martinu T (2025) Detailed cellular and spatial characterization of chronic lung allograft dysfunction using imaging mass cytometry. J Heart Lung Transplant 44(1):118–124. https://doi.org/10.1016/j.healun.2024.09.023

Renkema K, Jameson S, Hamilton S (2017) Unique properties of long-lived effector CD8 T cells. J Immunol 198(Supplement_1):151.20. https://doi.org/10.4049/jimmunol.198.Supp.151.20

Richter K, Agnellini P, Oxenius A (2010) On the role of the inhibitory receptor LAG-3 in acute and chronic LCMV infection. Int Immunol 22(1):13–23. https://doi.org/10.1093/intimm/dxp107

Righi I, Vaira V, Morlacchi LC, Croci GA, Rossetti V, Blasi F, Ferrero S, Nosotti M, Rosso L, Clerici M (2021) Immune checkpoints expression in chronic lung allograft rejection. Front Immunol 12(August):714132. https://doi.org/10.3389/fimmu.2021.714132

Righi I, Vaira V, Morlacchi LC, Croci GA, Rossetti V, Blasi F, Ferrero S, Nosotti M, Rosso L, Clerici M (2023) PD-1 expression in transbronchial biopsies of lung transplant recipients is a possible early predictor of rejection. Front Immunol 13(January):1024021. https://doi.org/10.3389/fimmu.2022.1024021

Righi I, Barone I, Rosso L, Morlacchi LC, Rossetti V, Caffarena G, Limanaqi F, Palleschi A, Clerici M, Trabattoni D (2024) Immunopathology of lung transplantation: from infection to rejection and vice versa. Front Immunol 15:1433469. https://doi.org/10.3389/fimmu.2024.1433469

Riquelme P, Haarer J, Kammler A, Walter L, Tomiuk S, Ahrens N, Wege AK et al (2018) TIGIT(+) iTregs elicited by human regulatory macrophages control T cell immunity. Nat Commun 9(1):2858. https://doi.org/10.1038/s41467-018-05167-8

Sabins NC, Chornoguz O, Leander K, Kaplan F, Carter R, Kinder M, Bachman K et al (2017) TIM-3 engagement promotes effector memory T cell differentiation of human antigen-specific CD8 T cells by activating mTORC1. J Immunol 199(12):4091–4102. https://doi.org/10.4049/jimmunol.1701030

Sakuishi K, Apetoh L, Sullivan JM, Blazar BR, Kuchroo VK, Anderson AC (2010) Targeting Tim-3 and PD-1 pathways to reverse T cell exhaustion and restore anti-tumor immunity. J Exp Med 207(10):2187–2194. https://doi.org/10.1084/jem.20100643

Samanta P, Clancy CJ, Hong Nguyen M (2021) Fungal infections in lung transplantation. J Thorac Dis 13(11):6695–6707. https://doi.org/10.21037/jtd-2021-26

Sato M, Waddell TK, Wagnetz U, Roberts HC, Hwang DM, Haroon A, Wagnetz D et al (2011) Restrictive allograft syndrome (RAS): a novel form of chronic lung allograft dysfunction. J Heart Lung Transplant 30(7):735–742. https://doi.org/10.1016/j.healun.2011.01.712

Sato Y, Silina K, van den Broek M, Hirahara K, Yanagita M (2023) The roles of tertiary lymphoid structures in chronic diseases. Nat Rev Nephrol 19(8):525–537. https://doi.org/10.1038/s41581-023-00706-z

Sayah DM, Mallavia B, Liu F, Ortiz-Muñoz G, Caudrillier A, DerHovanessian A, Ross DJ et al (2015) Neutrophil extracellular traps are pathogenic in primary graft dysfunction after lung transplantation. Am J Respir Crit Care Med 191(4):455–463. https://doi.org/10.1164/rccm.201406-1086OC

Selin LK, Lin MY, Kraemer KA, Pardoll DM, Schneck JP, Varga SM, Santolucito PA, Pinto AK, Welsh RM (1999) Attrition of T cell memory. Immunity 11(6):733–742. https://doi.org/10.1016/S1074-7613(00)80147-8

Shah RJ, Diamond JM, Lederer DJ, Arcasoy SM, Cantu EM, Demissie EJ, Kawut SM et al (2012) Plasma monocyte chemotactic protein-1 levels at 24 hours are a biomarker of primary graft dysfunction after lung transplantation. Transl Res 160(6):435–442. https://doi.org/10.1016/j.trsl.2012.08.003

Shen X-D, Ke B, Zhai Y, Gao F, Tsuchihashi S-I, Lassman CR, Busuttil RW, Kupiec-Weglinski JW (2007) Absence of toll-like receptor 4 (TLR4) signaling in the donor organ reduces ischemia and reperfusion injury in a murine liver transplantation model. Liver Transpl 13(10):1435–1443. https://doi.org/10.1002/lt.21251

Smirnova NF, Riemondy K, Bueno M, Collins S, Suresh P, Wang X, Patel KN et al (2022) Single-cell transcriptome mapping identifies a local, innate B cell population driving chronic rejection after lung transplantation. JCI Insight 7(18):e156648. https://doi.org/10.1172/jci.insight.156648

Snyder ME, Sembrat J, Noda K, Myerburg MM, Craig A, Mitash N, Harano T et al (2021) Human lung-resident macrophages colocalize with and provide costimulation to PD1hi tissue-resident memory T cells. Am J Respir Crit Care Med 203(10):1230–1244. https://doi.org/10.1164/rccm.202006-2403OC

Snyder ME, Moghbeli K, Bondonese A, Craig A, Popescu I, Fan L, Tabib T et al (2022) Modulation of tissue resident memory T cells by glucocorticoids after acute cellular rejection in lung transplantation. J Exp Med 219(4):e20212059. https://doi.org/10.1084/jem.20212059

Takahashi T, Hsiao HM, Tanaka S, Li W, Higashikubo R, Scozzi D, Bharat A et al (2018) PD-1 expression on CD8+ T cells regulates their differentiation within lung allografts and is critical for tolerance induction. Am J Transplant 18(1):216–225. https://doi.org/10.1111/ajt.14437

Trachuk P, Bartash R, Abbasi M, Keene A (2020) Infectious complications in lung transplant recipients. Lung 198(6):879–887. https://doi.org/10.1007/s00408-020-00403-9

Valentine VG, Gupta MR, James E Jr, Walker LS, Bonvillain RW, Lombard GA, Weill D, Dhillon GS (2009) Effect of etiology and timing of respiratory tract infections on development of bronchiolitis obliterans syndrome. J Heart Lung Transplant 28(2):163–169. https://doi.org/10.1016/j.healun.2008.11.907

Valenzuela NM, Mulder A, Reed EF (2013) HLA class I antibodies trigger increased adherence of monocytes to endothelial cells by eliciting an increase in endothelial P-selectin and, depending on subclass, by engaging FcγRs. J Immunol 190(12):6635–6650. https://doi.org/10.4049/jimmunol.1201434

van den Berg SPH, Pardieck IN, Lanfermeijer J, Sauce D, Klenerman P, van Baarle D, Arens R (2019) The hallmarks of CMV-specific CD8 T-cell differentiation. Med Microbiol Immunol 208(3–4):365–373. https://doi.org/10.1007/s00430-019-00608-7

Verleden GM, Glanville AR, Lease ED, Fisher AJ, Calabrese F, Corris PA, Ensor CR et al (2019) Chronic lung allograft dysfunction: definition, diagnostic criteria, and approaches to treatment—a consensus report from the Pulmonary Council of the ISHLT. J Heart Lung Transplant 38(5):493–503. https://doi.org/10.1016/j.healun.2019.03.009

Wagnetz D, Sato M, Hirayama S, Matsuda Y, Juvet SC, Yeung JC, Guan Z et al (2012) Rejection of tracheal allograft by intrapulmonary lymphoid neogenesis in the absence of secondary lymphoid organs. Transplantation 93(12):1212–1220. https://doi.org/10.1097/TP.0b013e318250fbf5

Wang X, Zhou J, Li X, Liu C, Liu L, Cui H (2024) The role of macrophages in lung fibrosis and the signaling pathway. Cell Biochem Biophys 82(2):479–488. https://doi.org/10.1007/s12013-024-01253-5

Watanabe T, Martinu T, Chruscinski A, Boonstra K, Joe B, Horie M, Guan Z et al (2019) A B cell–dependent pathway drives chronic lung allograft rejection after ischemia–reperfusion injury in mice. Am J Transplant 19(12):3377–3389. https://doi.org/10.1111/AJT.15550

Watanabe T, Juvet SC, Berra G, Havlin J, Zhong W, Boonstra K, Daigneault T et al (2023) Donor IL-17 receptor A regulates LPS-potentiated acute and chronic murine lung allograft rejection. JCI Insight 8(21):e158002. https://doi.org/10.1172/jci.insight.158002

Wei X, Valenzuela NM, Rossetti M, Sosa RA, Nevarez-Mejia J, Fishbein GA, Mulder A et al (2020) Antibody-induced vascular inflammation skews infiltrating macrophages to a novel remodeling phenotype in a model of transplant rejection. Am J Transplant 20(10):2686–2702. https://doi.org/10.1111/ajt.15934

Weigt SS, Wang X, Palchevskiy V, Patel N, Derhovanessian A, Shino MY, Sayah DM et al (2018) Gene expression profiling of bronchoalveolar lavage cells during aspergillus colonization of the lung allograft. Transplantation 102(6):986–993. https://doi.org/10.1097/TP.0000000000002058

Wherry EJ (2011) T cell exhaustion. Nat Immunol 12(6):492–499. https://doi.org/10.1038/ni.2035

Yamada Y, Brüstle K, Jungraithmayr W (2019) T helper cell subsets in experimental lung allograft rejection. J Surg Res 233(January):74–81. https://doi.org/10.1016/j.jss.2018.07.073

Zamora MR (2004) Cytomegalovirus and lung transplantation. Am J Transplant 4(8):1219–1226. https://doi.org/10.1111/j.1600-6143.2004.00505.x

Chapter 10
Chronic Rejection in Facial Vascularized Composite Allotransplantation (fVCA)

Felix J. Klimitz, Leonard Knoedler, Samuel Knoedler, Bohdan Pomahac, and Martin Kauke-Navarro

Abstract Chronic rejection (CR) remains a major barrier to long-term success in facial vascularized composite allotransplantation (fVCA), particularly in facial transplants. Unlike acute rejection, CR is a gradual, progressive process marked by vasculopathy, fibrosis, and functional graft decline. Histopathologic features include intimal hyperplasia, dermal sclerosis, adnexal atrophy, and telangiectasia. Immune mechanisms driving CR involve persistent activation of Th1 and Th17 T cells, macrophages, and, to a lesser extent, B cells, which form tertiary lymphoid structures. Dysregulated cytokines and chemokines, such as IFN-γ, IL-17, and IL-6, perpetuate inflammation and fibrosis, whereas the downregulation of regulatory mediators like IL-10 impairs immune resolution. Chronic antigen exposure, complement activation, and the presence of tissue-resident memory T cells further sustain graft injury. Clinically, CR presents with tightening of facial tissue, pigmentary changes, pain, and impaired oral and facial function, which can sometimes progress to necrosis and graft failure. Diagnostic challenges persist due to heterogeneous presentation and a lack of standardized criteria. Early detection through protocol biopsies,

F. J. Klimitz
Division of Plastic Surgery, Department of Surgery, Yale School of Medicine, New Haven, CT, USA

Department of Hand, Replantation, and Microsurgery, Burns and Plastic Surgery, BG Trauma Center Berlin, Chair of Hand, Replantation, and Microsurgery, Charité, Germany, Universitätsmedizin Berlin, Berlin, Germany

L. Knoedler
Division of Plastic Surgery, Department of Surgery, Yale School of Medicine, New Haven, CT, USA

Charité – Universitätsmedizin Berlin, corporate member of Freie Universität Berlin, Humboldt-Universität zu Berlin, and Berlin Institute of Health, Department of Oral and Maxillofacial Surgery, Berlin, Germany

S. Knoedler · B. Pomahac · M. Kauke-Navarro (✉)
Division of Plastic Surgery, Department of Surgery, Yale School of Medicine, New Haven, CT, USA
e-mail: kauke-navarro.martin@yale.edu

M. Kloc, R. M. Ghobrial (eds.), *Cellular and Molecular Mechanisms of Chronic Rejection in Organ Transplantation*, Results and Problems in Cell Differentiation 77, https://doi.org/10.1007/978-3-032-07686-1_10

mucosal sampling, and vascular imaging is essential. Future directions emphasize the need for molecular diagnostics, targeted immunomodulation, and antifibrotic therapies. A deeper understanding of CR pathophysiology is critical to improving graft longevity and patient quality of life in fVCA.

10.1 Introduction

Facial vascularized composite allotransplantation has emerged as a groundbreaking reconstructive option for patients with severe facial disfigurement, offering superior functional and aesthetic outcomes compared to conventional techniques such as autologous tissue transfer or prosthetics. This is supported by systematic reviews and case series demonstrating improved quality of life, restoration of facial function, and social reintegration in long-term survivors of fVCA (Tchiloemba et al. 2021; Govshievich et al. 2021).

Acute rejection episodes are common in fVCA, but are generally manageable with current immunosuppressive regimens, most often based on tacrolimus, mycophenolate mofetil, and corticosteroids (Huelsboemer et al. 2024a). However, chronic rejection remains a major long-term challenge, threatening graft survival and function. Unlike solid organ transplantation (SOT), where chronic rejection is well characterized and classified, the understanding of chronic rejection in vascularized composite allotransplantation is still evolving, and there is currently no standardized classification or universally accepted diagnostic criteria (Homsy et al. 2024; Moktefi et al. 2021; Petruzzo et al. 2015; Ng et al. 2019).

Chronic rejection in fVCA is characterized by progressive graft deterioration, with clinical manifestations such as sclerotic skin changes, telangiectasia, pigmentary alterations, and functional decline (e.g., reduced mouth opening and impaired facial movement) (Krezdorn et al. 2019; Kanitakis et al. 2016). Histopathological findings include vascular remodeling, dermal fibrosis, adnexal atrophy, and features resembling autoimmune cutaneous disorders. The pathophysiology involves a complex interplay of innate and adaptive immune responses, endothelial injury, persistent inflammatory signaling, and fibrotic remodeling, with both T cell- and macrophage-mediated mechanisms implicated (Krezdorn et al. 2019; Lee et al. 2023; Kollar et al. 2019).

10.2 Onset and Monitoring of Chronic Rejection over Time

The onset of chronic rejection typically occurs months to years after transplantation and is characterized by progressive graft dysfunction and histopathological changes. Chronic rejection in face transplants has been increasingly recognized as a late complication, with clinical onset generally reported from approximately 1 to 8 years post-transplantation. The clinical manifestations include progressive graft dysfunction and chronic disfiguring cutaneous changes, such as sclerosis and dyschromia, telangiectasia, and, in some cases, premature aging or alopecia. Functional decline, such as reduced mouth opening or speech changes, may also occur as the process advances (Moktefi et al. 2021; Petruzzo et al. 2015; Krezdorn et al. 2019; Huelsboemer et al. 2024b; Akdogan et al. 2021).

Histopathologically, chronic rejection is characterized by features such as dermal sclerosis, epidermal thinning, vascular changes (including vasculopathy and intimal hyperplasia), adnexal atrophy, and, in some cases, lymphocyte-mediated cytotoxicity and collagen deposition (Moktefi et al. 2021; Krezdorn et al. 2019; Kanitakis et al. 2016; Mundinger et al. 2013). These changes can be detected on protocol or clinically indicated biopsies and may resemble patterns seen in autoimmune cutaneous disorders or chronic graft-versus-host disease (Moktefi et al. 2021; Petruzzo et al. 2015; Krezdorn et al. 2019). Chronic rejection may progress despite control of acute rejection episodes and is a leading cause of late graft loss in face transplantation (Homsy et al. 2024; Huelsboemer et al. 2024b).

Given the lack of universally accepted diagnostic criteria or guidelines (Homsy et al. 2024; Moktefi et al. 2021; Krezdorn et al. 2019; Kanitakis et al. 2016), the most effective monitoring strategies for the early detection of chronic rejection in face transplant recipients are regular protocol skin and mucosal biopsies, combined with clinical examination and noninvasive vascular imaging (Moktefi et al. 2021; Krezdorn et al. 2019; Kanitakis et al. 2016).

Protocol skin biopsies remain the cornerstone for surveillance, as histopathological changes such as dermal sclerosis, epidermal thinning, adnexal atrophy, and vascular alterations are the earliest and most specific indicators of chronic rejection (Moktefi et al. 2021). However, mucosal biopsies (e.g., from the oral or nasal mucosa) should be integrated into routine monitoring, as they can detect immune rejection events that may be missed by skin biopsies alone and may provide earlier or more sensitive detection of subclinical rejection activity. Discordance between skin and mucosal biopsy findings is common, and mucosal tissue may reveal latent alloimmune activity that precedes overt graft dysfunction (Kauke-Navarro et al. 2025a). Serological monitoring for donor-specific antibodies (DSA) and C4d staining in biopsies may provide adjunctive information, but their roles in chronic rejection surveillance are not yet standardized in vascularized composite allotransplantation (Homsy et al. 2024; Huelsboemer et al. 2025).

Clinical examination is essential for identifying progressive graft dysfunction, including sclerotic skin changes, pigmentary alterations, telangiectasia, and functional decline (e.g., reduced mouth opening or speech changes). These findings should prompt targeted biopsies even outside of protocol intervals (Petruzzo et al. 2015; Krezdorn et al. 2019; Kanitakis et al. 2016).

Noninvasive vascular imaging (such as Doppler ultrasound, CT angiography, or MR angiography) can detect graft vasculopathy, which is a hallmark of chronic rejection and may not be apparent on superficial biopsies. Imaging is particularly valuable for assessing deep vessel involvement and monitoring for accelerated arteriosclerosis (Kanitakis et al. 2016; Lee et al. 2023).

10.3 Involved Immune Cells

Chronic rejection in fVCA involves both cellular and humoral immune responses, with a complex interplay of immune cell populations such as T cells, macrophages, and organized lymphoid structures playing central roles. Human studies consistently demonstrate that CD8+ T cells and macrophages are prominent in chronically rejected tissue, particularly infiltrating deep arteries and associating with endothelial injury and arteriosclerotic changes, supporting a T cell-/macrophage-driven process. Immunoproteomic profiling of chronically rejected face allografts shows upregulation of effector memory T cells and proinflammatory monocytes, further supporting an innate-adaptive synergy (Lee et al. 2023).

CD8+ T cells are frequently implicated as dominant mediators of chronic injury, with evidence of cytotoxic effects on basal keratinocytes and endothelial cells in human tissue. However, while CD8+ T cells are central, the overall immune infiltrate is heterogeneous and includes macrophages and, in some cases, organized lymphoid structures (Lee et al. 2023; Mundinger et al. 2013).

Nonhuman primate models corroborate these findings: chronic rejection manifests as transplant vasculopathy with CD3+ T-cell infiltrates in vessel walls and neointimal hyperplasia, often without overt skin rejection, and with the presence of tertiary lymphoid follicles. Importantly, both human and animal studies report that CD20+ B cells and tertiary lymphoid structures are observed, but their pathogenic role is less clear. Human studies show a lack of consistent correlation between donor-specific antibodies (DSA) or C4d deposition and chronic rejection, and primate studies similarly find no link between alloantibody presence or C4d and vasculopathy, suggesting a predominantly T cell-driven phenotype (Mundinger et al. 2013; Huelsboemer et al. 2025).

These findings support a model of chronic rejection in facial VCA that is primarily orchestrated by T cells and macrophages, with B-cell involvement largely limited to tertiary follicle formation rather than direct humoral injury (Lee et al. 2023; Mundinger et al. 2013; Huelsboemer et al. 2025).

10.4 Immune Pathways and Processes in Chronic Rejection in Vascularized Composite Allotransplantation

A complex interplay of immune mechanisms perpetuates low-grade inflammation, tissue remodeling, and fibrotic degeneration (Kanitakis et al. 2016). Unlike acute rejection, which is often abrupt and reversible, chronic rejection is sustained by persistent alloimmune responses, dysregulated cytokine networks, and insufficient regulatory control (Lee et al. 2023). Both adaptive and innate immune systems contribute—through T-cell and B-cell activation, auto- and alloantibody production, complement deposition, and chronic antigen exposure—leading to vascular injury, parenchymal fibrosis, and eventual graft dysfunction (Puscz et al. 2020). This section explores the vascular, cytokine, cellular, and fibrotic cascades that collectively drive chronic immune-mediated deterioration in fVCA grafts, highlighting the mechanistic parallels with and distinctions from solid organ rejection (Krezdorn et al. 2019). An overview of the key immune pathways, tissue changes, and clinical manifestations is shown in Fig. 10.1.

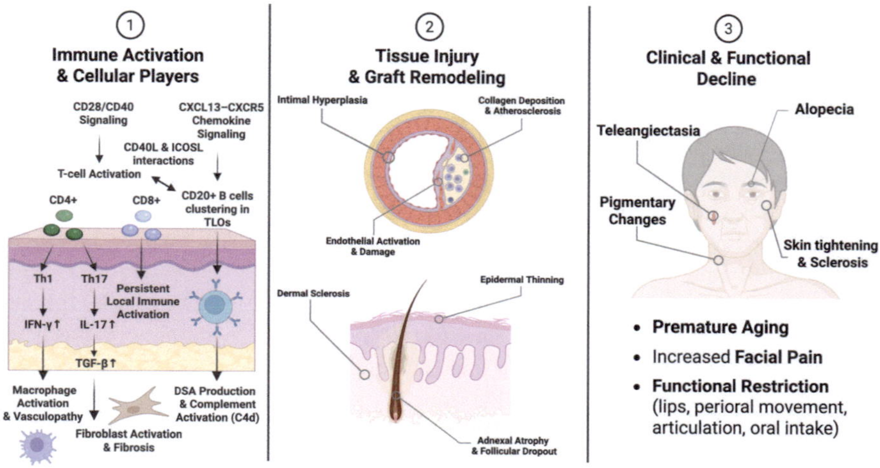

Fig. 10.1 Immunopathogenesis and clinical sequelae of chronic rejection in facial vascularized composite allotransplantation. Progressive stages of chronic rejection in facial vascularized composite allotransplantation (fVCA), from immune activation to clinical deterioration. *Panel 1*: Immune activation and cellular players—Chronic alloimmune responses are driven by T-cell activation via CD28/CD40 costimulatory signaling and reinforced by CD40L–ICOSL interactions. $CD4^+$ T cells polarize into Th1 and Th17 subsets, secreting IFN-γ and IL-17, respectively. These cytokines promote macrophage activation, vasculopathy, fibroblast activation, and fibrosis, in part through TGF-β upregulation. $CD8^+$ T cells and tissue-resident memory cells contribute to persistent local immune activation. B cells ($CD20^+$) are recruited via CXCL13–CXCR5 chemokine signaling and cluster in tertiary lymphoid organs (TLOs), where they drive donor-specific antibody (DSA) production and complement activation (marked by C4d deposition). *Panel 2*: Tissue injury and graft remodeling—Sustained immune responses lead to intimal hyperplasia, endothelial injury, collagen deposition, and atherosclerosis in graft vessels. In the skin, dermal sclerosis, epidermal thinning, adnexal atrophy, and follicular dropout reflect chronic tissue remodeling. *Panel 3*: Clinical and functional decline—Chronic rejection manifests with visible changes including telangiectasia, pigmentary alterations, alopecia, and skin tightening. Patients often experience premature aging, increased facial pain, and functional restrictions affecting lip movement, articulation, and oral intake. fVCA, Facial Vascularized Composite Allotransplantation. CD, Cluster of Differentiation. Th, T helper cell. IFN-γ, Interferon gamma. IL-17, Interleukin-17. TGF-β, Transforming Growth Factor Beta. TLOs, Tertiary Lymphoid Organs. CXCL13, C-X-C Motif Chemokine Ligand 13. CXCR5, C-X-C Motif Chemokine Receptor 5. CD40L, CD40 Ligand. ICOSL, Inducible T-cell Costimulator Ligand. DSA, Donor-Specific Antibody. C4d, Complement component 4d. (Created in BioRender. https://BioRender.com/dtsucek)

10.4.1 Vascular Changes

Chronic rejection is often marked by graft vasculopathy, characterized by intimal hyperplasia, vessel wall fibrosis, luminal occlusion, capillary thrombosis, and accelerated arteriosclerosis. These changes are driven by immune-mediated endothelial injury and subsequent reparative processes, affecting arteries more severely than veins (Ng et al. 2019; Kanitakis et al. 2016; Mundinger et al. 2013; Kaufman et al. 2020).

Graft vasculopathy is marked by the proliferation of α-SMA+ myofibroblasts, smooth muscle cells migrating from the media, and fibroblasts. This proliferation leads to luminal narrowing or occlusion (Petruzzo et al. 2015). Simultaneously, endothelial cell activation and dysfunction can be observed, together with cellular swelling, detachment, and reduced expression of vascular homeostasis markers such as CD31 and von Willebrand factor (Kaufman et al. 2020). Further, increased expression of adhesion molecules (e.g., ICAM-1, VCAM-1, and E-selectin) facilitates leukocyte adhesion and transmigration (Ng et al. 2019). Mononuclear inflammatory cell infiltration, primarily CD4+ T cells and CD68+ macrophages, has been shown to accumulate in the intima and adventitia, fueling chronic inflammation (Knoedler et al. 2023). Arterial walls frequently show fibrosis, elastin degradation, and neointimal thickening with collagen deposition, resembling transplant vasculopathy in solid organ transplantation, such as cardiac and renal transplants (Kaufman et al. 2020; Song et al. 2024; Pober et al. 2021). In smaller vessels, capillary leakage and perivascular fibrosis may be observed, which are associated with ischemic damage and impaired perfusion. Collectively, these vascular lesions disrupt microcirculation, leading to progressive tissue ischemia, fibrosis, and necrosis in skin and muscle tissue, among other components of the VCA allograft (Kanitakis et al. 2016; Knoedler et al. 2023).

Interestingly, graft vasculopathy and chronic skin changes (e.g., telangiectasis, epidermal thinning, and papillary dermal sclerosis) can occur independently of each other. Our group reported histopathological findings of one facial VCA that was removed after 10 years following the patient's death due to unrelated reasons. There was no graft vasculopathy, while skin tissue showed hallmarks of chronic rejection (e.g., telangiectasis, epidermal thinning, and papillary dermal sclerosis). Therefore, we hypothesized that two phenotypes of chronic rejection may exist, with one featuring chronic immune-mediated arteriosclerotic changes of the vasculature and one predominantly affecting the skin component (Kollar et al. 2020). Some face transplant patients can even develop chronic allograft rejection similar to autoimmune cutaneous disorders, such as lupus erythematosus and scleroderma/chronic sclerodermoid graft-versus-host disease (Krezdorn et al. 2019). Future research is warranted to decipher the interplay of graft vasculopathy and skin deterioration in chronic VCA rejection.

10.4.2 Inflammatory Cytokines and Chemokines

Dysregulation of cytokines and chemokines represents a hallmark of chronic rejection. Key pro-inflammatory cytokines of chronic VCA rejection range from IFN-γ, TNF-α, and IL-1β (Puscz et al. 2020). IFN-γ, produced primarily by alloreactive Th1 cells and NK cells, promotes MHC class II upregulation on endothelial and parenchymal cells, driving antigen presentation and immunogenicity of the graft. TNF-α and IL-1β contribute to endothelial activation, apoptosis, and upregulation of adhesion molecules such as ICAM-1 and VCAM-1, facilitating leukocyte extravasation (Jang et al. 2021; Lopez-Castejon and Brough 2011). IL-6 has also been found to be elevated in chronic VCA rejection (Knoedler et al. 2023). It promotes the differentiation of Th17 cells and the production of IL-17, which fuels neutrophilic inflammation and triggers tissue fibrosis. Chemokines such as CCL2, CXCL9, CXCL10, and CXCL11 are also upregulated during chronic VCA rejection (Rabbani et al. 2019). These chemokines bind to receptors such as CCR2 and CXCR3 on monocytes, macrophages, and T cells, navigating these cell types into the graft (Shroka et al. 2023). Interestingly, CXCL9, 10, and 11 are IFN-γ-inducible and can serve as potent chemoattractants for effector T cells, reinforcing Th1-mediated inflammatory responses and endothelial injury (Kak et al. 2018).

Conversely, chronic VCA rejection is also characterized by the downregulation of immunoregulatory and homeostatic cytokines and chemokines, which have been implicated in immune balance. For instance, IL-10 is a key anti-inflammatory cytokine produced by regulatory T cells, tolerogenic dendritic cells, and M2 macrophages, and has been demonstrated to be downregulated during chronic VCA rejection (Aviña et al. 2023; Anggelia et al. 2022). IL-10 suppresses the production of IFN-γ, TNF-α, and IL-1β and inhibits antigen-presenting cell maturation, ultimately mitigating alloreactive responses (Carlini et al. 2023).

The dysregulation of anti- and pro-inflammatory immunomodulatory mediators fosters a cascade of inflammation, vasculopathy, and fibrosis, ultimately contributing to chronic graft deterioration in VCA.

10.4.3 Immune Activation Pathways

Chronic rejection in VCA in general involves sustained alloimmune activation driven by both adaptive and innate pathways. Although the exact mechanisms underlying chronic allograft rejection remain poorly understood, data from preclinical models and human limb and facial VCAs have identified several potentially relevant pathways. Persistent T-cell responses against donor antigens are likely central to developing chronic rejection, supported by costimulatory signaling and helper T-cell polarization (Knoedler et al. 2023). T-cell costimulation through CD28–CD80/86 and CD40–CD154 (CD40L) pathways is critical for full T-cell activation (Lin et al. 2016). Experimental VCA models demonstrated that blockade of these signals markedly improves graft outcomes. For example, in a fully MHC-mismatched murine hindlimb VCA model, combined blockade of CD28–B7 (using CTLA4-Ig) and CD40–CD154 (using anti-CD154 antibody) prolonged graft survival (median >210 days with adjunctive radiation vs. 82 days without) (Oh et al. 2020). This combined costimulatory blockade promoted mixed chimerism, deletion of donor-reactive T cells, and donor-specific tolerance without depleting regulatory T cells (Tregs). Such findings imply that persistent CD28 and CD40 signals sustain chronic graft injury. Correspondingly, clinical and preclinical reports highlight the promise of belatacept (CTLA4-Ig)-based regimens in VCA, mirroring experience in solid organ transplant (Oh et al. 2020; Atia et al. 2020). Attempts to block other signals have also been explored: in a rhesus VCA model, adjunctive IL-17 pathway inhibitors (ustekinumab or secukinumab) reduced IL-17-producing T-cell infiltration but did not further extend graft survival beyond belatacept and steroids, suggesting that costimulatory blockade remains primary (Atia et al. 2020).

T helper cell polarization is also a key determinant of chronic allograft injury. Acute skin rejection in face and limb transplants often involves a codominant Th1/Th17 response, characterized by increases in IFN-γ- and IL-17-producing $CD4^+$ T cells (Borges et al. 2016). Although chronic rejection has been less well studied, murine models indicate that IL-17 is specifically required for fibrotic chronic rejection (Faust et al. 2009), as detailed in the fibrosis section. IFN-γ-secreting Th1 cells promote chronic vasculopathy and classical macrophage activation (Mitchell and Libby 2007). Indeed, infiltrating Th1 cells (and their effector cytokines) accumulate in chronically rejected VCA grafts, amplifying vascular injury and parenchymal damage by activating endothelial cells and fibroblasts (Borges et al. 2016; Faust et al. 2009). In contrast, the Th2 type has not been typically observed in VCA rejection (Borges et al. 2016). Together, these observations suggest that chronic VCA rejection involves a skewed Th1/Th17 response akin to that driving chronic rejection in solid organs.

In addition to T cell-driven mechanisms, B cells and humoral immunity also play an important role in chronic VCA rejection. Evidence from human, nonhuman primate, and rodent models revealed the presence of intragraft B-cell aggregates and the development of tertiary lymphoid organs (TLOs) in chronically rejecting VCA skin. These structures exhibit hallmarks of ectopic lymphoid neogenesis, including $CD20^+$ B-cell clusters and peripheral node addressin $(PNAd)^+$ high endothelial venules, which facilitate local lymphocyte recruitment and organization. Importantly, the presence of $PNAd^+$ vasculature correlated strongly with both $CD20^+$ infiltration and C4d deposition, suggesting that these B cell-rich niches contribute to local antibody-mediated rejection and complement activation (Hautz et al. 2014). A reported case of chronic rejection in a bilateral forearm transplant recipient further supports this mechanism, wherein B cell-rich TLOs were associated with de novo donor-specific antibodies (DSAs) and were responsive to B cell-depleting therapy (Weissenbacher et al. 2014). These findings indicate that B cells not only mediate systemic alloantibody production but also contribute to local immune amplification and chronic tissue injury in VCA.

Regulatory T cells (Tregs) typically temper alloimmunity, and deficits in Tregs number or function likely permit chronic rejection (Kauke-Navarro et al. 2023a). In VCA, insufficient Tregs infiltration has been implicated in pathologic rejection. Notably, experimentally enhancing Tregs recruitment can induce tolerance: a murine hindlimb VCA model used chemokine-releasing microparticles to attract Tregs locally. These Tregs recruitment microparticles induced long-term graft acceptance (>200 days) and donor-specific tolerance, significantly enriching FoxP3$^+$ Tregs in graft skin and draining lymph nodes (Fisher et al. 2020). This finding implies that chronically rejecting grafts may lack adequate Tregs regulation. In solid organ transplant, chronic rejection is associated with low Tregs/proliferating T-cell ratios, and similar deficits likely occur in VCA. Thus, a relative paucity of tolerogenic Tregs in the allograft microenvironment may allow persistent effector T-cell activation and fibrosis. Other regulatory subsets (e.g., myeloid-derived suppressors) have been less studied in VCA, and further research is warranted (Veglia et al. 2021).

Skin-resident memory T cells (Trm) represent a distinct reservoir of graft-reactive cells that may contribute to chronic inflammation. Donor skin harbors abundant Trm that remain in situ after transplant. In fact, human skin is densely populated by effector-memory T cells with a Th1 phenotype (Iske et al. 2019). In VCA recipients, persistent donor or host Trm could continuously respond to local antigens without needing recirculation (Beura et al. 2017). Supporting this, spatial proteomic analysis of a chronically rejected face transplant revealed dense aggregates of donor-derived CD8$^+$ T cells within deep graft vessels and arteriosclerotic lesions (Lee et al. 2023). These tissue-resident effector T cells, found in direct contact with chimeric endothelial cells, suggest that a cadre of local memory T cells may drive chronic graft injury. While specific markers of skin Trm (e.g., CD69/CD103) have not yet been characterized in VCA, by analogy to cutaneous diseases, they likely contribute to refractory inflammation.

Chronic antigen persistence is another driver of sustained rejection (Fisher et al. 2025; Sayegh and Carpenter 2004). VCA grafts continuously provide alloantigens (from skin, muscle, and bone), leading to ongoing immune stimulation. Chronic rejection models in rodents showed that prolonged alloantigen exposure leads to donor-specific humoral responses: in a murine hindlimb VCA chronic model, recipients developed significantly higher titers of IgG donor-specific antibodies and prominent C4d deposition compared to acute rejection groups (Fisher et al. 2025). These alloantibodies can form immune complexes and activate complement, exacerbating graft injury. Evidence of complement fixation in chronically rejecting VCA grafts indicated persistent antigen-driven humoral immunity. Thus, unlike a one-time insult, VCA permits continuous "re-exposure" to donor antigen, which can skew T cells toward an exhausted or chronically activated state. The net effect is a low-grade but relentless immune activation, maintained by residual antigen-presenting cells and graft-derived cytokines.

Innate immune sensors also participate in chronic VCA rejection. Recent profiling of a rejected human face transplant revealed upregulation of innate inflammatory pathways. Digital spatial proteomics highlighted marked expression of Indoleamine 2,3-dioxygenase 1 (IDO1) and STING (stimulator of interferon genes) in chronic rejection lesions (Lee et al. 2023). These molecules reflect innate sensing of danger signals: IDO1, typically induced by IFN-γ, can modulate tryptophan metabolism and dampen immune responses, but may also indicate a chronic IFN milieu; STING activation suggests cytosolic DNA or cGAS ligands in the graft microenvironment (perhaps from damaged cells). The presence of IDO1 and STING in arteriosclerotic graft vessels implies that innate immune "alarms" are persistently triggered in CR (Lee et al. 2023).

In sum, chronic fVCA rejection is likely the result of complex immune activation networks. T-cell costimulation through CD28 and CD40 pathways sustains effector responses, while Th1 and Th17 bias drive inflammation and fibrosis. Deficient Tregs activity and persistent tissue-resident memory T cells fail to constrain the response. Ongoing exposure to donor antigen, together with the formation of alloantibodies and complement activation, continually stimulates adaptive immunity. Concurrently, innate pathways, including STING and IDO1, maintain an inflammatory milieu in the graft. Importantly, chronic rejection in VCA does not appear to follow a singular pathogenic pathway. There seems to be a wide spectrum of clinical and histopathologic presentations, and our understanding of what defines chronic rejection—how it relates to prior acute rejection episodes, and how to distinguish chronic cellular from antibody-mediated rejection—remains incomplete. Current hypotheses suggest two main, potentially overlapping pathways: (1) a humoral-driven, antibody-mediated mechanism predominantly affecting the vasculature, often linked with complement activation and chronic vasculopathy; and (2) a cell-mediated process driven by repeated acute cellular rejection episodes, leading over time to chronic inflammation, tissue remodeling, and fibrosis—sometimes occurring even in the absence of significant vasculopathy (Kollar et al. 2020; Kauke et al. 2021a; Morelon et al. 2018). This variability between patients, with some experiencing early-onset chronic rejection and others maintaining long-term graft function despite multiple acute episodes, underscores the urgent need for longitudinal studies to define the molecular and clinical trajectories of chronic rejection in VCA.

10.4.4 Fibrosis and Sclerosis

Chronic VCA rejection is characterized by progressive fibrosis and sclerosis. Histologically, chronically rejected VCA grafts show epidermal thinning, dermal fibrosis, and loss of skin appendages (Knoedler et al. 2023). In a series of face-transplant biopsies, chronic rejection presented with epidermal thinning accompanied by focal lymphocyte-mediated cytotoxicity, hyperkeratosis, and plugging of hair follicles. Sclerotic dermal bands of dense type I collagen formed beneath the epidermis with associated follicular dropout (Krezdorn et al. 2019). Similarly, animal models of VCA chronic rejection demonstrated marked dermal fibrosis and adnexal atrophy. A murine hindlimb chronic model demonstrated significantly more dermal collagen deposition and follicular atrophy than acute rejection controls (Fisher et al. 2025). The Banff grading for chronic VCA highlights such changes:

graft biopsy findings include dermal fibrosis, follicular dropout, vascular intimal thickening and perivascular fibrosis, and increased dermal mast cells (Cendales et al. 2008, 2024). In chronically rejected rat VCA, dermal mast cells were significantly higher than in acutely rejecting grafts (Fisher et al. 2025). Mast cells likely contribute to fibrosis by releasing profibrotic mediators (e.g., TGF-β, histamine) and may be recruited by chronic inflammation (Artuc et al. 2002). Vascular changes also accompany fibrosis: chronic face-graft failure showed deep arteriosclerotic lesions with intimal proliferation and chimeric endothelium (Krezdorn et al. 2019). Thus, graft vasculopathy and cutaneous sclerosis co-occur, mirroring scleroderma and graft-versus-host disease phenotypes. Electron microscopy of sclerotic zones revealed myofibroblast-like spindle cells and dense collagen I bundles, consistent with extracellular matrix remodeling (Krezdorn et al. 2019).

At the transcriptional level, fibrogenesis in VCA chronic rejection is driven by key regulators. Notably, the activator protein-1 (AP-1) pathway has been implicated: chronically rejecting face grafts exhibited upregulation of AP-1 transcription factors JunB and c-Fos in fibrotic dermal regions. These same AP-1 components are known to drive collagen overproduction in systemic sclerosis (Ponticos et al. 2015). Transforming growth factor-β (TGF-β) is another regulator of fibrosis. Although direct VCA data are lacking, extrapolating from chronic allograft rejection in solid organs suggests that TGF-β signaling is crucial. In cardiac allografts, intragraft TGF-β drives fibroblast activation and Th17-mediated fibrosis. Chronically rejected VCA specimens often stain positively for TGF-β and phosphorylated SMADs in the dermis (as reported in hand and face transplant case reports), indicating active TGF-β signaling. This is congruent with the endothelial intimal thickening and collagen deposition observed. Other transcription factors linked to fibrosis, such as the SMAD family and β-catenin, have yet to be examined in VCA.

Extracellular matrix remodeling is a hallmark of chronic rejection fibrosis (Zhao et al. 2022). Type I collagen accumulates prominently in the dermis and vessel intima. Fibronectin and other matrix proteins (collagen III and vimentin) also increase. Dysregulated matrix metalloproteinases (MMPs) and their inhibitors (TIMPs) mediate ECM turnover. In face transplant recipients, proteomic profiling identified serum MMP-3 as highly upregulated during severe rejection episodes, indicative of active matrix remodeling (Kollar et al. 2018). The presence of MMP-3 implies that chronic rejection likely involves a persistent imbalance of MMP/TIMP activity. In chronic rat VCA, upregulation of CXC chemokines (e.g., CXCL9–11) was reported (Puscz et al. 2020); these chemokines can recruit inflammatory cells that secrete TGF-β and MMPs, linking innate signals to fibrosis. TIMP-1 expression is often elevated in chronic allografts; in experimental aortic allografts, mTOR inhibition reduced collagen III and TIMP-1 transcription (Murphy et al. 2003). By analogy, chronic VCA likely features altered TIMP/MMP expression, promoting matrix accumulation.

Profibrotic cell types in the VCA graft include activated fibroblasts (myofibroblasts), macrophages, mast cells, and certain T-lymphocyte subsets. Fibroblasts resident in the dermis or recruited from circulation differentiate into contractile myofibroblasts under cytokine signals (TGF-β and IL-13) and mechanical stress.

These myofibroblasts are the primary source of collagen synthesis. In fibrotic VCA biopsies, spindle-shaped fibroblasts with α-smooth muscle actin positivity have been observed. Although no study has yet isolated fibroblasts from chronically rejecting grafts, gene expression data suggest their activation: fibrosis-related genes such as COL1A1, COL3A1, and ACTA2 (α-SMA) are upregulated in chronic rejection models. Macrophages are abundant in chronic lesions: $CD68^+$ macrophages were notably increased in a chronic limb-allograft model (Puscz et al. 2020). These macrophages often exhibit an M2-like (alternatively activated) phenotype, producing TGF-β and PDGF, further driving fibroblast activation. Indeed, in a rat model, $CD68^+$ infiltration paralleled collagen deposition. Mast cells are another profibrotic innate cell; their accumulation in chronically rejected skin suggests a role. Mast cell-derived chymase and tryptase can activate latent TGF-β and stimulate fibroblasts. In a murine model, mast-cell depletion ameliorated skin fibrosis, implying that a similar effect may occur in VCA chronic rejection (Everett et al. 1995). Among lymphocytes, $CD4^+$ T cells may produce fibrogenic cytokines (IL-17 and IL-13) and have been observed infiltrating chronic VCA grafts (Cheng et al. 2025). $CD8^+$ T cells are less associated with direct fibrosis but can injure capillaries, leading secondarily to ischemia and scarring (Jiang et al. 2024).

Autoimmune fibrotic analogs provide insights into VCA chronic rejection. The skin changes in chronic VCA mirror those in systemic sclerosis and chronic graft-versus-host disease (Barausse et al. 2010). Both settings feature excessive type I collagen, vascular rarefaction, and appendage loss. As noted, AP-1 dysregulation (JunB/c-Fos overexpression) links VCA chronic rejection to scleroderma fibrosis (Ponticos et al. 2015). Moreover, systemic sclerosis patients showed increased dermal mast cells and M2 macrophages in fibrotic lesions, paralleling VCA findings (Brown and O'Reilly 2019). Accordingly, drugs that inhibit TGF-β or IL-6 (trialed in scleroderma) might mitigate VCA fibrosis. Likewise, anti-CTGF or angiotensin receptor blockers (used in cardiac fibrosis) could be considered, albeit off-label (AlQudah et al. 2020). However, no VCA-specific antifibrotic therapy has yet been tested or validated (Fuster-Martínez and Calatayud 2024).

From a therapeutic perspective, antifibrotic and immunomodulatory strategies are under exploration. Rapamycin (mTOR inhibitor) has both immunosuppressive and antifibrotic effects (Panwar et al. 2023). In chronic cardiac allograft models, rapamycin markedly reduced intimal hyperplasia and medial fibrosis (Murphy et al. 2003). By contrast, calcineurin inhibitors lack these antifibrotic effects and may even promote fibrosis by upregulating TGF-β (Rivelli et al. 2015). In a VCA context, switching calcineurin to mTOR inhibitors could attenuate chronic changes; indeed, anecdotal case reports suggest better vascular histology in face grafts on sirolimus (Morelon et al. 2017). Targeting TGF-β signaling directly (e.g., with neutralizing antibodies) is an intriguing but challenging strategy due to TGF-β's pleiotropy (Wang et al. 2023). Blocking IL-17 (e.g., with anti-IL-17A antibodies) may dampen fibrotic signaling without impairing Treg function, an approach that has been shown to be effective in scleroderma models (Zeng et al. 2020). Finally, therapies enhancing regulatory immune cells (e.g., Tregs infusion or CCL22 recruitment) have shown promise in prolonging VCA survival and may indirectly limit fibrosis by controlling inflammation (Fisher et al. 2020). In the future, novel antifibrotic interventions should be investigated in VCA as adjuncts to immunosuppression.

10.5 Clinical Manifestations and Functional Consequences

Chronic rejection leads to a gradual yet significant decline in both the appearance and functional performance of the facial graft. Clinically, patients may initially notice subtle changes such as premature aging, skin tightening, and telangiectasia, which can evolve into more severe manifestations, including purpuric lesions, scaly plaques, skin necrosis, and mucosal dryness (Krezdorn et al. 2019). Affected patients often report a progressive sensation of tightness and stiffness in the face, resembling scleroderma, as well as increased facial pain, likely due to soft tissue contraction (Petruzzo et al. 2015; Krezdorn et al. 2019). These changes ultimately compromise key facial functions—lip and perioral movements become restricted, resulting in difficulties with articulation, eating, and drinking (Petruzzo et al. 2015). In advanced cases, chronic vasculopathy and pan-fibrosis can lead to ischemic tissue damage, blistering, and full-thickness necrosis, necessitating graft explantation and, in rare cases, retransplantation (Lantieri et al. 2020). The most well-documented case of facial retransplantation followed CAMR-induced graft necrosis, which progressed over 12 months and resulted in total loss of facial motion, severe pain, disfigurement, and infection, illustrating the life-threatening consequences of untreated or unresponsive CR (Lantieri et al. 2020).

Recent reports have also highlighted progressive alopecia—affecting the beard and eyebrows—as a previously underrecognized but clinically significant manifestation of CR. In one case, hair loss emerged as the earliest sign of rejection, preceding sclerosis, pigmentary changes, and vasculopathy, and progressed despite intensified immunosuppression. This suggests that follicular involvement may serve as an early clinical marker of CR and expand the spectrum of visible phenotypes associated with graft deterioration (Akdogan et al. 2021). While earlier reports documented preserved core functions like lip closure despite skin sclerosis (Petruzzo et al. 2015), more recent cases reveal that CR can culminate in near-total functional loss, including impaired speech and mastication (Krezdorn et al. 2019; Lantieri et al. 2020). This discrepancy may reflect variations in immunologic injury severity or treatment timing.

These observations are consistent with broader clinical reviews, which emphasize that CR in fVCA typically presents with a heterogeneous mix of cutaneous and vascular changes—such as psoriasiform plaques, purpuric lesions, and necrotic ulcerations—that gradually impair function and disfigure appearance (Kanitakis et al. 2016). Notably, some patients developed CR despite adherence to standard triple immunosuppressive regimens (Kanitakis et al. 2016). These findings were further corroborated by findings from the first retransplantation case in a female patient, who showed progressive soft tissue contraction, facial pain, and a scleroderma-like appearance culminating in severe functional impairment—including restricted speech, eating, and facial movement—preceding the need for retransplantation. Cyanotic discoloration, blistering, and mucosal necrosis marked the late stages of graft failure (Kauke et al. 2021a).

In summary, chronic rejection in face transplantation is a multifaceted clinical entity marked by progressive aesthetic degradation and severe functional compromise. Early recognition—potentially including features like alopecia—paired with vigilant monitoring and tailored immunosuppression remains critical to preserving graft integrity and patient quality of life.

10.6 Future Directions in Understanding and Preventing Chronic Rejection in fVCA

10.6.1 Understanding the Link Between Acute and Chronic Rejection in VCA

Acute and chronic rejections continue to pose significant challenges in fVCA (Huelsboemer et al. 2024b; Knoedler et al. 2023; Kauke et al. 2021b). Repeated acute cellular rejection episodes are common, especially during the first few years post-transplantation (Tchiloemba et al. 2021; Huelsboemer et al. 2024b). Over time, chronic rejection may develop, with manifestations ranging from graft fibrosis, chronic ulceration, and loss of function to contracture and dyschromia (Tchiloemba et al. 2021; Petruzzo et al. 2015; Krezdorn et al. 2019; Kanitakis et al. 2016; Huelsboemer et al. 2024b; Knoedler et al. 2023; Kauke et al. 2021a). In some cases, a more dramatic form of chronic rejection has been reported, involving progressive graft vasculopathy and ultimately vessel thrombosis with graft necrosis (Kanitakis et al. 2016; Kauke et al. 2021a). In these cases, donor-specific antibodies are typically identified and thus suggest a chronic antibody-mediated rejection pathway (Kauke et al. 2021a).

Importantly, there seems to be a wide spectrum of how chronic rejection can present, and our understanding of what truly presents chronic rejection, how it is related to acute rejection events, and how to distinguish chronic cellular and antibody-mediated rejection remains incomplete. The current hypothesis suggests that chronic rejection may evolve from repeated acute rejection episodes or from chronic vascular injury (i.e., graft vasculopathy), though chronic changes have also been observed in the absence of vasculopathy (Kollar et al. 2020; Kauke et al. 2021a; Morelon et al. 2018). The mechanisms remain unclear and may differ between patients—some experience early-onset chronic rejection, while others maintain long-term graft function despite repeated early acute cellular rejection episodes (Huelsboemer et al. 2024b). This variability underscores the need for longitudinal studies to define the molecular and clinical progression of rejection more precisely. Molecular investigations remain limited, but suggest overlapping pathways with other fibrosing conditions (Krezdorn et al. 2019). For example, gene expression analyses have identified upregulation of AP-1, JunB, and c-Fos in chronic rejection samples of skin, which are also observed in systemic sclerosis (Krezdorn et al. 2019). Immunoproteomic profiling has further suggested a T cell-/macrophage-driven process, wherein donor-derived T cells target recipient-derived endothelial cells within the graft vasculature and lead to accelerated arteriosclerosis of the graft vessels (Lee et al. 2023).

Once established, chronic rejection likely remains irreversible, especially in cases of chronic antibody-mediated rejection with graft vessel thrombosis and acute necrosis. Management strategies have stabilized progression in some cases but have, as expected, not reversed tissue damage (Kanitakis et al. 2016; Kauke et al. 2021a; Morelon et al. 2018). As such, prevention may be the only viable strategy.

Efforts to mitigate chronic rejection may depend on better diagnosis and early intervention for acute rejection. Subclinical immune activity—especially within mucosal tissue—often precedes changes in the skin (Knoedler et al. 2023; Kauke et al. 2020; Kauke-Navarro et al. 2021, 2025b; Kanitakis et al. 2006; Bergfeld et al. 2013). Therefore, it is important to improve our understanding of mucosal rejection and the utility of mucosal tissues as a sentinel tissue for skin rejection (Kauke-Navarro et al. 2025a). In the future, an updated grading system of both acute and chronic rejection is important, ideally inclusive of both skin and mucosa, and, possibly sentinel flaps. Furthermore, biomarkers, including immune cell population changes and inflammatory markers, may provide useful diagnostic adjuncts and should be integrated in an updated grading system of VCA rejection (Chinta et al. 2025; Kauke-Navarro et al. 2023b; Kiefer et al. 2025).

10.6.2 Toward Prevention and Tolerance: Future Strategies in Chronic Rejection

Preventing and slowing down chronic rejection will require a multifaceted approach. Longitudinal sampling of both skin and mucosa (in case of facial transplants)—paired with clinical correlation—may help define how acute changes evolve into chronic pathology. In parallel, mitigating risk factors such as CMV mismatch, mechanical trauma, and poor adherence to immunosuppression is crucial (Kauke-Navarro et al. 2022; Kauke et al. 2021c). Immunologic matching between donors and recipients plays an important role in minimizing genetic mismatch and thereby reducing the risk of rejection (Krezdorn et al. 2019; Mathes et al. 2014). Furthermore, improved understanding of rejection mechanisms may open the door to novel immunomodulatory strategies that are likely to play a central role in the future. While tolerance induction through, for example, donor bone marrow infusion—such as the Pittsburgh protocol—has been proposed and trialed, clinical success has been limited, and tolerance has not been achieved clinically (Kauke et al. 2022; Robbins et al. 2019).

Regulatory T-cell therapies represent a promising avenue currently under investigation in solid organ transplantation (SOT) and may define the next frontier in vascularized composite allotransplantation (VCA) (Kauke-Navarro et al. 2023a). In SOT, the ONE study demonstrated the safety of regulatory cell therapy in living-donor kidney transplant recipients, with the potential to reduce reliance on systemic maintenance immunosuppression (Sawitzki et al. 2020). More advanced strategies that may become clinically relevant include engineered CAR-T regulatory cells designed to recognize donor-specific HLA antigens with their engineered surface receptor and deliver targeted immunosuppression—such as rapamycin-releasing microparticles—directly to the sites of inflammation within the graft, potentially limiting systemic toxicity while controlling local inflammation and rejection (Kauke-Navarro et al. 2023a).

Aside from cell-based therapies, biologic agents such as CTLA4-Ig (e.g., belatacept), which block T-cell costimulation (CD28/CD80/86), are increasingly used in solid organ transplantation and select VCA patients, especially as part of CNI-sparing regimens (Huelsboemer et al. 2024b). However, their role in long-term graft preservation remains to be fully established.

Ultimately, the field must continue working toward an integrated understanding of acute and chronic rejection at the molecular, cellular, and tissue levels. Early diagnosis, precision immunosuppression, and novel tolerance-inducing approaches may eventually shift the paradigm from reactive treatment to proactive prevention, which may slow down the development of chronic graft rejection.

10.7 Conclusion

Chronic rejection in facial vascularized composite allotransplantation represents a distinct and multifactorial immunopathological entity, characterized by progressive graft deterioration involving a variable degree of vascular injury, immune-mediated inflammation, and fibrotic tissue remodeling (Krezdorn et al. 2019; Kanitakis et al. 2016). Histopathologically, it can manifest as graft vasculopathy, dermal sclerosis, adnexal atrophy, and epithelial thinning—often paralleling features seen in chronic graft-versus-host disease and systemic autoimmune conditions. Central drivers include persistent alloantigen exposure, Th1/Th17-skewed T-cell responses, macrophage activation, dysregulated cytokine and chemokine signaling, and an insufficient regulatory immune response (Mundinger et al. 2013; Knoedler et al. 2023; Fisher et al. 2025). Importantly, chronic rejection does not appear to follow a uniform pathway. Current evidence suggests at least two partially overlapping mechanisms: one dominated by likely antibody-mediated vascular injury and chronic vasculopathy, and another driven by repeated episodes of acute cellular rejection leading to progressive inflammation, fibrosis, and tissue sclerosis—even in the absence of overt vasculopathy (Kollar et al. 2020; Kauke et al. 2021a; Morelon et al. 2018). This heterogeneity highlights the complexity of chronic rejection in fVCA and underscores the need for additional longitudinal studies to elucidate its molecular and clinical trajectories. Chronic rejection is further sustained by tissue-resident memory T cells, tertiary lymphoid structures, and complement-activating alloantibodies, alongside innate immune triggers such as STING and IDO1 (Fisher et al. 2025). Clinically, chronic rejection leads to significant aesthetic and functional decline, with symptoms ranging from telangiectasia and sclerosis to ischemic necrosis and facial immobility (Krezdorn et al. 2019; Kanitakis et al. 2016; Huelsboemer et al. 2024b). While current immunosuppressive regimens manage acute rejection effectively, they fall short in preventing or reversing chronic rejection. As such, early diagnosis through vigilant clinical monitoring, protocol biopsies, and the inclusion of mucosal surveillance is essential. Future efforts must focus on identifying reliable biomarkers, refining diagnostic criteria, and developing targeted immunomodulatory and antifibrotic therapies to preserve graft function and improve long-term outcomes in fVCA recipients.

References

Akdogan N, Ersoy-Evans S, Gokoz O, Erdem Y, Nasir S (2021) Early recognition of chronic rejection in a face allotransplant patient with alopecia. J Cutan Pathol 48(10):1286–1297. https://doi.org/10.1111/cup.14069

AlQudah M, Hale TM, Czubryt MP (2020) Targeting the renin-angiotensin-aldosterone system in fibrosis. Matrix Biol 91-92:92–108. https://doi.org/10.1016/j.matbio.2020.04.005

Anggelia MR, Cheng HY, Lai PC, Hsieh YH, Lin CH, Lin CH (2022) Cell therapy in vascularized composite allotransplantation. Biom J 45(3):454–464. https://doi.org/10.1016/j.bj.2022.01.005

Artuc M, Steckelings UM, Henz BM (2002) Mast cell-fibroblast interactions: human mast cells as source and inducers of fibroblast and epithelial growth factors. J Invest Dermatol 118(3):391–395. https://doi.org/10.1046/j.0022-202x.2001.01705.x

Atia A, Moris D, McRae M et al (2020) Th17 cell inhibition in a costimulation blockade-based regimen for vascularized composite allotransplantation using a nonhuman primate model. Transpl Int 33(10):1294–1301. https://doi.org/10.1111/tri.13612

Aviña AE, De Paz D, Huang SC et al (2023) IL-10 modified mRNA monotherapy prolongs survival after composite facial allografting through the induction of mixed chimerism. Mol Ther Nucleic Acids 31:610–627. https://doi.org/10.1016/j.omtn.2023.02.016

Barausse G, Caramaschi P, Scambi C et al (2010) Clinical, serologic and instrumental data of ten patients affected by sclerodermatous chronic graft versus host disease: similarities and differences in respect to systemic sclerosis. Int J Immunopathol Pharmacol 23(1):373–377. https://doi.org/10.1177/039463201002300139

Bergfeld W, Klimczak A, Stratton JS, Siemionow MZ (2013) A four-year pathology review of the near total face transplant. Am J Transplant 13(10):2750–2764. https://doi.org/10.1111/ajt.12379

Beura LK, Rosato PC, Masopust D (2017) Implications of resident memory T cells for transplantation. Am J Transplant 17(5):1167–1175. https://doi.org/10.1111/ajt.14101

Borges TJ, O'Malley JT, Wo L et al (2016) Codominant role of interferon-γ- and interleukin-17-producing T cells during rejection in full facial transplant recipients. Am J Transplant 16(7):2158–2171. https://doi.org/10.1111/ajt.13705

Brown M, O'Reilly S (2019) The immunopathogenesis of fibrosis in systemic sclerosis. Clin Exp Immunol 195(3):310–321. https://doi.org/10.1111/cei.13238

Carlini V, Noonan DM, Abdalalem E et al (2023) The multifaceted nature of IL-10: regulation, role in immunological homeostasis and its relevance to cancer, COVID-19 and post-COVID conditions. Front Immunol 14:1161067. https://doi.org/10.3389/fimmu.2023.1161067

Cendales LC, Kanitakis J, Schneeberger S et al (2008) The Banff 2007 working classification of skin-containing composite tissue allograft pathology. Am J Transplant 8(7):1396–1400. https://doi.org/10.1111/j.1600-6143.2008.02243.x

Cendales LC, Farris AB, Rosales I et al (2024) Banff 2022 vascularized composite allotransplantation meeting report: diagnostic criteria for vascular changes. Am J Transplant 24(5):716–723. https://doi.org/10.1016/j.ajt.2023.12.023

Cheng HY, Anggelia MR, Lin CH (2025) Unraveling the roles of macrophages in vascularized composite allotransplantation. Biomedicines 13(6):1425. https://doi.org/10.3390/biomedicines13061425

Chinta SR, Shah AR, Tran DL et al (2025) New paradigms in rejection monitoring: lymphocyte subsets as noninvasive graft markers in vascularized composite allotransplantation. Plast Reconstr Surg Glob Open 13(3):e6598. https://doi.org/10.1097/gox.0000000000006598

Everett ET, Pablos JL, Harley RA, LeRoy EC, Norris JS (1995) The role of mast cells in the development of skin fibrosis in tight-skin mutant mice. Comp Biochem Physiol A Physiol 110(2):159–165. https://doi.org/10.1016/0300-9629(94)00127-f

Faust SM, Lu G, Marini BL et al (2009) Role of T cell TGFbeta signaling and IL-17 in allograft acceptance and fibrosis associated with chronic rejection. J Immunol 183(11):7297–7306. https://doi.org/10.4049/jimmunol.0902446

Fisher JD, Zhang W, Balmert SC et al (2020) In situ recruitment of regulatory T cells promotes donor-specific tolerance in vascularized composite allotransplantation. Sci Adv 6(11):eaax8429. https://doi.org/10.1126/sciadv.aax8429

Fisher DT, Mackey E, Kononov E et al (2025) Chronic rejection models for vascularized composite tissue allotransplantation. Sci Rep 15(1):16882. https://doi.org/10.1038/s41598-025-01803-8

Fuster-Martínez I, Calatayud S (2024) The current landscape of antifibrotic therapy across different organs: a systematic approach. Pharmacol Res 205:107245. https://doi.org/10.1016/j.phrs.2024.107245

Govshievich A, Saleh E, Boghossian E et al (2021) Face transplant: current update and first Canadian experience. Plast Reconstr Surg 147(5):1177–1188. https://doi.org/10.1097/prs.0000000000007890

Hautz T, Zelger BG, Nasr IW et al (2014) Lymphoid neogenesis in skin of human hand, nonhuman primate, and rat vascularized composite allografts. Transpl Int 27(9):966–976. https://doi.org/10.1111/tri.12358

Homsy P, Huelsboemer L, Barret JP et al (2024) An update on the survival of the first 50 face transplants worldwide. JAMA Surg 159:1339–1345. https://doi.org/10.1001/jamasurg.2024.3748

Huelsboemer L, Boroumand S, Kochen A et al (2024a) Immunosuppressive strategies in face and hand transplantation: a comprehensive systematic review of current therapy regimens and outcomes. Front Transplant 3:1366243. https://doi.org/10.3389/frtra.2024.1366243

Huelsboemer L, Kauke-Navarro M, Boroumand S et al (2024b) Ten-year follow-up after face transplantation—a single-center retrospective cohort study. Am J Transplant 25:611–622. https://doi.org/10.1016/j.ajt.2024.10.007

Huelsboemer L, Hosseini H, Klimitz FJ et al (2025) The prognostic relevance of donor-specific antibodies in facial transplantation – a retrospective cohort study. J Plast Reconstr Aesthet Surg 103:286–296. https://doi.org/10.1016/j.bjps.2025.01.028

Iske J, Nian Y, Maenosono R, Maurer M, Sauer IM, Tullius SG (2019) Composite tissue allotransplantation: opportunities and challenges. Cell Mol Immunol 16(4):343–349. https://doi.org/10.1038/s41423-019-0215-3

Jang DI, Lee AH, Shin HY et al (2021) The role of tumor necrosis factor alpha (TNF-α) in autoimmune disease and current TNF-α inhibitors in therapeutics. Int J Mol Sci 22(5):2719. https://doi.org/10.3390/ijms22052719

Jiang W, Tang TT, Zhang YL et al (2024) CD8 T cells induce the peritubular capillary rarefaction during AKI to CKD transition. Int J Biol Sci 20(8):2980–2993. https://doi.org/10.7150/ijbs.96812

Kak G, Raza M, Tiwari BK (2018) Interferon-gamma (IFN-γ): exploring its implications in infectious diseases. Biomol Concepts 9(1):64–79. https://doi.org/10.1515/bmc-2018-0007

Kanitakis J, Badet L, Petruzzo P et al (2006) Clinicopathologic monitoring of the skin and oral mucosa of the first human face allograft: report on the first eight months. Transplantation 82(12):1610–1615. https://doi.org/10.1097/01.tp.0000248780.55263.33

Kanitakis J, Petruzzo P, Badet L et al (2016) Chronic rejection in human vascularized composite allotransplantation (hand and face recipients): an update. Transplantation 100(10):2053–2061. https://doi.org/10.1097/tp.0000000000001248

Kaufman CL, Kanitakis J, Weissenbacher A et al (2020) Defining chronic rejection in vascularized composite allotransplantation-The American Society of Reconstructive Transplantation and International Society of Vascularized Composite Allotransplantation chronic rejection working group: 2018 American Society of Reconstructive Transplantation meeting report and white paper research goals in defining chronic rejection in vascularized composite allotransplantation. SAGE Open Med 8:2050312120940421. https://doi.org/10.1177/2050312120940421

Kauke M, Safi AF, Zhegibe A et al (2020) Mucosa and rejection in facial vascularized composite allotransplantation: a systematic review. Transplantation 104(12):2616–2624. https://doi.org/10.1097/tp.0000000000003171

Kauke M, Panayi AC, Safi AF et al (2021a) Full facial retransplantation in a female patient-Technical, immunologic, and clinical considerations. Am J Transplant 21(10):3472–3480. https://doi.org/10.1111/ajt.16696

Kauke M, Panayi AC, Tchiloemba B et al (2021b) Face transplantation in a black patient – racial considerations and early outcomes. N Engl J Med 384(11):1075–1076. https://doi.org/10.1056/NEJMc2033961

Kauke M, Haug V, Obed D et al (2021c) Donor and recipient matching in facial vascularized composite allotransplantation: a closer look at the donor pool. Plast Reconstr Surg 148(1):194–202. https://doi.org/10.1097/prs.0000000000008094

Kauke M, Safi AF, Panayi AC et al (2022) A systematic review of immunomodulatory strategies used in skin-containing preclinical vascularized composite allotransplant models. J Plast Reconstr Aesthet Surg 75(2):586–604. https://doi.org/10.1016/j.bjps.2021.11.003

Kauke-Navarro M, Tchiloemba B, Haug V et al (2021) Pathologies of oral and sinonasal mucosa following facial vascularized composite allotransplantation. J Plast Reconstr Aesthet Surg 74(7):1562–1571. https://doi.org/10.1016/j.bjps.2020.11.028

Kauke-Navarro M, Panayi AC, Formica R et al (2022) Cytomegalovirus-related complications and management in facial vascularized composite allotransplantation: an international multicenter retrospective cohort study. Transplantation 106(10):2031–2043. https://doi.org/10.1097/tp.0000000000004132

Kauke-Navarro M, Knoedler S, Panayi AC, Knoedler L, Noel OF, Pomahac B (2023a) Regulatory T cells: liquid and living precision medicine for the future of VCA. Transplantation 107(1):86–97. https://doi.org/10.1097/tp.0000000000004342

Kauke-Navarro M, Knoedler S, Panayi AC et al (2023b) Correlation between facial vascularized composite allotransplantation rejection and laboratory markers: insights from a retrospective study of eight patients. J Plast Reconstr Aesthet Surg 83:155–164. https://doi.org/10.1016/j.bjps.2023.04.050

Kauke-Navarro M, Huelsboemer L, Klimitz FJ et al (2025a) A comparative analysis of lesional skin, sentinel flap, and mucosal biopsies in assessing acute face transplant rejection. Front Immunol 16:1562024. https://doi.org/10.3389/fimmu.2025.1562024

Kauke-Navarro M, Crisler WJ, Younis N et al (2025b) B-cell infiltration distinguishes mucosal from skin patterns of rejection in facial vascularized composite allografts. Am J Transplant 25:1193–1207. https://doi.org/10.1016/j.ajt.2025.01.013

Kiefer J, Zeller J, Schneider L et al (2025) C-reactive protein orchestrates acute allograft rejection in vascularized composite allotransplantation via selective activation of monocyte subsets. J Adv Res 72:401–420. https://doi.org/10.1016/j.jare.2024.07.007

Knoedler L, Knoedler S, Panayi AC et al (2023) Cellular activation pathways and interaction networks in vascularized composite allotransplantation. Front Immunol 14:1179355. https://doi.org/10.3389/fimmu.2023.1179355

Kollar B, Shubin A, Borges TJ et al (2018) Increased levels of circulating MMP3 correlate with severe rejection in face transplantation. Sci Rep 8(1):14915. https://doi.org/10.1038/s41598-018-33272-7

Kollar B, Kamat P, Klein HJ, Waldner M, Schweizer R, Plock JA (2019) The significance of vascular alterations in acute and chronic rejection for vascularized composite allotransplantation. J Vasc Res 56(4):163–180. https://doi.org/10.1159/000500958

Kollar B, Rizzo NM, Borges TJ et al (2020) Accelerated chronic skin changes without allograft vasculopathy: a 10-year outcome report after face transplantation. Surgery 167(6):991–998. https://doi.org/10.1016/j.surg.2020.01.010

Krezdorn N, Lian CG, Wells M et al (2019) Chronic rejection of human face allografts. Am J Transplant 19(4):1168–1177. https://doi.org/10.1111/ajt.15143

Lantieri L, Cholley B, Lemogne C et al (2020) First human facial retransplantation: 30-month follow-up. Lancet 396(10264):1758–1765. https://doi.org/10.1016/s0140-6736(20)32438-7

Lee CAA, Wang D, Kauke-Navarro M et al (2023) Insights from immunoproteomic profiling of a rejected full-face transplant. Am J Transplant 23(7):1058–1061. https://doi.org/10.1016/j.ajt.2023.04.008

Lin CH, Wang YL, Anggelia MR et al (2016) Combined anti-CD154/CTLA4Ig costimulation blockade-based therapy induces donor-specific tolerance to vascularized osteomyocutaneous allografts. Am J Transplant 16(7):2030–2041. https://doi.org/10.1111/ajt.13694

Lopez-Castejon G, Brough D (2011) Understanding the mechanism of IL-1β secretion. Cytokine Growth Factor Rev 22(4):189–195. https://doi.org/10.1016/j.cytogfr.2011.10.001

Mathes DW, Solari MG, Gazelle GS et al (2014) Stable mixed hematopoietic chimerism permits tolerance of vascularized composite allografts across a full major histocompatibility mismatch in swine. Transpl Int 27(10):1086–1096. https://doi.org/10.1111/tri.12380

Mitchell RN, Libby P (2007) Vascular remodeling in transplant vasculopathy. Circ Res 100(7):967–978. https://doi.org/10.1161/01.Res.0000261982.76892.09

Moktefi A, Hivelin M, Grimbert P et al (2021) Face transplantation: a longitudinal histological study focusing on chronic active and mucosal rejection in a series with long-term follow-up. Am J Transplant 21(9):3088–3100. https://doi.org/10.1111/ajt.16489

Morelon E, Petruzzo P, Kanitakis J et al (2017) Face transplantation: partial graft loss of the first case 10 years later. Am J Transplant 17(7):1935–1940. https://doi.org/10.1111/ajt.14218

Morelon E, Petruzzo P, Kanitakis J (2018) Chronic rejection in vascularized composite allotransplantation. Curr Opin Organ Transplant 23(5):582–591. https://doi.org/10.1097/mot.0000000000000571

Mundinger GS, Munivenkatappa R, Drachenberg CB et al (2013) Histopathology of chronic rejection in a nonhuman primate model of vascularized composite allotransplantation. Transplantation 95(10):1204–1210. https://doi.org/10.1097/TP.0b013e31828d1528

Murphy GJ, Bicknell GR, Nicholson ML (2003) Rapamycin inhibits vascular remodeling in an experimental model of allograft vasculopathy and attenuates associated changes in fibrosis-associated gene expression. J Heart Lung Transplant 22(5):533–541. https://doi.org/10.1016/s1053-2498(02)00571-5

Ng ZY, Lellouch AG, Rosales IA et al (2019) Graft vasculopathy of vascularized composite allografts in humans: a literature review and retrospective study. Transpl Int 32(8):831–838. https://doi.org/10.1111/tri.13421

Oh BC, Furtmüller GJ, Fryer ML et al (2020) Vascularized composite allotransplantation combined with costimulation blockade induces mixed chimerism and reveals intrinsic tolerogenic potential. JCI Insight 5(7):e128560. https://doi.org/10.1172/jci.insight.128560

Panwar V, Singh A, Bhatt M et al (2023) Multifaceted role of mTOR (mammalian target of rapamycin) signaling pathway in human health and disease. Signal Transduct Target Ther 8(1):375. https://doi.org/10.1038/s41392-023-01608-z

Petruzzo P, Kanitakis J, Testelin S et al (2015) Clinicopathological findings of chronic rejection in a face grafted patient. Transplantation 99(12):2644–2650. https://doi.org/10.1097/tp.0000000000000765

Pober JS, Chih S, Kobashigawa J, Madsen JC, Tellides G (2021) Cardiac allograft vasculopathy: current review and future research directions. Cardiovasc Res 117(13):2624–2638. https://doi.org/10.1093/cvr/cvab259

Ponticos M, Papaioannou I, Xu S et al (2015) Failed degradation of JunB contributes to overproduction of type I collagen and development of dermal fibrosis in patients with systemic sclerosis. Arthritis Rheumatol 67(1):243–253. https://doi.org/10.1002/art.38897

Puscz F, Dadras M, Dermietzel A et al (2020) A chronic rejection model and potential biomarkers for vascularized composite allotransplantation. PLoS One 15(6):e0235266. https://doi.org/10.1371/journal.pone.0235266

Rabbani PS, Rifkin WJ, Kadle RL et al (2019) Noninvasive monitoring of allograft rejection using a novel epidermal sampling technique. Plast Reconstr Surg Glob Open 7(8):e2368. https://doi.org/10.1097/gox.0000000000002368

Rivelli RF, Gonçalves RT, Leite M Jr et al (2015) Early withdrawal of calcineurin inhibitor from a sirolimus-based immunosuppression stabilizes fibrosis and the transforming growth factor-β signalling pathway in kidney transplant. Nephrology (Carlton) 20(3):168–176. https://doi.org/10.1111/nep.12368

Robbins NL, Wordsworth MJ, Parida BK et al (2019) Is skin the most allogenic tissue in vascularized composite allotransplantation and a valid monitor of the deeper tissues? Plast Reconstr Surg 143(4):880e–886e. https://doi.org/10.1097/prs.0000000000005436

Sawitzki B, Harden PN, Reinke P et al (2020) Regulatory cell therapy in kidney transplantation (the ONE study): a harmonised design and analysis of seven non-randomised, single-arm, phase 1/2A trials. Lancet 395(10237):1627–1639. https://doi.org/10.1016/s0140-6736(20)30167-7

Sayegh MH, Carpenter CB (2004) Transplantation 50 years later—progress, challenges, and promises. N Engl J Med 351(26):2761–2766. https://doi.org/10.1056/NEJMon043418

Shroka TM, Kufareva I, Salanga CL, Handel TM (2023) The dual-function chemokine receptor CCR2 drives migration and chemokine scavenging through distinct mechanisms. Sci Signal 16(770):eabo4314. https://doi.org/10.1126/scisignal.abo4314

Song EY, Barrow BE, Cendales LC (2024) Vascular changes in vascularized composite allotransplantation. Curr Opin Organ Transplant 29(6):363–367. https://doi.org/10.1097/mot.0000000000001184

Tchiloemba B, Kauke M, Haug V et al (2021) Long-term outcomes after facial allotransplantation: systematic review of the literature. Transplantation 105(8):1869–1880. https://doi.org/10.1097/tp.0000000000003513

Veglia F, Sanseviero E, Gabrilovich DI (2021) Myeloid-derived suppressor cells in the era of increasing myeloid cell diversity. Nat Rev Immunol 21(8):485–498. https://doi.org/10.1038/s41577-020-00490-y

Wang J, Zhao X, Wan YY (2023) Intricacies of TGF-β signaling in Treg and Th17 cell biology. Cell Mol Immunol 20(9):1002–1022. https://doi.org/10.1038/s41423-023-01036-7

Weissenbacher A, Hautz T, Zelger B et al (2014) Antibody-mediated rejection in hand transplantation. Transpl Int 27(2):e13–e17. https://doi.org/10.1111/tri.12233

Zeng C, Kahlenberg JM, Gudjonsson JE (2020) IL-17A softens the skin: antifibrotic properties of IL-17A in systemic sclerosis. J Invest Dermatol 140(1):13–14. https://doi.org/10.1016/j.jid.2019.07.678

Zhao M, Wang L, Wang M et al (2022) Targeting fibrosis, mechanisms and cilinical trials. Signal Transduct Target Ther 7(1):206. https://doi.org/10.1038/s41392-022-01070-3

Chapter 11
Evolving Therapeutic Approaches for Treatment of Antibody-Mediated Rejection in Renal Allograft Recipients

Stanley C. Jordan, Noriko Ammerman, Edmund Huang, and Ashley Vo

Abstract Antibody-mediated rejection (AMR) is now recognized as the primary cause of the decline of functional integrity and ultimately, graft failure in kidney transplants. Evolving understanding of effectors responsible for B-cell activation and plasma cell biology has led to the informed design of new therapeutics. Currently, several novel agents with varying mechanisms of action have emerged. These include imlifidase, FcRn inhibitors, anti-cytokine (IL-6/IL-6R) therapies, and plasma cell-directed therapies, especially anti-CD38s. As our understanding of critical effector pathways responsible for ARM grows, we now know that no single agent can control all facets of antibody generation and injury. The future of antibody-directed therapeutics will require protocols consisting of multiple agents positioned in a logical sequence to rapidly and durably remove pathogenic DSAs and block their re-emergence.

Abbreviations

ADCC	Antibody-dependent cellular cytotoxicity
AE	Adverse event
AMR	Antibody-mediated rejection
Anti-IL-6R	Anti-interleukin-6 Receptor
ATG	Anti-thymocyte globulin
BPAR	Biopsy-proven acute rejection
c-aAMR	Chronic active ABMR
CAR-T	Chimeric antigen receptor T cell
CDC	Complement-dependent cytotoxicity
cg	Chronic glomerulopathy

S. C. Jordan (✉) · N. Ammerman · E. Huang · A. Vo
Comprehensive Transplant Center, Cedars-Sinai Medical Center, West Hollywood, CA, USA
e-mail: stan.jordan@cshs.org

M. Kloc, R. M. Ghobrial (eds.), *Cellular and Molecular Mechanisms of Chronic Rejection in Organ Transplantation*, Results and Problems in Cell Differentiation 77, https://doi.org/10.1007/978-3-032-07686-1_11

CMR	T Cell-mediated rejection
CNI	Calcineurin inhibitor
cPRA	Calculated panel reactive antibody
DD	Deceased donor
DDKT	Deceased donor kidney transplant
DES	Desensitization
DSA	Donor-specific HLA antibody
eGFR	Estimated Glomerular filtration rate
ESRD	End-stage renal disease
FcRn	Fc Neonatal Receptor
FCXM	Flow crossmatch
FDA	Food & Drug Administration
G	Gram
HLA	Human leukocyte antigen
HS	Highly HLA sensitized
IgG1	Immunoglobulin G1
IL-6	Interleukin-6
IL-6R	Interleukin-6 Receptor
IVIG	Intravenous Immunoglobulin
KAS	Kidney allocation system
kg	Kilogram
KPD	Kidney paired donation
KT	Kidney transplant
LD	Living donor
M	Month
mAb	Monoclonal antibody
MCS	Mean channel shift
MFI	Mean fluorescence intensity
mg	Milligram
MM	Multiple myeloma
MMF	Mycophenolate mofetil
NS	Normal saline
PC	Plasma cells
PLEX	Plasmapheresis
q	Every
RCT	Randomized controlled trial
SAE	Serious adverse event
SCr	Serum creatinine
SOC	Standard of Care
subQ	Subcutaneous
TCZ	Tocilizumab
T_{fh}	T follicular helper
T-reg	T regulatory cell
U	Units
US	United States of America

11.1 Introduction

The detection and management of antibody-mediated rejection (AMR) after kidney transplantation has emerged as one of the most vexing and, as yet, unsolved problems facing transplant physicians and immunologist (Diebold et al. 2025; Hart et al. 2021a). AMR in acute and chronic forms usually leads to loss of functional integrity of the allograft, which contributes to poor graft survival (Loupy and Lefaucheur 2018; Hart et al. 2021b). Importantly, current immunosuppressive protocols are ineffective in the management of AMR (Schinstock et al. 2020; Irish et al. 2021) and there are currently no approved therapies. This results in a significant economic and emotional burden on patients and their families (Gosset et al. 2014). This reinforces the prescient need to develop and implement more effective therapies (Heeger et al. 2024).

Donor-specific antibodies (DSAs) are recognized as critical mediators of AMR. Here, DSAs arise after alloantigen exposure, which initiates germinal center activation of B cells that develop into IL-6- producing plasmablasts and mature plasma cells producing pathogenic complement-activating DSAs. Pathogenic DSAs bind to HLA molecules expressed on allograft endothelium, inducing microvascular inflammation (MVI) and allograft injury (Hart et al. 2021a; Loupy and Lefaucheur 2018). In addition to complement- dependent cytotoxicity (CDC), DSAs can also mediate graft damage through antibody-dependent cellular cytotoxicity (ADCC), where innate immune cells (natural killer (NK) cells) and monocytes/macrophages engage the Fc portion of DSA to Fcγ receptors (FcγR) (Loupy and Lefaucheur 2018; Heeger et al. 2024).

The search for effective therapeutics aimed at the prevention and treatment of AMR has been difficult. Over the past 20 years, a limited number of agents and treatments have been available to reduce DSA levels and deplete B cells and plasma cells. As noted in an important assessment of current therapies available for the treatment of AMR published in 2019, summarizing the Transplantation Society Working Group assessments of all therapeutic options, they concluded that there were no enough data to support any treatment and that optimizing baseline immunosuppression is the primary recommendation for late AMR, while apheresis combined with intravenous immunoglobulin remains the standard for early active AMR, despite limited evidence (Schinstock et al. 2020). Despite limited early successes, AMR treatment protocols have progressed considerably benefiting from the introduction of therapies originally developed for autoimmunity and B-cell/Plasma cell (PC) malignancies and aimed to reduce deleterious IgG molecules, target critical cytokines, and eliminate B cells and PCs. In this review, we will discuss emerging therapeutic options for treatment and prevention of AMR.

11.2 Advances in B-Cell Immunology Inform Therapeutic Approaches to AMR

11.2.1 B Cells and Plasma Cells

Understanding the contributions of B cells, PCs, and antibodies to alloimmune activation and progression of pathological responses is critical for the development of rational therapies for the prevention and treatment of AMR. Critical considerations for the development of AMR therapy should demonstrate reductions in DSAs, prevention of complement-dependent cytotoxicity (CDC), antibody-dependent cellular cytotoxicity (ADCC), and prevention or limitation of deleterious antibody rebound responses. Immune activation pathways critical for alloantibody generation are shown in Fig. 11.1a. Figure 11.1b shows the importance of CDC and ADCC in the mediation of AMR and how this can be addressed with emerging therapeutics. Several new agents have emerged to address this vexing immunologic problem. These are summarized in Fig. 11.2a, b.

11.2.2 Approaches to Treatment of AMR

Conventional approaches to AMR treatment include plasmapheresis (PLEX) and IVIg. PLEX is somewhat effective in removing circulating antibodies (along with other plasma proteins). Although PLEX is often considered a first-line therapy for AMR treatment, the data to support this are limited and derived from small, single-center studies (Allen et al. 1983; Bonomini et al. 1985; Kirubakaran et al. 1981). In a study of AMR treatment comparing the efficacy of pathogenic HLA antibody removal by PLEX + IVIg v. Imlifidase (IgG endopeptidase) + IVIg, imlifidase showed 97% efficacy for DSA reduction for imlifidase compared to 42% for PLEX. Additionally, imlifidase reduced DSA to noncomplement fixing levels, whereas PLEX failed to do so (Halleck et al. 2024). This information is promising, but the study failed to show benefits for AMR treatment likely due to excessive chronicity in both treatment groups and strong rebound of DSAs after cessation of treatment.

11.2.2.1 Importance of IVIg in Transplant Therapeutics

IVIg is derived from pooled plasma of thousands of donors and can modify pathogenic antibodies and effector cell functions, including inhibition of B-cell and T-cell proliferation, enhanced clearance of pathogenic IgG by saturating the Fc neonatal receptor (FcRn) (Fernández-Cruz et al. 2009; Stapleton et al. 2015), significantly reducing antibody half-life, neutralizing pathogenic antibody effector functions via idiotypic or anti-idiotypic interactions, inhibiting or limiting complement

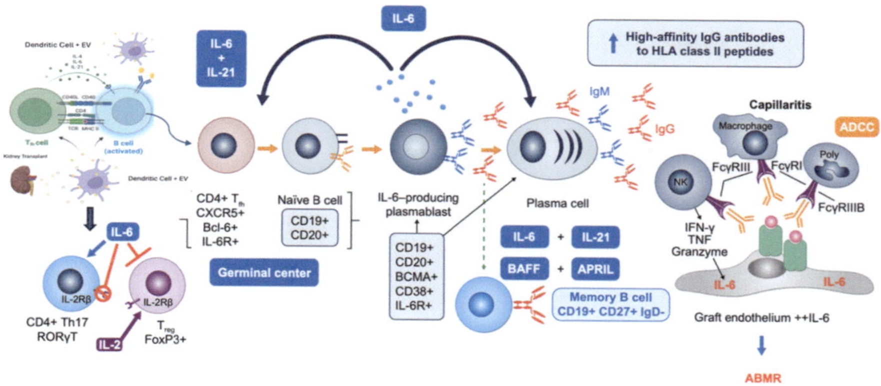

Fig. 11.1 (**a**) The immune activation pathways critical for the development of alloimmunity. Antigen-presenting cells (APCs) process and present alloantigens to naïve CD4+ T cells. Recent data suggest that extracellular vesicles released from damaged allografts, complement coated and carrying alloantigens are disseminated throughout the body and are processed by APCs in the spleen and lymph nodes. This process requires alloantigen binding to cognate T-cell receptor (TCR) complexes followed by costimulation through CD80/86 (APC) and CD28 (naïve T cell). Activated CD4+ cells then migrate to regional lymph nodes and spleen. Here, under the influence of CXCR5, they mature into T follicular (T_{FH}) cells that stimulate alloantigen responses in naïve B cells stimulated by the cytokines (IL-6 and IL-21). Activated B cells develop into B_{MEM} cells with specific cells evolving into plasmablasts and ultimately to antibody-producing plasma cells. BAFF and APRIL are also critical to this process, interacting with BCMA (B-cell maturation antigen on plasma cells and some B cells). Antibody affinity to graft alloantigens (DSAs) evolves from low-affinity IgG (purple) to high affinity, complement-activating alloantibodies (in red) binding to donor-specific HLA molecules in the allograft and initiates the clinical and pathologic features of AMR. The antigenic anatomy of each cell type is shown, which could represent relevant targets for therapeutic intervention. AMR, antibody-mediated rejection; APC, antigen-presenting cells; IL, interleukin; T_{FH}, T follicular helper; BAFF (B cell-activating factor of TNF family); APRIL (a proliferation-inducing ligand); BCMA (B-cell maturation antigen). (**b**) This figure demonstrates the importance of ADCC and CDC as critical antibody effector pathways responsible for endothelial cell injury and ultimately, antibody rejection. First, ADCC is mediated by NK cells, monocytes, and polymorphonuclear leukocytes through recognition of Fc fragments of pathogenic DSAs bound to endothelial cell targets (HLA molecules). Interaction of FcγR+ cells with Fc of target-bound DSAs initiates cell activation with the release of perforins and granzyme, in initiating target cell destruction. NK cells also release chemokines and cytokines that result in the recruitment of monocytes and macrophages that increase ADCC and cell adhesion to the target endothelium. CDC is mediated primarily by the activation of the classic complement pathway by pathogenic DSAs. This activation ultimately results in the formation of the C5b-C9 membrane attack complex (MAC), which destroys target cells. During complement activation, potent anaphylatoxins (C3a and C5a) are released and are critical for initiation of monocyte and NK cell recruitment. In addition C3a and C5a bind to receptors on B cells and T cells, resulting in activation of T cell-directed immune responses and enhancement of high-affinity DSA production. Important therapeutic agents to combat ADCC and CDC include imlifidase, which removes Fc from DSAs and thus inhibits their ability to mediate these important injurious pathways. This is also likely true for the new Fc receptor neonatal (FcRn) inhibitors, which rapidly reduce circulating IgG (DSA) by inhibiting IgG recycling mechanisms. This would also be true for high-dose IVIG. In addition, anti-complement therapies (anti-C5, anti-C1s, and C1INH) inhibit classic, mannose binding and alternative pathways, limiting CDC and activation of immune cells. ADCC can also be inhibited by the use of anti-CD38, which can deplete NK cells and monocytes/macrophages as shown. (Modified from: Vo et al. (2024) with permission)

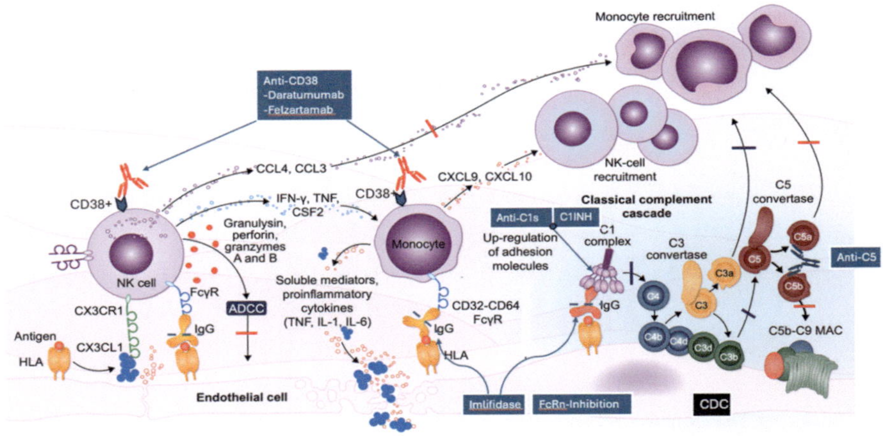

Fig. 11.1 (continued)

activation, and activating immune inhibitory signaling through FcγRIIb receptors present on certain immune cells (Perez et al. 2017). Thus, IVIg remains an essential element of AMR protocols. In addition to the immune regulatory pathways, IVIg also provides passive immunity against most infectious agents, including COVID-19 and variants (Jordan et al. 2022a). It will also be an important adjunct for patients treated with B cell- and plasma cell-depleting agents, as hypogammaglobulinemia is a common side effect of these therapies.

IVIg was initially evaluated for desensitization as reported in a multicenter randomized controlled trial (RCT) (Jordan et al. 2004). Here, IVIg, 2 g/kg monthly for 4 months before and 4 months after transplant, was effective in reducing HLA antibodies and increasing transplant rates (35% IVIg vs. 17% placebo; $p < 0.05$) in highly HLA-sensitized (HS) kidney transplant candidates (Jordan et al. 2004). Montgomery et al. demonstrated a significant survival benefit for HS patients who received living donor kidney transplantation (LDKT) after desensitization with low-dose IVIg (100 mg/kg) and plasma exchange (PLEX) compared to HS patients who were transplanted without desensitization or who remained on dialysis (8-year patient survival of 80.6%, 49.1%, and 30.5%, respectively; $p < 0.001$) (Montgomery et al. 2000). These initial studies led to the implementation of these basic protocols for desensitization (Gloor et al. 2003; Lim et al. 2019). In addition, these protocols were rapidly adopted for the treatment of AMR, as this was a frequent complication of desensitization protocols after transplantation. A better understanding of the IgG-Fc/FcRn interaction has allowed exploration of newer therapies with higher affinity to block FcRn and, therefore, decrease pathogenic anti-HLA antibodies or enhance the therapeutic efficacy of other monoclonals (Seijsing et al. 2018). These may find application in AMR treatment in the future (Jordan et al. 2020).

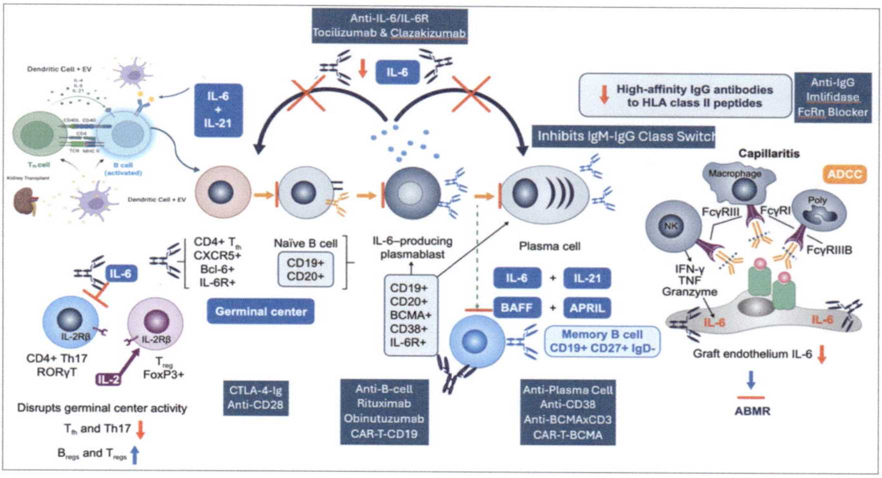

Fig. 11.2 (**a**) Figure 11.2a shows therapeutic approaches for the prevention and treatment of antibody-mediated rejection that are shown in conjunction with their targeted alloantigen activation and effector pathways. As discussed in the text, the major pathogenic factor is high-affinity IgG complement-activating DSAs. Initial alloactivation events leading to T-cell activation can be blocked by CTLA4Ig or anti-CD28. Importantly, these costimulatory blockers are also potent inhibitors of germinal center (T_{FH}) activity and can prevent primary immune events induced by alloantigens. New data also suggest that CTLA4-Ig may directly inhibit CD80/86+ PCs. Inebilizumab (anti-CD19) depletes B cells, PB, and ~50% of PC, anti–IL-6 (clazakizumab) or anti-IL-6R (tocilizumab) block activation of T_{FH} cell, Th17, and PB. Anti-PC therapies inhibit the production of complement-activating DSAs; these include daratumumab, isatuximab (anti-CD38), inebilizumab (anti-CD19), and REGN5459 (bispecific anti-BCMA × CD3). Once pathogenic DSAs are present in copious amounts, imlifidase can cleave all IgG molecules, inhibiting CDC/ADCC. FcRn is an HLA class 1-like molecule that is present in most cells and is responsible for recycling IgG molecules. Inhibition of FcRn with monoclonals, Fc fragments, or saturation with IVIg enhances pathogenic IgG degradation, limiting pathogenesis. Complement inhibitors (C1INH and anti-C5) can inactivate effector pathways critical to allograft injury (CDC). Combining antibody reduction therapies along with inhibition of B-cell activation and plasma cell DSA production represents an achievable pathway for the prevention and treatment of AMR. PC, plasma cells; PB, plasma blasts; IL-6R, interleukin-6-receptor; IL-6, interleukin-6-ligand. Figure 11.2b shows the mechanism of action of a novel bispecific antibody (Anti-BCMA × CD3) These engineered IgG monoclonals have one arm with binding site for the plasma cell-specific target, B-cell maturation antigen (BCMA), and the other arm specific for the pan-T-cell antigen CD3. Here, the monoclonal mimics CAR-T cells except that it brings naïve CD3+ T cells in proximity to BCMA+ plasma cells. The T cells are activated and destroy the plasma cells. The drug REGN 5459 is being investigated as a desensitization agent and could have utility in the treatment of c-aAMR (Fig. 11.2b developed from Biorender, with permission)

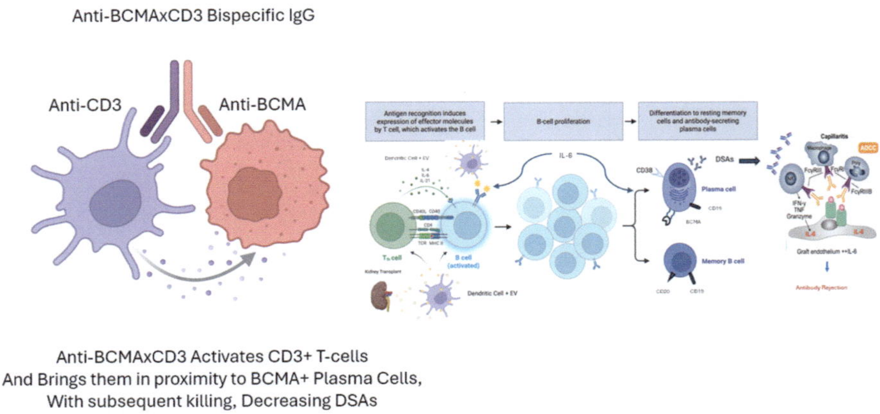

Fig. 11.2 (continued)

Toyoda et al. (2017) also demonstrated that IVIg had benefits in reducing the risk of opportunistic infections in patients undergoing desensitization. However, both high-dose IVIg and PLEX + low-dose IVIg cannot modulate B memory cells (B_{MEM}), plasmablast, and PC rebound activity, which limits efficacy and increases the risk for post-transplant AMR.

11.3 Importance of B-Cell Activation and Critical Effector Pathways (ADCC and CDC) in Antibody Rejection

11.3.1 Understanding Mechanisms Responsible for Allosensitization

Recent data by Chen et al. (2025) helped explain why allografts can be so immunogenic to recipients and result in intense immune activation, resulting in pathogenic DSA generation. In an experimental model of heart transplantation, the investigators showed that transplants release copious amounts of extracellular vesicles (EV) that disseminate donor alloantigens throughout the body, resulting in consumption by follicular dendritic cells with antigen presentation to T and B cells. Here, complement binding to the EVs is critical for opsonization by the follicular dendritic cells. This opens several important considerations, suggesting that early intervention with complement blockers or B-cell depletion could significantly alter the immune activation events and DSA generation.

11.3.2 Importance of CDC and ADCC in Mediating Allograft Injury

Once pathogenic DSAs are generated, they mediate allograft injury through ADCC and CDC, which are critical effector pathways responsible for injury and destruction of cognate antigen-binding targets. ADCC consists of innate immune cells (i.e., NK cells and macrophages/monocytes) that express IgG Fc receptors that seek FcγR of antibodies bound to alloantigen targets and, when engaged, activate the release of cytokines (IFN-γ and TNF-α) and initiate cytotoxic enzyme release (perforin and granzyme), destroying the endothelium and perpetuating graft injury. In kidney transplantation, the prime targets are HLA antigens on the donor vascular endothelium. For CDC-mediated injury, critical segments of the Fc region of DSA activate C1Q and initiate the classic complement pathway to form a C5b-C9 membrane attack complex (MAC), which rapidly destroys endothelial cells and results in graft injury. Both processes are significant and can arise in concert (or individually). However, in early severe graft rejection, CDC is more prevalent while in c-aAMR, ADCC likely plays a more prominent role (Heeger et al. 2024).

11.3.3 Importance of B Cell-Focused Therapies for AMR

The emergence of B cell-directed therapy for desensitization and AMR treatment leveraged the critical nexus between B-cell responses in AMR and decreased allograft survival (Lynch et al. 2013; Zachary et al. 2013; Jordan et al. 2011; Genberg et al. 2006). In AMR, these protocols include anti-CD20 (rituximab or biosimilar equivalents) as a critical component, depleting B cells and preventing recall antibody responses by B_{MEM} cell depletion (Gloor et al. 2003; Lynch et al. 2013; Jordan et al. 2011; Zhao et al. 2014; Lobashevsky et al. 2013).

CD20 is an antigen expressed on B cells and some plasmablast, but not plasma cells (Heeger et al. 2024). Rituximab is a first-in-class anti-CD20 originally developed for the treatment of B-cell malignancies that has found importance in the treatment of autoimmune diseases and AMR. Our group first decribed the importance of rituximab as a desensitization agent (Vo 2008, 2013). Loupy et al. (2010) showed a reduction in DSA, cAMR, and transplant glomerulopathy at 1 year in DDKT recipients (Loupy et al. 2010). In an RCT, a substantial decrease in rejection was seen in high-risk patients receiving rituximab induction (17.9% versus 38.2% placebo, $p = 0.004$) (van den Hoogen et al. 2015). Comparable results were observed in a meta-analysis including 589 HS patients; patients treated with rituximab pre-transplant had significantly fewer AMR episodes post-transplant and higher 1-year graft survival rates (Freeman and Sehn 2018). This is consistent with work reported by others as well (Zhao et al. 2014).

11.3.3.1 Advancement in Anti-CD20 Antibody Engineering for AMR Treatment

Important advancements have occurred in the development of genetically engineered anti-CD20 antibodies. The prototype is obinutuzumab (type II anti-CD20), which was clinically superior to rituximab in the treatment of B-cell lymphoma and systemic lupus erythematosus treatment (Freeman and Sehn 2018). In an open-label phase 1b study of obinutuzumab in HS patients with ESRD, IVIg + obinutuzumab for desensitization did not demonstrate efficacy in diminishing anti-HLA antibodies in HS patients (Redfield et al. 2019). However, obinutuzumab did demonstrate substantial reductions of B cells, $B_{MEM,}$ and plasmablasts. This suggests that obinutuzumab could be effective in modifying antibody rebound responses from B_{MEM} cells and plasmablasts. A recent report examining obinutuzumab treatment in HS patients undergoing desensitization or receiving AMR treatment showed that obinutuzumab had minimal effects on the B-cell complement-dependent cytotoxicity (CDC) crossmatch. This suggests that obinutuzumab depletes B cells predominantly via antibody-dependent cellular cytotoxicity (ADCC) and antibody-dependent phagocytosis as opposed to rituximab (NasrAllah et al. 2022). However, subsequent data from our HLA laboratory did not confirm these findings (Zhang et al. 2021). Obinutuzumab is very efficient in decreasing or eliminating autoantibodies vs. alloantibodies. This likely represents an important immunologic difference in B-cell immune responses in autoimmunity vs. alloantigen targeting in AMR. Here, autoantibodies are predominantly produced by plasmablasts, while HLA antibodies are generated by long-lived PCs, which are CD20-deficient. In addition, it has become apparent that HLA antibodies are present in very high titers in HS patients. This may represent a persistence of immune stimulation due to the presence of allogeneic material in the HS patients (i.e., rejected allografts). This remains a major obstacle to the advancement of desensitization therapies and treatment of AMR.

11.4 Advancements in Treatment of AMR

11.4.1 Inhibiting Interleukin-6 and Interleukin-6 Receptor

Interleukin-6(IL-6) is a multifunctional cytokine critical for T follicular/helper cells T_{FH} and adaptive immunity (Hunter and Jones 2015; Tanaka et al. 2014). A review of data regarding the benefits of inhibiting the IL-6/IL-6R signaling pathway suggests that it would likely result in lowering pathogenic IgG antibodies and T effector/memory responses while enhancing T_{REG} populations. In addition, recent data show that anti-HLA class II antibodies binding to human endothelial cells stimulate copious IL-6 production, leading to vascular injury and obliterative vasculopathy. Here, anti-IL6 treatment limited endothelial cell responses to anti-HLA antibodies, resulting in a reduction of AMR-mediated pathologic injury (Jordan et al. 2022b).

11.4.1.1 Tocilizumab (Anti-IL-6R)

Tocilizumab (TCZ) is a humanized anti-IL-6R monoclonal antibody (mAb). TCZ + IVIg was initially evaluated for desensitization in 10 HS kidney transplant recipients who previously failed standard desensitization (Vo et al. 2015a). No AMR was seen on 6-month protocol biopsies and DSA reductions were seen for some antibodies. Daligault et al. reported on TCZ administration to treatment-naïve HS patients ($N = 14$) (Daligault et al. 2021). TCZ significantly decreased dominant anti-HLA antibody sensitization. TCZ has not been formally evaluated in controlled clinical trials for the treatment of AMR. However, there are several significant reports of clinical experience with TCZ for the treatment of AMR and c-aAMR. Choi et al. from our group (2017) reported on the first study of TCZ as a rescue therapy for c-aAMR in 36 KT recipients who were resistant to the standard of care. Graft- and patient survival rates were excellent, (80% and 91%) at 6 years after AMR diagnosis. We also showed a significant decrease in DSA, C4d deposition, and microvascular infiltration after TCZ treatment, whereas the chronic glomerulopathy (cg) score and renal function remained stable. More recently, Noble et al. (2025) reported on the use of TCZ for the treatment of patients with DSA+ c-aAMR and patients with microvascular inflammation DSA (−)/C4d (−) suspicious for AMR. This is an important study since it involves a large number of patients with well-defined pathology and DSA status. Estimated glomerular filtration rate (eGFR) and DSA were assessed 1 year before and after TCZ initiation. The study included 64 patients who received TCZ between 2018 and 2023. Here, the eGFR pretreatment decline significantly decreased after TCZ treatment (-1.2 ± 0.2 vs. 0.03 ± 0.2 mL/min/1.73 m^2/month pre- vs. post-TCZ, respectively; $p = 0.001$). The proportion of patients with DSA decreased from 63.9% to 38.9% ($p < 0.001$), and the average MFI decreased from 9537 to 7250 ($p = 0.001$). Patient survival was 98.4%, and graft survival was 93.7% at 1 year. The authors conclude that first-line therapy with TCZ for c-aAMR or MVI + DSA-C4d- is associated with an improvement in eGFR trajectories, reduced DSA numbers, and MFI and histological inflammation in glomeruli. Pearl et al. (2022) have shown similar findings in the stabilization of eGFR in pediatric patients with c-aAMR treated with monthly TCZ. Other authors have shown the benefit of TCZ in kidney transplant patients with c-aAMR (Boonpheng et al. 2023) that is associated with reductions in cell-free donor-derived DNA (cf-dd-DNA) levels suggesting modification of injury to the allograft endothelium. Importantly, there are reports of beneficial effects on the treatment of AMR in cardiac and lung transplant recipients (Pottebaum et al. 2024; January et al. 2023). These studies have renewed interest in a more formal evaluation of TCZ for the treatment of c-aAMR. Pending studies of TCZ for the treatment of c-aAMR in kidney transplant patients include the (INTERCEPT) trial, which will enter 50 patients with c-aAMR randomized to receive placebo vs. TCZ with standard assessments as outlined above (Streichart et al. 2024). Another important trial that has completed enrollment and is awaiting assessment is an NIH randomized placebo-controlled trial of TCZ for the prevention of cardiac allograft vasculopathy.

This is a prospective, multicenter phase 2 clinical trial that included 200 primary heart transplant recipients randomized (1:1) to receive either TCZ or placebo plus standard triple maintenance immunosuppression (NCT03644667).

11.4.1.2 Clazakizumab (Anti-IL-6)

Clazakizumab, an IgG1 anti-IL-6 mAb (Vo et al. 2022), was evaluated for desensitization at our center. Here, 20 HS patients (cPRA >50%) received PLEX + IVIg, and then clazakizumab monthly × 6 M (Vo et al. 2022). If transplanted, patients received clazakizumab × 12 M. After 6 M, clazakizumab treatment was associated with significant declines in HLA antibodies, including strong-binding HLA antibodies (>10,000 MFI) with 18 patients transplanted. After transplantation, only one patient remained DSA-positive at 12 M. At 12 M, patient and graft survival rates were 100% and 94%, and mean eGFR was 58 ± 29 ml/min/1.73m^2. Clazakizumab was also evaluated in two phase II studies in patients with c-aAMR (Doberer et al. 2021; Jordan et al. 2022c), which showed diminished DSA levels, pathologic and molecular features of cAMR, and, importantly, eGFR stabilization. However, interim data analysis from a phase 3 randomized placebo-controlled trial of clazakizumab for c-aAMR treatment in kidney transplant patients (IMAGINE) showed no benefit at 1 year (Nickerson 2022). The reasons for this are unclear; however, only 50% of the doses used in the desensitization and c-aAMR phase II trial were used in the IMAGINE study. Nonetheless, anti-IL-6/IL-6R treatments will likely remain important agents in desensitization, DSA rebound prevention, and treatment of c-aAMR in certain populations.

11.5 Evolution of Plasma Cell-Directed Therapies for Treatment of AMR

Since long-lived PCs are thought to be the prime producers of HLA antibodies, it seems rational to develop PC-directed therapies. Here, we will discuss important advances in plasma cell-directed therapies.

11.5.1 Proteosome Inhibitors

11.5.1.1 Bortezomib

Bortezomib, a proteosome inhibitor (PI) used to treat multiple myeloma (MM), was studied for AMR treatment. Unfortunately, data from a well-conducted, placebo-controlled trial exhibited no meaningful ability to reduce DSAs, stabilize renal function, or impact patient/graft survival (Eskandary et al. 2018a). Additionally, studies showed considerable AE/SAEs with bortezomib administration (Moreno Gonzales et al. 2017).

11.5.1.2 Carfilzomib

Carfilzomib is a second-generation proteosome inhibitor (PI) that is more robust in depleting PCs with fewer neuropathic side effects. A study of carfilzomib in lung transplant recipients with AMR appeared to reduce DSAs and improve lung function (Tremblay et al. 2020). One-year graft survival for patients responsive to carfilzomib treatment was 78% vs. 20% for nonresponders. Another study of lung transplant recipients demonstrated similar results (Pham et al. 2021). No deaths occurred during treatment, but at 1 year, 7 of 14 patients died from graft failure. Although encouraging, it appears that additional therapy with agents aimed at preventing DSA rebound and allograft injury will likely be necessary to sustain the initial benefits seen with carfilzomib (Ensor et al. 2017).

11.6 Plasma Cell-Directed Monoclonal Antibodies

11.6.1 Daratumumab (Anti-CD38)

Daratumumab is a humanized IgG1κ mAb with specificity for CD38 transmembrane glycoproteins. Daratumumab was originally developed as a plasma cell/plasmablast-depleting agent, but it also can deplete several other cell types that express CD38. This includes natural killer (NK) cells, monocytes/macrophages, some B cells, and T_{REG} cells (Krejcik et al. 2016). Importantly, subsets of B_{MEM} cells are CD38-. The use of daratumumab in a primate model of allosensitization reduced DSAs and marginally improved renal allograft survival (Kwun et al. 2019), but a swift antibody rebound with extreme T cell-mediated rejection (CMR) was seen at 1 M. Our group reported using daratumumab for desensitization in an HS heart transplant candidate and resistant AMR treatment in a kidney transplant recipient (Jordan et al. 2019). HLA antibodies diminished in both patients. AMR findings improved in the kidney transplant patient post-daratumumab, but CMR developed on subsequent biopsy. These findings were also reported by others (Joher et al. 2021; Scalzo et al. 2023). Although promising, daratumumab exhibits numerous off-target immune effects (i.e., CD38+ T_{REG} and B_{REG} cell depletion), which could increase CMR risk (van de Donk 2018). The ability to deplete NK cells and monocytes may account for improvements in glomerulitis and PTCitis observed after daratumumab treatment of AMR.

11.6.2 Felzartamab

Recently, Mayer et al. (2024) presented data from a phase II trial of the anti-CD38 monoclonal felzartamab for the treatment of AMR/c-aAMR patients. This was a double-blind, randomized, placebo-controlled trial, examining the safety and limited efficacy of felzartamab in 22 adult patients with AMR. Patients participating in the trial had a biopsy-proven AMR and were at least 180 days post-transplantation. All had HLA donor-specific antibodies, and the majority had c-aAMR at presentation. Patients received nine infusions of felzartamab (16 mg/kg) or placebo for 6 months, followed by a 6-month observation. Important secondary outcomes included renal biopsy results at 6 M and 12 M, DSA levels, peripheral NK-cell counts, and dd-cf-DNA levels. Serious adverse events (SAEs) occurred in 1 patient in the felzartamab group and 4 patients in the placebo group. The treatments were overall well tolerated without infectious events.

After 6 M treatment, with no meaningful DSA reductions, resolution of morphologic and molecular features of AMR was seen more frequently in the felzartamab-treated patients, 9 of 11 [82%], while placebo showed resolution in only 2 of 10 [20%] patients. Another important observation was the value of the noninvasive biomarker dd-cf-DNA in predicting responses to AMR treatment with felzartamab. The level of dd-cf-DNA was significantly lower in patients treated with felzartamab. At 12 M, (6 M after no therapy), AMR recurrence was seen in 3 of 9 patients, but most patients showed increased molecular scores of AMR on repeat biopsy. Despite this, the 1-year eGFR slope was −0.39 ml/min/1.73m^2 in the felzartamab group and −4.53 ml/min/1.73m^2 in the placebo group. Thus, felzartamab treatment resulted in a reduction of histologic and molecular features of AMR, dd-cf-DNA, and stabilization of eGFR compared to placebo without DSA reductions. No cases of CMR were seen. These observations are encouraging but certainly lead to some confusion regarding the role of DSAs in the mediation of c-aAMR. How can this be scientifically accounted for?

11.7 NK Cells: Important in Mediation of AMR

Human NK cells are cytotoxic effector cells that represent a critical component of innate immunity and are primarily responsible for cancer surveillance and elimination of virally infected cells (Heeger et al. 2024). NK cells have been increasingly implicated as effector cells in AMR. Lamarthe'e et al (2023) used transcriptional and spatial profiling of kidney allografts with AMR to investigate the role of FcγRIII+ NK and FcγRIII+ nonclassical monocytes (also CD38+) in AMR, with specificity to the glomerular area (Lamarthée et al. 2023). The authors concluded that these results highlight a critical role of innate cellular immunity in the pathogenesis of AMR and identify potential targets for the prevention and treatment of AMR that could improve allograft outcomes. This study could support the thesis that FcγRIII+ NK and nonclassical monocytes are primary mediators of injury in AMR. Thus, the use of anti-CD38 could eliminate these effector cells that are critical to the mediation of ADCC.

One of the most important and unique features of the felzartamab study was the measurement of circulating $CD16^{bright}$ natural killer (NK) cells in the peripheral blood of patients treated with felzartamab vs. placebo. Here, the authors found a significant reduction of $CD16^{bright}$ NK cells after felzartamab treatment that was sustained at 12 M. Thus, peripheral blood $CD16^{bright}$ NK cell assessments appear to be an important indicator of treatment efficacy, even more so than DSAs. Since DSAs were not reduced, we can postulate that the depletion of CD38+ $CD16^{bright}$ NK cells and possibly other ADCC-inducing cells was responsible for the reduction of inflammation and improved eGFR in treated patients. This concept is quite revolutionary, and like so many things in medicine, a therapy aimed at a putative target (i.e., DSAs) ends up working by another mechanism (i.e., NK-cell depletion). The serendipity here is that eliminating ADCC effectors appears to be an effective treatment of AMR. These concepts are now being evaluated in a Phase III trial (TRANSCEND, NCT06685757), which could result in the first anti-CD-38 monoclonal approval for the treatment of c-aAMR.

11.8 Costimulatory Blockade and Combination Therapy

Costimulatory blockade primarily inhibits naïve CD4+ T-cell activation by impeding CD28/B7-1/B7-2 costimulation (Vincenti et al. 2016). The fusion protein (CTLA4-Ig) interferes with signal 2 costimulatory events. CTLA4-Ig is a potent inhibitor of naïve B-cell activation by T_{FH} cells in germinal centers. This likely explains the advantages of CTLA4-Ig in preventing dnDSA development (Kim et al. 2016). However, CTLA4-Ig has little or no impact on B_{MEM} or plasma cell function (Kim et al. 2016). Importantly, CTLA4-Ig does not inhibit CD8+ T effector/T memory, which likely accounts for the high rates of rejection seen in patients receiving CTLA4-Ig without CNI immediately post-transplant. Recent studies in a primate model of kidney transplant demonstrated the inability of CTLA4-Ig to modify CD8+ T-effector/memory cells in animals maintained on CTLA4-Ig alone. However, this was reversed with tocilizumab co-administration (Kubo et al. 2023) with significant reductions in CD8+ T-effector/memory cell populations.

Jain et al. evaluated the efficacy of bortezomib + CTLA4-Ig in 6 kidney transplant patients as rescue therapy for acute AMR or mixed rejection (Jain et al. 2020). DSA (classes I and II) reductions were seen at 25 days of treatment. DSA levels remained low or undetectable for 10–30 M. Thus, the combination of costimulatory blockade and anti-plasma cell therapies may be promising therapeutic agents for the treatment of AMR.

11.9 Bispecific Antibody Targeting BCMA × CD3

REGN5459/5458 are anti-B-cell maturation antigen (anti-BCMA) × anti-CD3 bispecific antibodies, binding BCMA on PCs and CD3 on cytotoxic T lymphocytes. Bispecific monoclonals represent a novel approach for B-cell/plasma cell depletion that in many ways would mimic CAR-T-cell therapy. The bispecific approach brings CD3+ cells in proximity to PCs with subsequent induction of PC death by the activated CD3+ cells. A dose escalation, multicenter, phase 1/2, open-label study of REGN5459 to desensitize HS ESRD patients is currently ongoing (Study for Desensitization of Chronic Kidney Disease in Adult Patients in Need of a Kidney Transplant Who Are Highly Sensitized to Human Leukocyte Antigen n.d.). Patients receive weekly infusions X 3 and, if transplanted, are followed up in a noninterventional, observational study to collect safety and data outcomes. REGN5459/5458 may provide advantages over anti-CD38, which targets multiple cell types, since BCMA is primarily located on PCs. Here, it is important to discern the antigenic anatomy of PCs responsible for HLA DSA generation. If they explicitly express BCMA, this method could be promising. The question remains regarding the possibility that BCMA- PCs and/or BCMA- B_{MEM} cells could continue producing HLA antibodies, limiting therapeutic efficacy. Regardless, important information will be obtained about the biology of HLA antibody production and effector functions. The mechanism of action associated with bispecific antibodies to BCMA × CD3 depleting plasma cells is depicted in Fig. 11.2b.

11.10 CAR-T-Cells CAR-T (CD19) and CAR-T(BCMA)

In another important attempt to translate lessons learned from cancer therapeutics, investigators have developed studies to assess chimeric antigen receptor (CAR) T cells (CD19 & BCMA) as potential desensitization agents (NCT06056102). However, there is little information on this approach. Jackson et al. reported a retrospective analysis of HLA antibodies in HS patients receiving CAR-T(CD19) treatment for MM (Hill et al. 2023). However, the investigators saw no significant reductions in HLA antibodies after CAR-T(CD19) treatment. CAR-T(BCMA) impact on HLA antibodies was not analyzed since patients were not sensitized. Investigators determined that HLA antibody-producing cells are likely $CD19^{NEG}$. CD19 is expressed on all B cells and ~50% of PCs. This is similar to the experience with obinutuzumab (anti-CD20), which diminishes plasmablasts and some PCs, without affecting HLA antibodies in HS patients (Redfield et al. 2019; Looney et al. 2023). However, Zhang et al. (2023) recently reported on the use of CAR-T (CD19) and CAR-T (BCMA) in animal models of desensitization and in highly HLA-sensitized patients. Here, investigators evaluated the ability of (CAR) T cells targeting CD19 and BCMA to eliminate alloantibodies in a skin pre-sensitized murine model of islet allotransplantation. They reported that treatment of allo-sensitized mice with CAR T cells targeting B_{MEM} and long-lived plasma cells (LLPCs) eliminates DSAs and prevents hyperacute rejection of subsequent islet allografts. They next assessed the clinical efficacy of CAR T therapy for desensitization in patients with multiple myeloma (MM) who were known to have pre-existing HLA antibodies. Here, patients were treated with a combination of CART-BCMA and CART-19. In three subjects who received treatment with CAR-T (CD19) and CAR-T (BCMA), HLA antibodies declined by 47–97% within 12 months following the CAR T-cell infusion (ClinicalTrials.gov: NCT03549442). These reductions certainly are clinically meaningful and could improve transplant rates for HS patients. However, to date, there are no reports of HS patients receiving kidney transplantation after CAR-T-cell desensitization. It is important to decide if there are therapeutic differences in efficacy of CAR-T cells compared to bispecific antibodies, which mimic CAR-T-cell actions (Zhang et al. 2023). In addition, recent data have demonstrated that anti-CD19 monoclonal antibodies (inebilizumab) are effective in modifying autoimmune diseases and reducing pathogenic antibodies (Nowak et al. 2025). To our knowledge, there are no reports of anti-CD19 or anti-CD19 × CD3 monoclonals in desensitization or treatment of c-aAMR.

11.11 Therapies That Target Antibody Reduction: Imlifidase

Imlifidase, or IdeS (IgG-degrading enzyme derived from *Streptococcus pyogenes*), is a novel IgG endopeptidase with promise as a desensitizing agent for HS patients (Huang et al. 2022; Jordan et al. 2017, 2021). Imlifidase cleaves IgG antibodies at the IgG hinge region, resulting in two fragments, F(ab′)2 and Fc. This enzymatic digestion of the intact IgG components transiently eliminates IgG-mediated ADCC and CDC. Imlifidase cleaves the B-cell receptor from B cells and selectively cleaves IgG antibodies without affecting other immunoglobulin molecules. This prevents pan-hypogammaglobulinemia and sustains the beneficial immune responses rendered by IgM, IgE, and IgA. An in-depth review of imlifidase in transplantation was recently published (de Weerd et al. 2024).

In a phase I/II trial, we reported the combined experience from two open-label studies (USA and Sweden) administering imlifidase to HS patients before transplantation ($N = 25$). We observed imlifidase administration completely removed IgG HLA antibodies with 24 patients transplanted (Jordan et al. 2017). Kjellman et al. published the combined 3-year data for the 39 FCXM+ patients receiving imlifidase before kidney transplant in the four open-label, phase 2 studies (Kjellman et al. 2021). At 3 years, for patients who were AMR+ vs. AMR−, patient survival was 85% vs. 94%, and mean eGFR was 49 ml/min/1.73m^2 vs. 61 ml/min/1.73m^2, but death-censored allograft survival was 93% vs. 77%, respectively. Briefly, 5-year follow-up of crossmatch + HS patients transplanted with imlifidase treatment showed that patient survival was 90% and graft survival 82%. The mean eGFR was 50.1 ml/min/1.73m^2. Most patients had no DSAs or low-level DSAs and there were no reported AMR episodes between the 3-year and 5-year analyses (Jordan et al.

2024). We feel these data continue to demonstrate reasonable patient and graft survival with stable renal function and indicate that imlifidase is a viable treatment option for HS patients whose breadth and depth of HLA sensitization prevent their access to life-saving kidney transplantation. Data from this and other studies resulted in conditional approval of imlifidase by the European Medicines Agency for desensitization of crossmatch-positive individuals awaiting DDKT (Al-Salama 2020). Imlifidase is being evaluated in a phase III trial conducted in the USA (ConfIdeS) for desensitization in HS (>99.90%) kidney transplant candidates (NCT05369975) (Renal Function in Highly Sensitized Patients 1 Year After Desensitization With Imlifidase Prior to DD Kidney Tx (ConfIdeS) n.d.).

A multi-center, international RCT administering imlifidase for AMR treatment vs. PLEX has also been completed ($N = 30$) (Halleck et al. 2024). The primary study endpoint was DSA reduction. Here, imlifidase showed superiority, as a 97% reduction in DSAs was seen with imlifidase compared to 42% for PLEX. Additionally, imlifidase reduced DSAs to noncomplement fixing levels, whereas PLEX failed to do so. However, there were no meaningful improvements in biopsy findings or renal function. Here, the heterogeneity of patients entered with most showing eGFRs of ~20–30 ml/min at entry probably had a limited capacity to recover after any treatment. This also demonstrates that the removal of antibodies alone will not be sufficient to treat AMR/c-aAMR patients. It will be important to consider other agents that will address rebound DSA generation and/or inhibition of ADCC/CDC.

Important concerns limit imlifidase use. Notably, it can only be used once, due to immunogenicity resulting in anti-imlifidase-neutralizing antibody formation. This has prompted efforts to develop a less immunogenic imlifidase. Furthermore, imlifidase requires accompaniment with other treatments to decrease rebound IgG production. In this regard, recent data by Schrezenmeier et al. (2024) showed that imlifidase desensitization followed by daratumumab (anti-CD38) resulted in long-term rejection-free survival of a highly HLA-sensitized patient with crossmatch positivity at the time of transplant. Using therapeutically relevant combinations of agents with imlifidase may help overcome the antibody rebound problem that currently complicates its use as a single agent.

11.12 Novel Approaches: FcRn Inhibitors

Another emerging approach that could be important in the treatment of c-aAMR is the proposed use of agents that block pathogenic IgG recycling by saturating or inhibiting the Fc neonatal receptor (FcRn). The FcRn is an HLA class I-like molecule that is responsible for recycling two important proteins (IgG and albumin). Importantly, FcRn recycling extends IgG half-life (~28 days), while the absence of FcRn recycling in FcRn (−/−) animals results in an IgG half-life of approximately 3 days. In-depth reviews of FcRn potential applications in transplantation are available (Baldwin 3rd et al. 2019; Hoy 2023). A better understanding of the IgG-Fc/FcRn interaction has resulted in a class of novel therapies with high affinity for the

FcRn, which out-competes pathogenic IgG resulting in rapid declines in their levels. Other efforts have been made to design therapeutic monoclonal antibodies that have Fc mutations that increase affinity for the FcRn. This allows the half-life of these therapeutic agents to be increased (Seijsing et al. 2018). FcRn, originally discovered as a transport mechanism for IgG from maternal circulation to fetus, is broadly distributed and prolongs IgG half-life (please also refer to our in-depth review) (Jordan et al. 2020). FcRn is closely related to nonclassical MHC Class I and is a novel target to endorse IgG degradation and reduce half-life of pathogenic auto- or alloantibodies or extend therapeutic monoclonal half-life. Knechtle et al. evaluated the effectiveness of anti-FcRn monoclonals administered pre- and post-kidney transplantation in rhesus macaque monkeys, sensitized by skin grafting (Manook et al. 2021). The authors showed that perioperative anti-FcRn monoclonal administration did not substantially reduce total circulating IgG via FcRn-mediated recycling possibly because of ongoing DSA produced by PCs. They also determined that anti-FcRn treatment transiently decreased DSAs without impact on DSA synthesis.

The first FDA-approved FcRn inhibitor, efgartigimod, is an Fc fragment with high affinity for the FcRn, resulting in diminished circulating IgG half-life. Efgartigimod was FDA approved for generalized myasthenia gravis treatment, in intravenous and subcutaneous formulations (Howard Jr. et al. 2019; Heo 2023). Clinical trials are now underway evaluating the efficacy of efgartigomod for the treatment of AMR/c-aAMR (ShAMRock, NCT06503731) in kidney transplant patients.

11.13 Complement Inhibition for Treatment of AMR

11.13.1 Eculizumab (Anti-C5)

Eculizumab is a monoclonal antibody aimed at complement activation initiated by C5 convertase, resulting in the formation of the membrane attack complex (C5b-C9MAC). Treatment results in terminal complement inhibitions by preventing C5b-C9MAC formation. Stegall et al. demonstrated that eculizumab significantly reduced AMR incidence compared to a historical control of patients desensitized with PLEX+IVIG (7.7% (2/26) vs. 41.2% (21/51); $p = 0.003$) (Stegall et al. 2011). Although the results were promising, patients receiving long-term eculizumab with DSA positivity did not display a difference in c-aAMR rates vs. controls. This is an important demonstration regarding the need for antibody reduction therapies in combination with CDC inhibition to improve outcomes. Without DSA depletion, ADCC continues and results in c-aAMR.

Lefaucheur et al. (2018) reported on 931 kidney transplant recipients who exhibited complement-activating DSAs, biopsy evaluations, and molecular scores. The investigators reported that eculizumab improved graft survival and was specifically effective against rejection phenotypes associated with complement-activating

DSAs. Assessment of C1Q-activating DSAs is not routinely done in most US centers. However, one can typically assume that DSAs of >10,000 MFI are complement-activating (Wiebe et al. 2017).

Ravulizumab, an Fc-engineered anti-C5, developed from eculizumab, has enhanced binding to the FcRn. IgG-Fc/FcRn interactions extend therapeutic antibody half-life. Additionally, alterations in F(ab) amino acid structure allow C5 release at pH 6.5 when anti-C5 is bound to FcRn, permitting ligand (C5) degradation in endosomes. This allows refreshed anti-C5 to be released into the circulation, permitting continuous C5 shuttling to the endosomes for degradation. In summary, anti-C5 remains a valuable therapy for AMR where clear evidence of rapid graft decline relates to complement (C4d) deposition and the presence of complement-activating anti-HLA antibodies (Roth et al. 2018).

11.13.2 C1 Esterase Inhibitor (C1-INH)

C1-INH blocks complement activation via the classical and mannose- binding lectin (MBL) pathways. C1INH was investigated as an adjunct to desensitization. Data from animal models (Eerhart et al. 2022; Chun et al. 2017) and human clinical trials by our team suggest that C1-INH treatment ameliorates ischemia/reperfusion injury (IRI) (Huang et al. 2020; Jordan et al. 2018). Here, C1-INH's prevention of IRI likely prevents immune activation events within the allograft, leading to B-cell activation and alloantibody production (Heeger et al. 2024; Chen et al. 2025). Vo et al. conducted a phase 1/2, blinded, RCT using C1-INH (Berinert) in 20 HS patients (cPRA >50%, DSA+, and FCXM+) for AMR prevention, with acceptable safety outcomes (Vo et al. 2015b). After desensitization with IVIG + rituximab ± PLEX, patients received C1-INH or placebo intraoperatively, and then, twice weekly ×7. At 6 months, no patients in the C1-INH group developed AMR vs. 1 in placebo. C1-INH treatment significantly reduced $C1q^+$ HLA antibodies. Additionally, post-transplant, C1-INH-treated patients had lower DSA rebound rates and *dn*DSA development vs. placebo. Viglietti et al. published their experience using C1INH in an open-label trial with IVIg for acute AMR treatment unresponsive to SOC therapy (Viglietti et al. 2016). All patients ($N = 6$) showed eGFR improvement. No changes in histological features were observed, except a decrease in the C4d deposition rate.

Humanized mAb-targeting C1s were shown to inhibit alloantibody- initiated classical pathway complement activation in vitro (Wahrmann et al. 2017). In a small clinical trial investigating the efficacy of anti-C1s for treatment of AMR, (NCT02502903) (Eskandary et al. 2018b), investigators showed that there were no severe AEs, and anti-C1s eliminated DSA-triggered classic pathway complement activation. Also, $C4d^+$ staining was significantly reduced in most patients. However, there were no meaningful reductions in microcirculation inflammation, gene expression patterns, DSA levels, or kidney function improvement. Again, selective CDC inhibition, without DSA reductions, likely has limited benefits. Figure 11.1b summarizes the effector pathways of CDC and ADCC and their importance in mediating injury to allograft endothelial cells. This demonstrates how therapies targeting inhibition of CDC/ADCC are crucial to inhibit damage to the graft endothelium (Heeger et al. 2024).

11.14 Conclusions

Today, there is a rapid resurgence of new therapeutic agents aimed at the treatment of AMR/c-aAMR. Most have emerged from therapeutic advances in the management of autoimmune diseases and B-cell/plasma cell malignancies. Evolving understanding of effectors responsible for B-cell activation and plasma cell biology has led to the informed design of new therapeutics. Currently, several novel agents with varying mechanisms of action have emerged. These include imlifidase, FcRn inhibitors, anti-cytokine (IL-6/IL-6R) therapies, and PC- directed therapies, especially anti-CD38s. With the evolving enlightenment in transplant-directed therapeutics, it is important to realize that no single agent can control all facets of antibody generation and injury. The future of antibody-directed therapeutics will require protocols consisting of multiple agents positioned in a logical sequence to rapidly and durably remove pathogenic DSAs and block their re-emergence. In addition, the use of anti-CD38 monoclonals will have the additional benefit of inhibiting critical effector pathways (ADCC) in the prevention and treatment of AMR. In summary, we are entering a new era of transplant therapeutics that will transcend our current understanding of immunosuppression and usher in an era of immune modulation.

Acknowledgments We gratefully acknowledge the patients who participated in clinical trials evaluating novel desensitization and AMR therapies for their contributions to advance our understanding of alloimmunity. We are thankful for the members of the Transplant Immunotherapy Program, the HLA & Immunogenetics Lab, Transplant Immunology Lab at Cedars-Sinai Medical Center for their commitment to improving patient lives.

Financial Support and Sponsorship None.

Conflict of Interest Disclosures S.C.J. received grant support from Hansa Medical, CSL-Behring, Regeneron, and argenx and is a consultant for Vera therapeutics, Regeneron, Hansa Biopharma, argenx. A.V. is a consultant for Hansa Biopharma and argenx. N.A. and E.H. have no disclosure to report.

References

Allen NH, Dyer P, Geoghegan T, Harris K, Lee HA, Slapak M (1983) Plasma exchange in acute renal allograft rejection. A controlled trial. Transplantation 35(5):425–428. https://doi.org/10.1097/00007890-198305000-00006

Al-Salama ZT (2020) Imlifidase: first approval. Drugs 80(17):1859–1864. https://doi.org/10.1007/s40265-020-01418-5

Baldwin WM 3rd, Valujskikh A, Fairchild RL (2019) The neonatal Fc receptor: key to homeostasic control of IgG and IgG-related biopharmaceuticals. Am J Transplant 19(7):1881–1887. https://doi.org/10.1111/ajt.15366

Bonomini V, Vangelista A, Frascà GM, Di Felice A, Liviano D'AG (1985) Effects of plasmapheresis in renal transplant rejection. A controlled study. Trans Am Soc Artif Intern Organs 31:698–703

Boonpheng B, De Castro ICC, Ng YH, Blosser C, Bakthavatsalam R, Gimferrer I, Smith K, Leca N (2023) Tocilizumab for treatment of chronic active antibody-mediated rejection in kidney transplant recipients. Clin Transpl 37(5):e14936. https://doi.org/10.1111/ctr.14936. Epub 2023 Feb 24

Chen R, Powell JS, Shufesky WJ, Bardhi E, Sullivan MLG, Tkacheva OA, Camirand G, Goncalves Z, Stolz DB, Chalasani G, Catz SD, Watkins SC, Larregina AT, Morelli AE (2025) Transplants foster B cell alloimmunity by relaying extracellular vesicles to follicular dendritic cells. Cell Rep 44(6):115832. https://doi.org/10.1016/j.celrep.2025.115832

Choi J, Aubert O, Vo A, Loupy A, Haas M, Puliyanda D, Kim I, Louie S, Kang A, Peng A, Kahwaji J, Reinsmoen N, Toyoda M, Jordan SC (2017) Assessment of Tocilizumab (anti-Interleukin-6 receptor monoclonal) as a potential treatment for chronic antibody-mediated rejection and transplant Glomerulopathy in HLA-sensitized renal allograft recipients. Am J Transplant 17(9):2381–2389. https://doi.org/10.1111/ajt.14228. Epub 2017 Mar 10

Chun N, Fairchild RL, Li Y et al (2017) Complement dependence of murine costimulatory blockade-resistant cellular cardiac allograft rejection. Am J Transplant 17(11):2810–2819. https://doi.org/10.1111/ajt.14328

Daligault M, Bardy B, Noble J et al (2021) Marginal impact of Tocilizumab monotherapy on anti-HLA alloantibodies in highly sensitized kidney transplant candidates. Transplant Direct 7(5):e690. https://doi.org/10.1097/txd.0000000000001139

de Weerd AE, Roelen DL, van de Wetering J, Betjes HS, Reinders MEJ (2024) Imlifidase desensitization in HLA-incompatible kidney transplantation: finding the sweet spot. Transplantation 108(2):335–345. https://doi.org/10.1097/tp.0000000000004689

Diebold M, Mayer KA, Hidalgo L, Kozakowski N, Budde K, Böhmig GA (2025) Chronic rejection after kidney transplantation. Transplantation 109(4):610–621. https://doi.org/10.1097/TP.0000000000005187. Epub 2024 Aug 28

Doberer K, Duerr M, Halloran PF et al (2021) A randomized clinical trial of anti–IL-6 antibody Clazakizumab in late antibody-mediated kidney transplant rejection. J Am Soc Nephrol 32(3):708–722. https://doi.org/10.1681/asn.2020071106

Eerhart MJ, Reyes JA, Blanton CL et al (2022) Complement blockade in recipients prevents delayed graft function and delays antibody-mediated rejection in a nonhuman primate model of kidney transplantation. Transplantation 106(1):60–71. https://doi.org/10.1097/TP.0000000000003754

Ensor CR, Yousem SA, Marrari M et al (2017) Proteasome inhibitor Carfilzomib-based therapy for antibody-mediated rejection of the pulmonary allograft: use and short-term findings. Am J Transplant 17(5):1380–1388. https://doi.org/10.1111/ajt.14222

Eskandary F, Regele H, Baumann L et al (2018a) A randomized trial of Bortezomib in late antibody-mediated kidney transplant rejection. J Am Soc Nephrol 29(2):591–605. https://doi.org/10.1681/asn.2017070818

Eskandary F, Jilma B, Muhlbacher J et al (2018b) Anti-C1s monoclonal antibody BIVV009 in late antibody-mediated kidney allograft rejection-results from a first-in-patient phase 1 trial. Am J Transplant 18(4):916–926. https://doi.org/10.1111/ajt.14528

Fernández-Cruz E, Alecsandru D, Sánchez Ramón S, Ramón SS (2009) Mechanisms of action of immune globulin. Clin Exp Immunol 157(Suppl 1):1–2. https://doi.org/10.1111/j.1365-2249.2009.03955.x

Freeman CL, Sehn LH (2018) A tale of two antibodies: obinutuzumab versus rituximab. Br J Haematol 182(1):29–45. https://doi.org/10.1111/bjh.15232

Genberg H, Hansson A, Wernerson A, Wennberg L, Tydén G (2006) Pharmacodynamics of rituximab in kidney allotransplantation. Am J Transplant 6(10):2418–2428. https://doi.org/10.1111/j.1600-6143.2006.01497

Gloor JM, DeGoey SR, Pineda AA et al (2003) Overcoming a positive crossmatch in living-donor kidney transplantation. Am J Transplant 3(8):1017–1023. https://doi.org/10.1034/j.1600-6143.2003.00180.x

Gosset C, Lefaucheur C, Glotz D (2014) New insights in antibody-mediated rejection. Curr Opin Nephrol Hypertens 23(6):597–604. https://doi.org/10.1097/MNH.000000000000

Halleck F, Böhmig GA, Couzi L et al (2024) A randomized trial comparing Imlifidase to plasmapheresis in kidney transplant recipients with antibody-mediated rejection. Clin Transpl 38(7):e15383. https://doi.org/10.1111/ctr.15383

Hart A, Schladt DP, Matas AJ et al (2021a) Incidence, risk factors, and long-term outcomes associated with antibody-mediated rejection—the long-term deterioration of kidney allograft function (DeKAF) prospective cohort study. Clin Transpl 35:e14337

Hart A, Zaun D, Itzler R et al (2021b) Cost, healthcare utilization, and outcomes of antibody-mediated rejection in kidney transplant recipients in the US. J Med Econ 24:1011–1017

Heeger PS, Haro MC, Jordan SC (2024) Translating B cell immunology to the treatment of antibody-mediated allograft rejection. Nat Rev Nephrol 20:218–232. https://doi.org/10.1038/s41581-023-00791-030069

Heo YA (2023) Efgartigimod Alfa in generalised myasthenia gravis: a profile of its use. CNS Drugs 37(5):467–473. https://doi.org/10.1007/s40263-023-01000-z

Hill JA, Kiem ES, Bhatti A et al (2023) Anti-HLA antibodies in recipients of CD19 versus BCMA-targeted CAR T-cell therapy. Am J Transplant 23(3):416–422. https://doi.org/10.1016/j.ajt.2022.11.001

Howard JF Jr, Bril V, Burns TM et al (2019) Randomized phase 2 study of FcRn antagonist efgartigimod in generalized myasthenia gravis. Neurology 92(23):e2661–e2673. https://doi.org/10.1212/wnl.0000000000007600

Hoy SM (2023) Rozanolixizumab: first approval. Drugs 83(14):1341–1347. https://doi.org/10.1007/s40265-023-01933-1

Huang E, Vo A, Choi J et al (2020) Three-year outcomes of a randomized, double-blind, placebo-controlled study assessing safety and efficacy of C1 esterase inhibitor for prevention of delayed graft function in deceased donor kidney transplant recipients. Clin J Am Soc Nephrol 15(1):109–116. https://doi.org/10.2215/CJN.04840419

Huang E, Maldonado AQ, Kjellman C, Jordan SC (2022) Imlifidase for the treatment of anti-HLA antibody-mediated processes in kidney transplantation. Am J Transplant 22(3):691–697. https://doi.org/10.1111/ajt.16828

Hunter CA, Jones SA (2015) IL-6 as a keystone cytokine in health and disease. Nat Immunol 16(5):448–457. https://doi.org/10.1038/ni.3153

Irish W, Nickerson P, Astor BC et al (2021) Change in estimated GFR and risk of allograft failure in patients diagnosed with late active antibody-mediated rejection following kidney transplantation. Transplantation 105:648–659

Jain D, Rajab A, Young JS et al (2020) Reversing donor-specific antibody responses and antibody-mediated rejection with bortezomib and belatacept in mice and kidney transplant recipients. Am J Transplant 20(10):2675–2685. https://doi.org/10.1111/ajt.15881

January SE, Fester KA, Halverson LP, Witt CA, Byers DE, Vazquez-Guillamet R, Alexander-Brett J, Tague LK, Kreisel D, Gelman A, Puri V, Bahena RN, Takahashi T, Hachem RR, Kulkarni HS (2023) Tocilizumab for antibody-mediated rejection treatment in lung transplantation. J Heart Lung Transplant 42(10):1353–1357. https://doi.org/10.1016/j.healun.2023.05.012. Epub 2023 Jun 1. PMID: 37268051; PMCID: PMC10529998

Joher N, Matignon M, Grimbert P (2021) HLA desensitization in solid organ transplantation: anti-CD38 to across the immunological barriers. Front Immunol 12:688301. https://doi.org/10.3389/fimmu.2021.688301

Jordan SC, Tyan D, Stablein D et al (2004) Evaluation of intravenous immunoglobulin as an agent to lower allosensitization and improve transplantation in highly sensitized adult patients with end-stage renal disease: report of the NIH IG02 trial. J Am Soc Nephrol 15(12):3256–3262. https://doi.org/10.1097/01.ASN.0000145878.92906.9F

Jordan SC, Kahwaji J, Toyoda M, Vo A (2011) B-cell immunotherapeutics: emerging roles in solid organ transplantation. Curr Opin Organ Transplant 16(4):416–424. https://doi.org/10.1097/MOT.0b013e32834874f7

Jordan SC, Lorant T, Choi J (2017) IgG endopeptidase in highly sensitized patients undergoing transplantation. N Engl J Med 377(17):1693–1694. https://doi.org/10.1056/NEJMc1711335

Jordan SC, Choi J, Aubert O et al (2018) A phase I/II, double-blind, placebo-controlled study assessing safety and efficacy of C1 esterase inhibitor for prevention of delayed graft function in deceased donor kidney transplant recipients. Am J Transplant 18(12):2955–2964. https://doi.org/10.1111/ajt.14767

Jordan SC, Toyoda M, Ammerman N, Huang E, Peng A, Sethi S, Najjar R, Lim K, Vo A (2019) Daratumumab for treatment of antibody-mediated rejection in a kidney transplant recipient. Am J Transplant 19(suppl 3):1062

Jordan SC, Ammerman N, Vo A (2020) Implications of Fc neonatal receptor (FcRn) manipulations for transplant Immunotherapeutics. Transplantation 104(1):17–23. https://doi.org/10.1097/TP.0000000000002912

Jordan SC, Legendre C, Desai NM et al (2021) Imlifidase desensitization in Crossmatch-positive, highly sensitized kidney transplant recipients: results of an international phase 2 trial (Highdes). Transplantation 105(8):1808–1817. https://doi.org/10.1097/TP.0000000000003496

Jordan SC, Berg A, Shin B, Vo A, Ammerman N, Zhang R (2022a) Intravenous immunoglobulin contains high-titer neutralizing IgG antibodies to SARS-CoV-2. Am J Transplant 10:2484–2485

Jordan SC, Ammerman N, Huang E, Vo A (2022b) Importance of IL-6 inhibition in prevention and treatment of antibody-mediated rejection in kidney allografts. Am J Transplant 22(Suppl 4):28–37. https://doi.org/10.1111/ajt.17207

Jordan SC, Ammerman N, Choi J et al (2022c) Evaluation of Clazakizumab (anti-Interleukin-6) in patients with treatment-resistant chronic active antibody-mediated rejection of kidney allografts. Kidney Int Rep 7(4):720–731. https://doi.org/10.1016/j.ekir.2022.01.1074

Jordan SC, Maldonado A et al (2024) Long-term follow up of imlifidase desensitized kidney transplant recipients: 5 year pooled analysis. Am J Transplant 24(6):1

Kim I, Wu G, Chai NN, Klein AS, Jordan SC (2016) Immunological characterization of de novo and recall alloantibody suppression by CTLA4Ig in a mouse model of allosensitization. Transpl Immunol 09(38):84–92. https://doi.org/10.1016/j.trim.2016.08.001

Kirubakaran MG, Disney AP, Norman J, Pugsley DJ, Mathew TH (1981) A controlled trial of plasmapheresis in the treatment of renal allograft rejection. Transplantation 32(2):164–165. https://doi.org/10.1097/00007890-198108000-00019

Kjellman C, Maldonado AQ, Sjöholm K et al (2021) Outcomes at 3 years posttransplant in imlifidase-desensitized kidney transplant patients. Am J Transplant 21(12):3907–3918. https://doi.org/10.1111/ajt.16754

Krejcik J, Casneuf T, Nijhof IS et al (2016) Daratumumab depletes CD38+ immune regulatory cells, promotes T-cell expansion, and skews T-cell repertoire in multiple myeloma. Blood 128(3):384–394. https://doi.org/10.1182/blood-2015-12-687749

Kubo MSK, Lu L, Vujevich V, Gunabushanam V, Ganoza A, Wijkstrom M, Humar A, Thomson A, Ezzelarab M (2023) IL-6 receptor blockade attenuates effector T cell function and significantly prolongs kidney allograft survival in CTLA4Ig-treated rhesus monkeys. AJT 23(6, Supplement 1):S765

Kwun J, Matignon M, Manook M et al (2019) Daratumumab in sensitized kidney transplantation: potentials and limitations of experimental and clinical use. J Am Soc Nephrol 30(7):1206–1219. https://doi.org/10.1681/ASN.2018121254

Lamarthée B, Callemeyn J, Van Herck Y et al (2023) Transcriptional and spatial profiling of the kidney allograft unravels a central role for FcyRIII+ innate immune cells in rejection. Nat Commun 14(1):4359. https://doi.org/10.1038/s41467-023-39859-7

Lefaucheur C, Viglietti D, Hidalgo LG et al (2018) Complement-activating anti-HLA antibodies in kidney transplantation: allograft gene expression profiling and response to treatment. J Am Soc Nephrol 29(2):620–635. https://doi.org/10.1681/ASN.2017050589

Lim JH, Cho JH, Jung HY et al (2019) Excellent outcome after desensitization in high immunologic risk kidney transplantation. PLoS One 14(9):e0222537. https://doi.org/10.1371/journal.pone.0222537

Lobashevsky AL, Higgins NG, Rosner KM, Mujtaba MA, Goggins WC, Taber TE (2013) Analysis of anti-HLA antibodies in sensitized kidney transplant candidates subjected to desensitization with intravenous immunoglobulin and rituximab. Transplantation 96(2):182–190. https://doi.org/10.1097/TP.0b013e3182962c84

Looney CM, Schroeder A, Tavares E et al (2023) Obinutuzumab effectively depletes key B-cell subsets in blood and tissue in end-stage renal disease patients. Transplant Direct 9(2):e1436. https://doi.org/10.1097/txd.0000000000001436

Loupy A, Lefaucheur C (2018) Antibody-mediated rejection of solid-organ allografts. N Engl J Med 379:1150–1160

Loupy A, Suberbielle-Boissel C, Zuber J et al (2010) Combined posttransplant prophylactic IVIg/anti-CD 20/plasmapheresis in kidney recipients with preformed donor-specific antibodies: a pilot study. Transplantation 89(11):1403–1410. https://doi.org/10.1097/TP.0b013e3181da1cc3

Lynch RJ, Silva IA, Chen BJ, Punch JD, Cascalho M, Platt JL (2013) Cryptic B cell response to renal transplantation. Am J Transplant 13(7):1713–1723. https://doi.org/10.1111/ajt.12308

Manook M, Flores WJ, Schmitz R et al (2021) Measuring the impact of targeting FcRn-mediated IgG recycling on donor-specific alloantibodies in a sensitized NHP model. Front Immunol 12:660900. https://doi.org/10.3389/fimmu.2021.660900

Mayer KA, Schrezenmeier E, Diebold M et al (2024) A randomized phase 2 trial of Felzartamab in antibody-mediated rejection. N Engl J Med 391:122–132. https://doi.org/10.1056/NEJMoa2400763

Montgomery RA, Zachary AA, Racusen LC et al (2000) Plasmapheresis and intravenous immune globulin provides effective rescue therapy for refractory humoral rejection and allows kidneys to be successfully transplanted into cross-match-positive recipients. Transplantation 70(6):887–895. https://doi.org/10.1097/00007890-200009270-00006

Moreno Gonzales MA, Gandhi MJ, Schinstock CA et al (2017) 32 doses of Bortezomib for desensitization is not well tolerated and is associated with only modest reductions in anti-HLA antibody. Transplantation 101(6):1222–1227. https://doi.org/10.1097/tp.0000000000001330

NasrAllah MM, Elalfy M, El Ansary M et al (2022) Obinutuzumab in kidney transplantation: effect on B-cell counts and crossmatch tests. Transplantation 106:369–372. https://doi.org/10.1097/TP.0000000000003686

Nickerson PW (2022) Rationale for the IMAGINE study for chronic active antibody-mediated rejection (caAMR) in kidney transplantation. Am J Transplant 22:38–44. https://doi.org/10.1111/ajt.17210

Noble J, Comai G, Corredetti V, Laamech R, Dard C, Jouve T, Giovannini D, Le Gouellec A, Wadnerkar S, Cravedi P, Apuzzo D, Vetrano D, Busutti M, Abenavoli C, Malvezzi P, Rostaing LP, Lamanna G (2025) Tocilizumab-based treatment of microvascular inflammation in kidney transplant recipients: a retrospective study. Transpl Int 16(38):14502. https://doi.org/10.3389/ti.2025.14502. PMID: 40454296; PMCID: PMC12124137

Nowak RJ, Benatar M, Ciafaloni E, Howard JF Jr, Leite MI, Utsugisawa K, Vissing J, Rojavin M, Li Q, Tang F, Wu Y, Rampal N, Cheng S, Investigators MINT (2025) A phase 3 trial of Inebilizumab in generalized myasthenia gravis. N Engl J Med 392(23):2309–2320. https://doi.org/10.1056/NEJMoa2501561. Epub 2025 Apr 8

Pearl M, Weng PL, Chen L, Dokras A, Pizzo H, Garrison J, Butler C, Zhang J, Reed EF, Kim IK, Choi J, Haas M, Zhang X, Vo A, Chambers ET, Ettenger R, Jordan S, Puliyanda D (2022)

Long term tolerability and clinical outcomes associated with tocilizumab in the treatment of refractory antibody mediated rejection (AMR) in pediatric renal transplant recipients. Clin Transpl 36(8):e14734. https://doi.org/10.1111/ctr.14734. Epub 2022 Jun 12. PMID: 35657013; PMCID: PMC9378624

Perez EE, Orange JS, Bonilla F et al (2017) Update on the use of immunoglobulin in human disease: a review of evidence. J Allergy Clin Immunol 139(3s):S1–s46. https://doi.org/10.1016/j.jaci.2016.09.023

Pham C, Pierce BJ, Nguyen DT, Graviss EA, Huang HJ (2021) Assessment of Carfilzomib treatment response in lung transplant recipients with antibody-mediated rejection. Transplant Direct 7(4):e680. https://doi.org/10.1097/txd.0000000000001131

Pottebaum AA, January SE, Liu C, Lavine S, Schilling JD, Lavine KJ (2024) Feasibility of Interleukin-6 receptor blockade in cardiac antibody-mediated rejection. Transplantation 108(2):539–544. https://doi.org/10.1097/TP.0000000000004784. Epub 2023 Aug 29. PMID: 37638881; PMCID: PMC10798586

Redfield RR, Jordan SC, Busque S et al (2019) Safety, pharmacokinetics, and pharmacodynamic activity of obinutuzumab, a type 2 anti-CD20 monoclonal antibody for the desensitization of candidates for renal transplant. Am J Transplant 19(11):3035–3045. https://doi.org/10.1111/ajt.15514

Renal Function in Highly Sensitized Patients 1 Year After Desensitization With Imlifidase Prior to DD Kidney Tx (ConfIdeS) (n.d.). https://classic.clinicaltrials.gov/ct2/show/NCT04935177

Roth A, Rottinghaus ST, Hill A et al (2018) Ravulizumab (ALXN1210) in patients with paroxysmal nocturnal hemoglobinuria: results of 2 phase 1b/2 studies. Blood Adv 2(17):2176–2185. https://doi.org/10.1182/bloodadvances.2018020644

Scalzo RE, Sanoff SL, Rege AS et al (2023) Daratumumab use prior to kidney transplant and T cell-mediated rejection: a case report. Am J Kidney Dis 81(5):616–620. https://doi.org/10.1053/j.ajkd.2022.11.010

Schinstock CA, Mannon RB et al (2020) Recommended treatment for antibody-mediated rejection after kidney transplantation: the 2019 expert consensus from the transplantation society working group. Transplantation 104(5):911–922. https://doi.org/10.1097/TP.0000000000003095

Schrezenmeier E, Choi M, Globke B et al (2024) Successful desensitization with Imlifidase and Daratumumab in a highly immunized, crossmatch positive, blood group-incompatible living-donor re-transplant recipient with systemic lupus erythematosus and antiphospholipid syndrome. Transfus Med Hemother 51(3):158–163. https://doi.org/10.1159/000538513

Seijsing J, Yu S, Frejd FY, Höiden-Guthenberg I, Gräslund T (2018) In vivo depletion of serum IgG by an affibody molecule binding the neonatal Fc receptor. Sci Rep 8(1):5141. https://doi.org/10.1038/s41598-018-23481-5

Stapleton NM, Einarsdóttir HK, Stemerding AM, Vidarsson G (2015) The multiple facets of FcRn in immunity. Immunol Rev 268(1):253–268. https://doi.org/10.1111/imr.12331

Stegall MD, Diwan T, Raghavaiah S et al (2011) Terminal complement inhibition decreases antibody-mediated rejection in sensitized renal transplant recipients. Am J Transplant 11(11):2405–2413. https://doi.org/10.1111/j.1600-6143.2011.03757.x

Streichart L, Felldin M, Ekberg J, Mjörnstedt L, Lindnér P, Lennerling A, Bröcker V, Mölne J, Holgersson J, Daenen K, Wennberg L, Lorant T, Baid-Agrawal S (2024) Tocilizumab in chronic active antibody-mediated rejection: rationale and protocol of an in-progress randomized controlled open-label multi-center trial (INTERCEPT study). Trials 25(1):213. https://doi.org/10.1186/s13063-024-08020-0. PMID: 38519988; PMCID: PMC10958896

Study for Desensitization of Chronic Kidney Disease in Adult Patients in Need of a Kidney Transplant Who Are Highly Sensitized to Human Leukocyte Antigen (n.d.). https://classic.clinicaltrials.gov/ct2/show/NCT05092347

Tanaka T, Narazaki M, Kishimoto T (2014) IL-6 in inflammation, immunity, and disease. Cold Spring Harb Perspect Biol 6(10):a016295. https://doi.org/10.1101/cshperspect.a016295

Toyoda M, Shin BH, Ge S et al (2017) Impact of desensitization on antiviral immunity in HLA-sensitized kidney transplant recipients. J Immunol Res 2017:5672523. https://doi.org/10.1155/2017/5672523

Tremblay S, Driscoll JJ, Rike-Shields A et al (2020) A prospective, iterative, adaptive trial of carfilzomib-based desensitization. Am J Transplant 20(2):411–421. https://doi.org/10.1111/ajt.15613

van de Donk NWCJ (2018) Immunomodulatory effects of CD38-targeting antibodies. Immunol Lett 199:16–22. https://doi.org/10.1016/j.imlet.2018.04.005

van den Hoogen MW, Kamburova EG, Baas MC et al (2015) Rituximab as induction therapy after renal transplantation: a randomized, double-blind, placebo-controlled study of efficacy and safety. Am J Transplant 15(2):407–416. https://doi.org/10.1111/ajt.13052

Viglietti D, Gosset C, Loupy A et al (2016) C1 inhibitor in acute antibody-mediated rejection nonresponsive to conventional therapy in kidney transplant recipients: a pilot study. Am J Transplant 16(5):1596–1603. https://doi.org/10.1111/ajt.13663

Vincenti F, Rostaing L, Grinyo J et al (2016) Belatacept and long-term outcomes in kidney transplantation. N Engl J Med 374(4):333–343. https://doi.org/10.1056/NEJMoa1506027

Vo AA, Lukovsky M, Toyoda M et al (2008) Rituximab and intravenous immune globulin for desensitization during renal transplantation. N Engl J Med 359(3):242–251. https://doi.org/10.1056/NEJMoa0707894

Vo AA, Petrozzino J, Yeung K et al (2013) Efficacy, outcomes, and cost-effectiveness of desensitization using IVIG and rituximab. Transplantation 95(6):852–858. https://doi.org/10.1097/TP.0b013e3182802f88

Vo AA, Choi J, Kim I et al (2015a) A phase I/II trial of the Interleukin-6 receptor-specific humanized monoclonal (Tocilizumab) + intravenous immunoglobulin in difficult to desensitize patients. Transplantation 99(11):2356–2363. https://doi.org/10.1097/TP.0000000000000741

Vo AA, Zeevi A, Choi J et al (2015b) A phase I/II placebo-controlled trial of C1-inhibitor for prevention of antibody-mediated rejection in HLA sensitized patients. Transplantation 99(2):299–308. https://doi.org/10.1097/TP.0000000000000592

Vo AA, Huang E, Ammerman N et al (2022) Clazakizumab for desensitization in highly sensitized patients awaiting transplantation. Am J Transplant 22(4):1133–1144. https://doi.org/10.1111/ajt.16926

Vo A, Ammerman N, Jordan SC (2024) New therapies for highly sensitized patients on the waiting list. Kidney360 5(8):1207–1225. https://doi.org/10.34067/KID.0000000000000509

Wahrmann M, Muhlbacher J, Marinova L et al (2017) Effect of the anti-C1s humanized antibody TNT009 and its parental mouse variant TNT003 on HLA antibody-induced complement activation-a preclinical in vitro study. Am J Transplant 17(9):2300–2311. https://doi.org/10.1111/ajt.14256

Wiebe C, Gareau AJ, Pochinco D et al (2017) Evaluation of C1q status and titer of De Novo donor-specific antibodies as predictors of allograft survival. Am J Transplant 17(3):703–711. https://doi.org/10.1111/ajt.14015

Zachary AA, Lucas DP, Montgomery RA, Leffell MS (2013) Rituximab prevents an anamnestic response in patients with cryptic sensitization to HLA. Transplantation 95(5):701–704. https://doi.org/10.1097/TP.0b013e31827be3c1

Zhang X, Li F, Jordan SC (2021) Obinutuzumab in kidney transplantation: effect on B-cell counts and crossmatch tests. Transplantation 105(11):e272–e273. https://doi.org/10.1097/TP.0000000000003849. PMID: 34709221

Zhang Z, Markmann C, Yu M et al (2023) Immunotherapy targeting B cells and long-lived plasma cells effectively eliminates pre-existing donor-specific allo-antibodies. Cell Rep Med 4(12):101336. https://doi.org/10.1016/j.xcrm.2023.101336

Zhao YG, Shi BY, Qian YY, Bai HW, Xiao L, He XY (2014) Clinical efficacy of rituximab for acute rejection in kidney transplantation: a meta-analysis. Int Urol Nephrol 46(6):1225–1230. https://doi.org/10.1007/s11255-013-0599-4

Chapter 12
Chronic Liver Allograft Rejection: Histopathological Insights and Future Directions

Samar Seeman, Joshua Wong, Adib Edilbi, Ashton A. Connor, John M. Vierling, Mary R. Schwartz, and Sadhna Dhingra

Abstract Chronic liver allograft rejection (CR) remains a significant cause of late graft dysfunction and failure despite advances in surgical techniques and immunosuppression. This chapter reviews the current understanding of CR, integrating diagnostic histopathological findings, atypical histological patterns of late rejection that are frequently associated with CR, and emerging diagnostic tools. CR is driven by overlapping T cell-mediated and antibody-mediated mechanisms, leading to progressive bile duct loss, vascular injury, and structural alterations. Diagnostic histopathological criteria include interlobular bile duct atrophy or loss, foam cell obliterative arteriopathy, and portal/periportal, perisinusoidal, and perivenular fibrosis. Chronic antibody-mediated rejection (cAMR) and mixed phenotypes exhibit additional vascular and sinusoidal injury, often with subtle biochemical changes. Structural sequelae such as nodular regenerative hyperplasia and variable parenchymal fibrosis underscore the heterogeneity of late graft injury. This chapter also examines late-onset rejection phenotypes, i.e., plasma cell-rich rejection, idiopathic post-transplant hepatitis, and isolated central perivenulitis and their potential to progress to chronic rejection. Future directions emphasize the integration of molecular diagnostics, digital pathology, and noninvasive biomarkers, such as donor-specific antibodies, liver stiffness measurement, and donor-derived cell-free DNA (dd-cfDNA), to complement or replace protocol biopsies, to guide personalized

S. Seeman · A. Edilbi · A. A. Connor
Department of Surgery, JC Walter Jr. Transplant Center Houston Methodist Hospital, Houston, TX, USA

J. Wong · M. R. Schwartz · S. Dhingra (✉)
Department of Pathology and Genomic Medicine, Houston Methodist Hospital, Houston, TX, USA
e-mail: sdhingra@houstonmethodist.org

J. M. Vierling
Margaret M. and Albert B. Alkek Department of Medicine, Section of Gastroenterology and Hepatology and Michael E. DeBakey Department of Surgery, Division of Abdominal Transplantation, Baylor College of Medicine, Houston, TX, USA

M. Kloc, R. M. Ghobrial (eds.), *Cellular and Molecular Mechanisms of Chronic Rejection in Organ Transplantation*, Results and Problems in Cell Differentiation 77, https://doi.org/10.1007/978-3-032-07686-1_12

immunosuppression, and to enable earlier intervention. Advances in regenerative medicine and antifibrotic therapies have the potential to alter the natural history of CR, shifting toward precision medicine approaches that may extend long-term graft survival after liver transplantation.

12.1 Introduction

Liver transplantation (LT) remains the only curative option for patients with liver failure occurring acutely or as a result of cirrhosis, with or without associated primary liver carcinomas. The underlying causes of liver diseases leading to transplantation are evolving. With the introduction of highly effective treatments for hepatitis C virus (HCV) infection, the implementation of birth-cohort screening programs, and the development of new medications achieving sustained virological response rates of 99%, chronic viral hepatitis C may become a rare disease in the future (Yi et al. 2017). There is a notable shift toward LT for alcohol-associated liver disease (ALD) and metabolic dysfunction-associated steatohepatitis (MASH), now the two most common indications for LT among patients without hepatocellular carcinoma (HCC). At the same time, MASH is the leading etiology among those with HCC (Younossi et al. 2024). Additionally, cholangiocarcinoma (both perihilar and intrahepatic), metastatic neuroendocrine neoplasms, and oligometastatic colorectal carcinoma have emerged as relatively new oncologic indications for LT, significantly altering the landscape of transplant oncology (Dasari et al. 2025; Ros et al. 2023; Sapisochin et al. 2020).

LT outcomes have improved markedly, with current 1-, 5-, and 10-year survival rates of approximately 85–95%, 68–80%, and 50–70%, respectively (Kaltenmeier et al. 2021). These improvements reflect advances in immunosuppression, surgical techniques, intra- and postoperative care, and standardized protocols for infection prophylaxis and optimal patient selection, especially for those with hepatocellular and cholangiocarcinomas. Tacrolimus-based regimens significantly reduce the incidence of acute rejection episodes and chronic rejection compared to those with cyclosporine (McAlister et al. 2006). Despite these successes, long-term graft survival remains a main challenge. While cardiovascular and renal diseases are leading nonhepatic causes of morbidity and mortality post-transplant, chronic graft failure is emerging as a leading cause of morbidity and the need for re-transplantation.

According to data from the 2020 Organ Procurement and Transplant Network/Scientific Registry of Transplant Recipients (OPTN/SRTR) report, allograft failure occurred in 5.9% of deceased donor liver transplant recipients at 6 months and 7.9% at 1 year for LTs performed in 2019, in 14.7% at 3 years for LTs in 2017, in 20.7% at 5 years for LTs in 2015, and in 40.6% at 10 years for LTs in 2010 (Kwong et al. 2022). Chronic allograft failure is attributable to recurrent or de novo diseases, chronic rejection, biliary strictures, and vascular disease (Lucey et al. 2013). Among these causes, chronic rejection (CR) accounts for approximately 5–7% of long-term allograft failures, though this may be underreported due to challenges in diagnosis and overlap with other chronic pathologic conditions (Angelico et al. 2021; Zarrinpar and Busuttil 2013).

Chronic rejection is a slow progressive alloimmune insult to liver allograft parenchyma, leading to irreversible bile duct and vascular injury. The mechanisms underlying chronic liver allograft rejection are not yet fully understood, particularly given the liver's distinctive capacity for immune tolerance (Angelico et al. 2021). Following liver transplantation, the allograft develops partial tolerance to immune-mediated injury, allowing for a reduction in immunosuppression after the first 90 days. Although a small subset of recipients may achieve operational tolerance, defined as stable allograft function without ongoing immunosuppression, this remains a rare occurrence. Most patients require lifelong immunosuppressive therapy to maintain allograft function. Chronic rejection develops over months to years, with subtle clinical signs and an insidious course (Yilmaz et al. 2006). When clinically apparent, chronic rejection typically manifests as elevated liver enzymes, prompting a liver biopsy to investigate the underlying etiology. Protocol biopsies in asymptomatic pediatric LT recipients at 1, 5, and 10 years have shown that idiopathic chronic hepatitis is a common finding at 10 years post-LT and is associated with progressive fibrosis and risk of cirrhosis (Evans et al. 2006; Sebagh et al. 2003). Liver biopsy, whether performed "for cause" in a symptomatic patient or as part of a surveillance protocol, remains the gold standard for assessing liver allograft health and etiology of dysfunction.

Chronic rejection in liver allografts arises from both T cell-mediated and antibody-mediated immune responses, each contributing through distinct, yet often-overlapping, mechanisms. In chronic T cell-mediated rejection (TCMR), recipient CD8+ cytotoxic T lymphocytes recognize donor-derived class I and class II HLA antigens presented on both hepatocytes and biliary epithelial cells; paradoxically, however, it is the intrahepatic bile ducts that undergo preferential destruction, suggesting a selective vulnerability of cholangiocytes despite shared antigenicity (Demirci et al. 1996; Vierling and Howell 1990). Ductular reaction is unusual in early-onset chronic rejection (Demetris et al. 2000). However, progressive bile duct loss and chronic cholestasis in late-onset chronic rejection can induce more periportal ductular reaction and fibrosis, which may progress to end-stage biliary cirrhosis over time (Sebagh et al. 2002). In contrast, chronic antibody-mediated rejection (AMR) is driven by donor-specific antibodies (DSAs) that bind to endothelial antigens, activating the complement cascade and causing sustained vascular injury. The resulting lesions, such as foam cell arteriopathy, sub-sinusoidal fibrosis, and perivenular sclerosis, reflect this humoral insult. While TCMR primarily targets the bile ducts through CD8+ cytotoxic T lymphocytes targeting class II HLA antigens expressed by bile duct epithelium, AMR predominantly affects the vasculature and may secondarily induce ischemic bile duct injury. In practice, chronic rejection often represents an amalgamation of both TCMR and AMR, complicating both diagnosis and management. Mixed phenotypes are frequently more resistant to treatment and are associated with worse long-term outcomes (Leonard et al. 2023). Recognizing the contribution of both immune pathways is essential for accurate diagnosis and for tailoring immunosuppressive strategies.

This chapter focuses on the histopathology of chronic rejection (CR) after LT. It also explores atypical histological patterns observed in late rejection and their possible relationship to CR. The final section discusses emerging noninvasive biomarkers that may help monitor long-term liver allograft health and guide immunosuppression strategies to reduce the incidence of CR beyond 10 years post-transplant.

12.2 Typical Histopathology of Chronic Rejection

CR is defined by progressive, irreversible injury to interlobular bile ducts, hepatic arteries and arterioles, terminal hepatic veins, and the surrounding parenchyma. As a result, both the portal tracts and the perivenular areas are affected in CR. The diagnostic criteria for CR include any one of the following: diffuse interlobular bile duct atrophy/senescent changes with or without bile duct loss (early CR), foam cell obliterative arteriopathy involving medium-sized arteries, or diffuse bile duct loss in more than 50% of portal tracts (Demetris 2000). These findings are typically seen in failed liver allografts removed during re-transplantation. However, in needle biopsies, the diagnosis often depends on identifying severe interlobular bile duct injury and bile duct loss, since medium-sized arteries are usually not sampled. Early CR shows ongoing low-grade necroinflammatory changes in portal tracts and perivenular regions in addition to biliary epithelial senescent changes. At this stage, the patient may benefit from the augmented immunosuppression, with the potential to reverse the damage to liver structure and function. In contrast, late CR is less inflammatory but more fibrotic, often associated with architectural distortion and irreversible functional decline.

12.2.1 Bile Duct Injury

Interlobular bile ducts (size <60-micron diameter) undergo progressive damage characterized by distinctive epithelial changes referred to as "senescent" changes (Fig. 12.1a). In early CR, there is eosinophilic transformation of biliary cytoplasm, nuclear pyknosis, uneven spacing of nuclei, and multinucleation. As injury progresses, bile ducts may only be partially lined by epithelial cells or shed epithelial cells into the lumen (Wight 1996). With progressive damage, multiple small bile ducts are destroyed, leading to ductopenia, also referred to as "vanishing bile duct syndrome" (O'Grady et al. 1988). Histologically, ductopenia is defined by the absence of bile ducts in greater than 50% of portal tracts (Fig. 12.1b). In early stages, bile duct loss is best appreciated as hepatic artery unaccompanied by a bile duct (Wiesner et al. 1991). Immunohistochemical staining for cytokeratin 7 (CK7) helps in quantifying bile duct loss (Fig. 12.1d). Destruction of bile ducts is not accompanied by ductular reaction in CR. Other features of late CR include

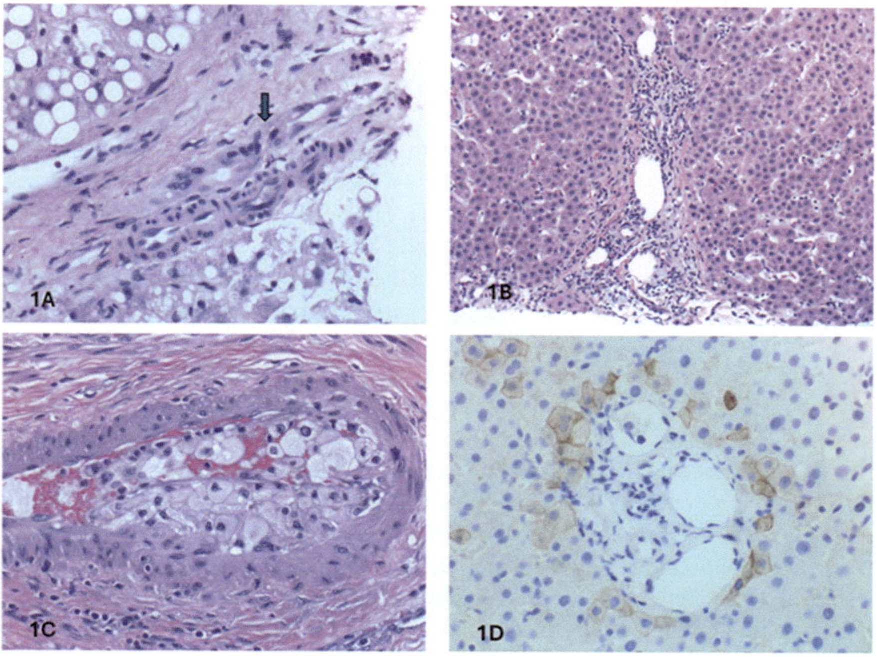

Fig. 12.1 Chronic liver allograft rejection. (**a**) Portal tract with bile duct showing senescent changes (arrow). Hematoxylin and eosin stain 60x. (**b**) Portal tract with mild lymphocytic inflammation and loss of interlobular bile duct. Hematoxylin and eosin stain 40x. (**c**) Hepatic artery with foam cell arteriopathy. Hematoxylin and eosin stain 40x. (**d**) Portal tract with loss of bile duct. Immunohistochemical stain for CK7 40x

perivenular hepatocyte swelling, hepatocytic and canalicular cholestasis, and intrasinusoidal foam cell clusters.

Clinically, bile duct loss correlates with progressive cholestasis, elevated alkaline phosphatase levels, and eventual biliary cirrhosis. The mechanisms driving bile duct injury in chronic rejection are both immune-mediated and ischemic. Allograft T lymphocytes recognize and attack cholangiocytes, triggering apoptosis or necrosis of bile duct epithelial cells (Hubscher et al. 1991). Electron microscopic studies and TUNEL assays have demonstrated programmed cell death in rejected bile ducts, indicating that cytotoxic T cells induce apoptosis in these cells (Afford et al. 1999; Nawaz and Fennell 1994). Bile duct epithelia derive the majority of their blood supply from the peribiliary capillary plexi supplied by the accompanying arteries in the portal triads, and thus suffer ischemic injury due to rejection-related foam cell arteriopathy (Demetris et al. 1997, 1998).

12.2.2 Inflammation

CR often has a relatively mild, patchy portal mixed inflammatory cell infiltrate (Hayashi et al. 1995; McCaughan et al. 1990) composed of a majority of T lymphocytes (both CD4+ helper and CD8+ cytotoxic T cells), accompanied by smaller numbers of CD20+ B lymphocytes, CD68+ macrophages (histiocytes), and an occasional plasma cell (Lunz 3rd et al. 2001). The portal inflammatory cell infiltrate is centered on the interlobular bile ducts and portal veins, manifesting as lymphocytic cholangitis and portal venulitis. Early CR is characterized by inflammation of terminal hepatic venules and surrounding perivenular parenchyma referred to as "central perivenulitis" and spotty necrosis of perivenular hepatocytes with the presence of ceroid-laden macrophages and early perivenular fibrosis. Central perivenulitis can occur in acute TCMR, late-onset TCMR, and CR.

Chronic ductopenic rejection is primarily a chronic cytotoxic (CD8+) T cell-mediated process directed against donor-derived antigens expressed on hepatocytes, biliary epithelial cells, and endothelium, initiating an inflammatory cascade that injures these structures. Paradoxically, however, the brunt of this injury is borne by the biliary epithelium, with cholangiocytes undergoing preferential destruction despite their shared antigenic profile with hepatocytes. (Demirci et al. 1996). Persistent alloimmune cytotoxicity driven by inadequate regulatory mechanisms that normally promote allograft tolerance leads to cumulative structural damage and worsening allograft function (Vierling 1999). Ultimately, this leads to a fibrotic and functionally compromised liver allograft.

12.2.3 Vascular Changes

The principal vascular manifestation of CR is foam cell obliterative arteriopathy. Histologically, medium-sized muscular arteries in the allograft develop a distinctive concentric intimal fibrosis with disruption of the internal elastic lamina secondary to infiltration by foamy macrophages (Fig. 12.1c). These foamy macrophages are monocytes that migrate from blood to intima after endothelial injury and acquire a foamy appearance secondary to phagocytosis of lipids and myointimal proliferation (Miyagawa-Hayashino et al. 2004). In early/evolving CR, lymphocytic inflammation accompanies the macrophage infiltration, causing arteritis. Progressive CR leads to luminal narrowing or even complete obliteration of the arterial lumen (Miyagawa-Hayashino et al. 2004). The Banff schema for allograft pathology recognizes foam cell obliterative arteriopathy as a defining lesion of CR, even preceding advanced ductopenia (Bellamy et al. 2024). In late CR, arteries of all calibers are involved, resulting in destruction of small intrahepatic arterioles and peribiliary capillary plexi in smaller interlobular portal tracts. Loss of hepatic arterioles is significant when it affects greater than 25% of portal tracts. CR frequently creates a vicious cycle where concurrent immune-mediated duct injury and arterial obliteration exacerbate other, resulting in accelerated bile duct loss (Demetris et al. 1998; Oguma et al. 1989a, b). When both bile ducts and hepatic arterioles are lost, portal tracts become the fibrous remnant, creating a challenge for the pathologists to identify portal tracts and quantify bile duct loss in assessment for ductopenia. Clinically, the extent of arterial narrowing correlates with allograft dysfunction and is associated with a poor prognosis for the allograft (Haas et al. 2014). Once foam cell arteriopathy is established, CR often progresses despite intensified immunosuppressive therapy and frequently culminates in allograft failure (Demetris 2000).

In addition to obliterative arteriopathy, intimal fibrosis has also been shown to develop in small terminal hepatic veins, medium hepatic veins, and portal veins as a manifestation of CR (Nakazawa et al. 2000a). The obliterative intimal fibrosis of terminal hepatic veins, referred to as veno-occlusive disease (VOD), is the earliest immune-mediated lesion of chronic rejection, which leads to progressive thrombosis of medium hepatic veins and portal veins. It is associated with progressive centrilobular fibrosis (Nakazawa et al. 2000a). Although morphologically similar to chemotherapy-related veno-occlusive disease, this lesion is immune-mediated in liver allografts.

Together, the triad of progressive interlobular bile duct loss, perivenular and portal inflammation, and obliterative vasculopathy forms the histologic backbone of chronic liver allograft rejection, with each component variably expressed across disease stages and influencing clinical management.

12.3 Chronic Antibody-Mediated Rejection (cAMR)

For decades, antibody-mediated rejection (AMR) in liver allografts was considered rare and of limited clinical relevance. The liver's unique immunologic properties were useful in combined liver–kidney transplantation for protecting renal allografts from alloantibody injury, particularly in highly sensitized recipients. This resistance is attributed to several complementary mechanisms (Demetris et al. 2016). The liver's dual blood supply creates a low-pressure, sinusoidal microenvironment that facilitates tolerogenic antigen presentation. Fenestrated liver sinusoidal endothelial cells and Kupffer cells efficiently clear immune complexes via Fc receptor-mediated phagocytosis, and the vast endothelial surface allows extensive antibody interaction normally without triggering injury. Additionally, hepatocytes and other parenchymal cells express HLA class I molecules diffusely, while HLA class II molecule expression remains low and is largely restricted to activated antigen-presenting cells, thereby limiting CD4+ T cell activation and downstream alloimmune amplification. The liver's robust regenerative capacity further enables repair and reversal of fibrosis following injury. Collectively, these features dampen humoral immune responses, rendering AMR less frequent and often subclinical (Haas et al. 2018).

Studies, however, have demonstrated a strong association between donor-specific antibodies (DSAs), chronic rejection, and histologic features such as inflammation, fibrosis, and architectural distortion in long-term liver allograft recipients (Demetris et al. 2016; Miyagawa-Hayashino et al. 2004; O'Leary et al. 2014; Schotters et al. 2021). In the cyclosporine era, DSA positivity was consistently linked to classical ductopenic forms of chronic AMR (Demetris et al. 1997; Nakazawa et al. 2000a). With the widespread use of tacrolimus-based immunosuppression, the incidence of overt ductopenic rejection has declined, while subtler histologic features suggestive of cAMR have emerged. These include low-grade portal, periportal and perivenular lymphoplasmacytic inflammation, low-grade interface hepatitis, perivenular lymphoplasmacytic inflammation, and fibrosis involving the portal/periportal, perivenular, or sinusoidal compartments (Fig. 12.2a–c) (Demetris et al. 2016; O'Leary et al. 2014). To identify at-risk patients, a cAMR histology score has been proposed, incorporating interface activity, lobular inflammation, portal venopathy, sinusoidal fibrosis, dense portal collagenization, and HCV status (O'Leary et al. 2016). Notably, patients with elevated cAMR scores may have near-normal liver enzymes and yet remain at high risk for progressive graft injury. In such cases, serologic evaluation for DSAs and C4d immunostaining is warranted.

C4d immunostaining serves as a surrogate marker for complement activation, providing supportive evidence of antibody-mediated injury. C4d, a stable degradation product of complement component C4, indicates DSA-mediated complement

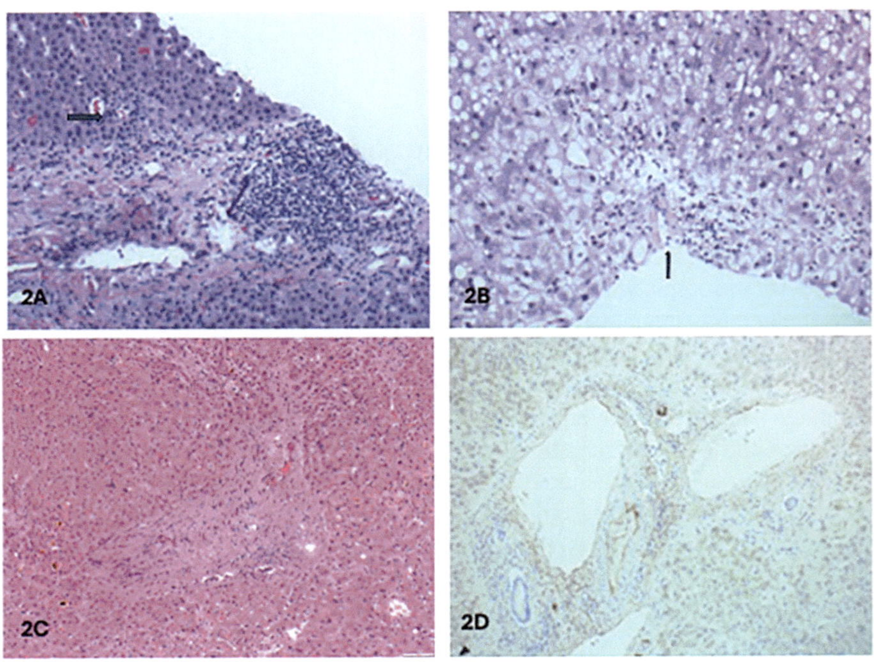

Fig. 12.2 Chronic antibody-mediated rejection. (**a**) Mild portal inflammation with mild interface hepatitis (arrow) Hematoxylin and eosin stain 20x. (**b**) Central perivenulitis (arrow showing central vein). Hematoxylin and eosin stain 20x. (**c**) Densely collagenized portal tract. Hematoxylin and eosin stain 20x. (**d**) C4d staining of the portal stroma, portal veins, hepatic arteries, and peribiliary capillaries. C4d immunostain 20x

fixation. While linear and granular microvascular endothelial cell C4d staining is accepted evidence of acute AMR in liver allografts, its role in cAMR is more limited. It can be observed in sinusoids, portal capillaries, portal veins, central veins, arterial endothelium, or portal stroma, but does not consistently correlate with histologic features of rejection. Furthermore, C4d staining can be present in nonrejection contexts, including ischemic or recurrent disease, and its absence does not exclude AMR. C4d staining of portal veins, peribiliary capillaries, and sinusoids seems to be specific for acute AMR. In cAMR, C4d staining is often less intense and variable, or can be negative (Fig. 12.2d) (Miyagawa-Hayashino et al. 2004; O'Leary et al. 2014). C4d staining does not always correlate with cAMR histology, such as foam cell arteriopathy, bile duct loss, or sinusoidal fibrosis. Recognizing these limitations, the 2016 Banff Working Group recommended that C4d be interpreted as a supportive rather than essential feature in diagnosing cAMR, emphasizing the importance of integrating histology, DSA status, and clinical graft function (Demetris et al. 2016).

In summary, a high cAMR histology score may indicate antibody-mediated injury, but does not reliably predict C4d positivity, and C4d positivity alone is not diagnostic of cAMR. Thus, the diagnostic value of C4d for cAMR diagnosis in LT remains limited, and its interpretation must be correlated with the presence of DSAs, both HLA and non-HLA DSAs. Serologic testing for DSAs remains essential when histologic features raise concern for chronic AMR, even in clinically silent cases.

12.4 Structural Alterations

12.4.1 Parenchymal Fibrosis

Progressive fibrosis and architectural remodeling are the key outcomes of CR with prognostic implications. In early CR, the fibrosis begins around the terminal hepatic veins, precipitating perivenular fibrosis accompanying hepatocyte dropout in zone 3 (Fig. 12.3a) (Demetris 2000; Nakazawa et al. 2000b). This corresponds to regions of ischemic injury due to arteriolar loss, as the centrilobular reticulin framework collapses and is, gradually, replaced by collagen. In addition, the development of VOD in terminal hepatic venules has been associated with thrombosis of medium-sized hepatic veins and portal veins, resulting in the loss of both inflow and outflow venous systems. This disruption compromises blood flow to the hepatic acinus, leading to parenchymal extinction and fibrosis (Nakazawa et al. 2000b). As rejection progresses and parenchymal extinction advances, bridging fibrosis between portal and central zones becomes more prominent, disrupting normal lobular architecture (Fig. 12.3b, c) (Nakazawa et al. 2000a). Fibrotic bands may even evolve to encircle regenerative nodules, meeting the definition of cirrhosis (Tariq and Nasir 2015).

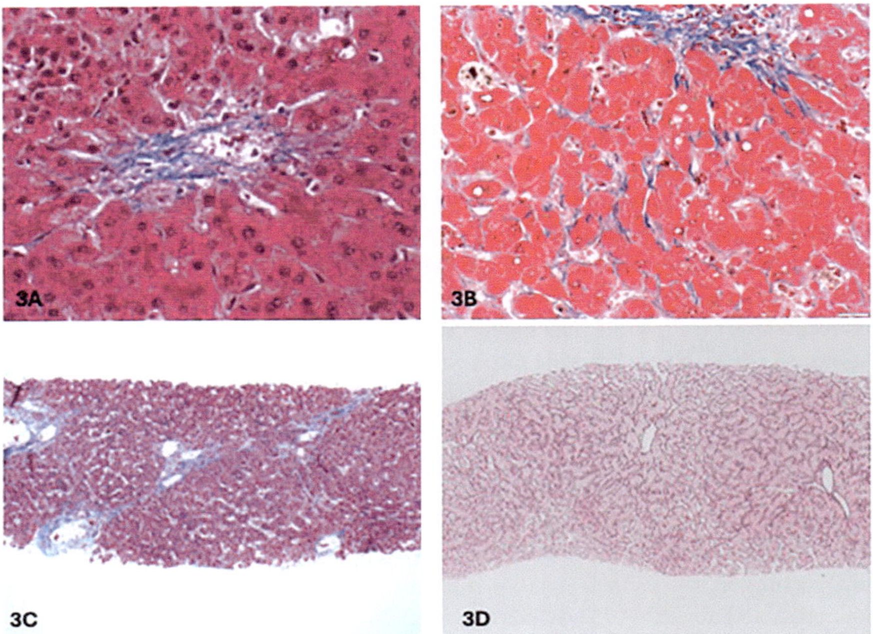

Fig. 12.3 Parenchymal alterations. (**a**) Perivenular fibrosis. Trichrome stain 40x. (**b**) Periportal fibrosis and perisinusoidal fibrosis. Trichrome stain 40x. (**c**) Central to portal bridging fibrosis. Trichrome stain 20x. (**d**) Nodular regenerative hyperplasia. Reticulin stain 20x

Traditional liver fibrosis scoring systems, such as METAVIR and Ishak staging, were developed for native liver diseases and do not fully capture the unique patterns of fibrosis seen in liver allografts. Idiopathic fibrosis in transplanted livers is considered an immune- mediated process and attributed primarily to chronic antibody-mediated rejection (CAMR) or a mixed phenotype of CTCMR and CAMR. Based on pathophysiology, the fibrosis often follows a heterogeneous distribution involving portal/periportal, sinusoidal, and/or centrilobular zones. To address the limitations of traditional fibrosis staging in liver allografts, Venturi et al. (2012) proposed a novel semiquantitative histologic scoring system known as the liver allograft fibrosis score (LAFSc), which has since been adopted and recommended by the Banff Working Group (Demetris et al. 2016). This system evaluates fibrosis at three distinct anatomical sites—portal, sinusoidal, and perivenular—and assigns a score from 0 to 3 for each, yielding a total possible score of 9. Rather than relying solely on severity, the score incorporates the pattern and distribution of fibrosis to better reflect the underlying pathophysiology. This approach provides a more nuanced assessment and has important clinical implications for guiding immunosuppression and monitoring long-term graft health. It also facilitates standardized reporting and research across transplant centers.

12.4.2 Nodular Regenerative Hyperplasia

Nodular regenerative hyperplasia (NRH) has emerged as a significant structural alteration contributing to long-term liver allograft dysfunction, particularly in the context of chronic antibody-mediated rejection (cAMR). NRH is a histopathological entity defined by the diffuse transformation of hepatic parenchyma into innumerable small regenerative nodules that involve the entire liver, without significant accompanying fibrosis (Fig. 12.3d). Originally described by Ranstrom in 1953 as "miliary hepatocellular adenomatosis" and renamed "nodular regenerative hyperplasia" by Steiner in 1959, these nodules are typically the size of a hepatic lobule and develop as an adaptive response to localized lobular injury diffusely distributed across the liver (Ludwig 1989). Histologically, the nodules reflect hepatocellular hypertrophy in zones 1 and 2, bordered by atrophy in zone 3. This transformation is not driven by fibrosis, but rather represents a regenerative response to subtle, widespread obliterative portal venopathy involving small portal veins, resulting in heterogeneous hepatic perfusion. Although frequently subclinical and underrecognized, NRH can manifest clinically as noncirrhotic portal hypertension, which has been reported in approximately 70% of affected patients (Bakshi et al. 2020).

Emerging data suggest that de novo NRH can develop in both pediatric and adult liver transplant recipients, even in the absence of overt biochemical abnormalities (González et al. 2019; Sebagh et al. 2023). A systematic review by Biesterveld et al. (2023) reported a mean incidence of 2.9% for de novo NRH following LT. Key risk factors identified include exposure to azathioprine and the presence of vascular abnormalities. NRH is often discovered incidentally on protocol surveillance biopsies in asymptomatic patients, though symptomatic individuals typically present with signs of noncirrhotic portal hypertension. Outcomes are variable, with graft failure rates ranging from 6% to 44%, and patient survival rates between 63% and 93%. While historically considered a nonrejection-related vascular lesion, recent studies suggest an association between NRH and chronic immune-mediated injury, particularly in donor-specific antibody (DSA)-positive recipients with subtle, progressive graft dysfunction (Sebagh et al. 2023). The Banff Working Group has recognized NRH as a potential long-term sequela of alloimmune injury, wherein vascular insult, including sinusoidal and portal venopathy, results in parenchymal extinction and compensatory nodular regeneration (Bellamy et al. 2024; Demetris et al. 2016).

Sebagh et al. (2023) provided compelling histopathological evidence that de novo NRH is not merely incidental in long-term liver allografts, but may be a structural endpoint of chronic antibody-mediated rejection (cAMR) or mixed T cell- and antibody-mediated CR. In their study of 18 adult LT recipients with biopsy-proven de novo NRH, DSAs were detected in 33.3% of cases. Features considered to represent cAMR included perisinusoidal fibrosis (33%), sinusoidal microvasculitis (33%), and C4d positivity in sinusoids and portal venules (44.4%) in index biopsies. Follow-up histology demonstrated progression, with portal venopathy in 14%, C4d positivity in 71.4%, and sinusoidal microvasculitis in 50% of patients. There was also increasing CD34 expression in sinusoidal endothelial cells, diffuse α-smooth muscle actin (α-SMA) expression, and enhanced HLA class II expression, particularly in the sinusoidal compartment. These findings reinforce the hypothesis that a subset of post-transplant NRH cases arise from ongoing, active antibody-mediated rejection. This aligns with other studies proposing NRH as a manifestation of chronic vascular injury in the context of alloimmunity (Bellamy et al. 2024; Demetris et al. 2016). Collectively, these insights underscore that NRH, especially when accompanied by sinusoidal fibrosis and perivenular changes, should not be regarded as a benign or incidental lesion, but rather as a potential histologic endpoint of chronic rejection. It warrants careful consideration, particularly in patients with DSAs or subtle evidence of allograft dysfunction.

12.5 Atypical Patterns of Late-Onset Rejection Related to Chronic Rejection

12.5.1 Idiopathic Post-transplant Hepatitis

Protocol liver biopsies in pediatric LT recipients procured more than 12 months post-transplant show mild inflammatory changes and fibrosis (Hübscher 2009). Such inflammatory changes in the absence of recurrent disease have been designated as "idiopathic chronic post-transplant hepatitis (IPTH)." Histologically, the portal tracts have a mild mononuclear inflammatory infiltrate comprised of lymphocytes, monocytes, and a few plasma cells, which is associated with mild interface hepatitis; there is also a noticeable component of lobular hepatitis with or without central perivenulitis (Fig. 12.4a–c). Portal inflammation is localized more prominently at the periphery of portal tracts, with spilling into the periportal hepatocytes in association with hepatocyte injury (interface hepatitis). The bile ducts and portal veins are relatively spared. This contrasts with acute TCMR, where the portal inflammatory infiltrate is mixed, comprising lymphocytes, eosinophils, and

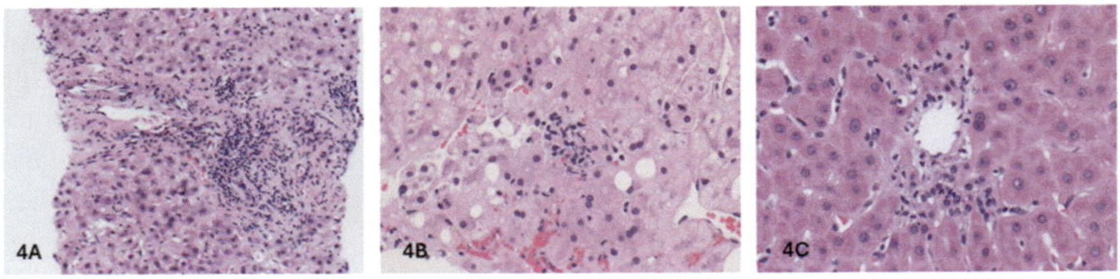

Fig. 12.4 Idiopathic post-transplant hepatitis. (**a**) Portal lymphocyte predominant inflammation with mild interface hepatitis. (**b**) Lobular hepatitis. (**c**) Central perivenulitis. Hematoxylin and Eosin stain 40x

neutrophils, and is centered on the bile ducts and portal veins, causing lymphocytic cholangitis and portal venulitis. The prevalence of idiopathic chronic hepatitis has been reported to increase over time from 22% to 24% at 1–2 years, rising to 64% at 10 years in the pediatric population, while it ranges between 10% and 30% in adults (Evans et al. 2006; Maor-Kendler et al. 2000). IPTH is often subclinical with normal liver enzymes, but can also be associated with elevated liver enzymes. IPTH is seen more frequently in patients who are on a low dose of immunosuppression or have compliance issues. Moreover, molecular profiling of IPTH most closely resembles that of TCMR, indicating an alloimmune-mediated hepatitis (Londoño et al. 2018). IPTH is a form of rejection associated with the presence of autoantibodies, de novo class II donor-specific HLA antibodies, and previous episodes of rejection that can improve with augmented immunosuppression (Kelly et al. 2016; Miyagawa-Hayashino et al. 2009). IPTH has a strong association with progressive graft fibrosis that results in cirrhosis over time (Evans et al. 2006; Herzog et al. 2008). IPTH has been reported to be the most common cause of graft cirrhosis in the absence of recurrent disease (Seyam et al. 2007).

12.5.2 *Plasma Cell-Rich Rejection*

Plasma cell-rich rejection (PCRR) is a morphologic variant of late (greater than 6 months) allograft rejection that can lead to graft loss in both pediatric and adult LT recipients. PCRR has emerged over the past two decades as a topic of growing clinical and pathological interest, with ongoing debates regarding its underlying pathophysiology and treatment challenges. It accounts for approximately 3–5% of cases of late liver allograft dysfunction (Kerkar et al. 1998; Spada et al. 2001). Historically, this entity was referred to as de novo autoimmune hepatitis or plasma cell hepatitis, reflecting its histologic and clinical resemblance to autoimmune hepatitis in the native liver (Kerkar et al. 1998; Ward et al. 2009). However, increasing evidence suggests that PCRR represents a distinct alloimmune process, particularly in cases with plasma cell-rich portal inflammation with significant interface activity, sometimes accompanied by IgG4-positive plasma cells and features overlapping with acute and chronic rejection.

In 2006, the Banff Liver Working Group proposed minimal criteria for the diagnosis of PCRR, then referred to as de novo AIH in an allograft. These criteria included interface hepatitis with portal lymphocytic infiltrates, high serologic titers (greater than 1:160) of antinuclear antibodies (ANA), antismooth muscle antibodies (SMA), or antibodies to liver kidney microsomal type 1 (LKM-1), hypergammaglobulinemia, and exclusion of viral hepatitis, drug-induced injury, and late acute or chronic rejection (Banff Working Group et al. 2006). The terminology was updated to "plasma cell-rich rejection" in the 2016 Banff update, which also revised the morphologic features supportive of diagnosis (Demetris et al. 2016).

The diagnosis of PCRR in a liver allograft requires evidence of allograft dysfunction in the absence of a history of autoimmune hepatitis in the native liver, and exclusion of other potential causes of autoimmunity, such as drug-induced liver injury or viral hepatitis. Morphologic features supportive of PCRR include the presence of plasma cell-rich (>30%) portal and perivenular infiltrates (Fig. 12.5a, b); frequent and severe lymphocytic cholangitis, a significant component of IgG4-positive plasma cells, microvascular C4d staining, and concurrent features of T cell-mediated rejection and chronic rejection (Demetris et al. 2016). Notably, non-organ-specific autoantibodies such as ANA and SMA are no longer included in the diagnostic criteria. Current serologic markers supportive of PCRR include LKM-1 antibodies and the presence of donor-specific antibodies (DSAs), particularly Class II (Wozniak et al. 2015). Given the strong association of anti-glutathione S-transferase T1 (GSTT1) antibodies with de novo AIH, antibodies to GSTT1 in GSTT1-null recipients of GSTT1-positive donors are considered contributory, but nonessential for the diagnosis of PCRR (Aguilera et al. 2018).

Despite its recurring presence in clinical practice, PCRR has not been formally defined as a distinct entity within the Banff classification. Instead, it is currently included under T cell-mediated rejection (TCMR). Some PCRR cases exhibit features suggestive of antibody-mediated rejection (AMR), such as donor-specific antibodies and/or microvascular C4d deposition (Aguilera et al. 2011; Wozniak et al. 2015). However, many cases lack serologic or histologic evidence of AMR, reinforcing the view that PCRR primarily represents a TCMR phenotype. The precise pathophysiology, whether purely TCMR or a TCMR–AMR overlap, is uncertain.

PCRR remains a diagnostically and therapeutically challenging entity, highlighting the limitations of current classification systems and underscoring the need for future Banff updates to more explicitly incorporate plasma cell-rich phenotypes into diagnostic criteria and treatment algorithms. Implementing serum- and tissue-based biomarkers in diagnosis has the potential for personalized management of PCRR (Saravanan et al. 2025).

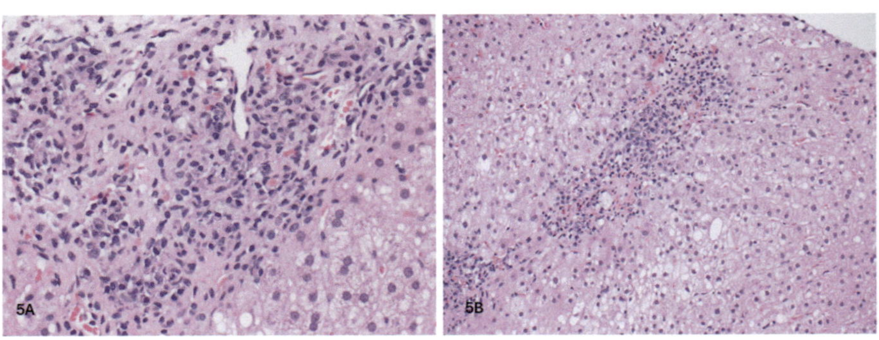

Fig. 12.5 Plasma cell-rich rejection. (**a**) Portal tracts with plasma cell-rich inflammation. Hematoxylin and eosin stain 40x. (**b**) Central perivenulitis with clusters of plasma cells. Hematoxylin and eosin stain 20x

12.5.3 Isolated Central Perivenulitis

Central perivenulitis (CP) is characterized histologically by zone 3 perivenular necro-inflammatory lesions and is considered a feature of late-onset T cell-mediated rejection. It can also be seen in acute TCMR and early CR. CP is a common histological finding in severe acute rejection occurring in about 40% of patients during the first rejection episode (Lovell et al. 2004). CP persists in cases of partially resolved acute rejection that progress to chronic rejection, suggesting this to be a poor prognostic lesion (Neil and Hubscher 2002). Isolated central perivenulitis (ICP) is defined as "central perivenulitis without portal changes of rejection." It is considered rejection-associated when secondary etiologies such as recurrent AIH and drug-induced liver injury are clinically excluded. It is seen more commonly in late allograft biopsies and has been reported to occur in 27% of protocol biopsies in pediatric liver transplant recipients (Krasinskas et al. 2008). The frequency of ICP increases over time and predisposes to the development of zone 3 fibrosis (25%) and chronic ductopenic rejection (15%) in a subset of patients (Abraham et al. 2008).

12.6 Future Directions in the Diagnosis of Liver Allograft Health and Rejection: Beyond Histomorphology

Emerging evidence from pediatric 10-year protocol biopsies showed that low-grade chronic inflammatory, fibrotic, and microvascular lesions reflect an ongoing injury to the liver allograft and have the potential to cause allograft failure over time (Sebagh et al. 2003, 2023). These lesions result from a complex interplay between alterations in the vascular flow, donor-specific antibodies (DSAs), and T cell-mediated alloimmune responses, often occurring in the context of suboptimal immunosuppression. This creates a well-recognized conundrum in LT management to minimize total immunosuppression because of the toxicities associated with excessive immunosuppression, which include chronic kidney disease (CKD), end-stage renal disease (ESRD), serious infections, post-transplant lymphoproliferative disorder (PTLD), and an increased risk of tumor recurrence, or de novo malignancy (Garcia et al. 2000). Conversely, suboptimal immunosuppression increases the risk of sensitization, immune-mediated injury, and progressive structural damage to the allograft. The limited sensitivity and specificity of routine liver function tests in detecting parenchymal injury underscore the importance of protocol biopsies, particularly at the 10-year post-transplant mark (Sebagh et al. 2003, 2023). These biopsies enable optimization of immunosuppression and detection of subclinical pathology that might otherwise go unnoticed. As such, protocol liver biopsies are instrumental in guiding personalized immunosuppressive strategies aimed at achieving long-term graft survival while minimizing treatment-related morbidity (Banff Working Group on Liver Allograft Pathology 2012).

Although histomorphological assessment of liver biopsies remains the gold standard for diagnosing different forms of allograft rejection, it is limited by sampling variation, interobserver variability, overlap of features with other entities, and reduced sensitivity for detecting early or subtle injury. Recent advancements in molecular diagnostics, including transcriptomic and proteomic profiling, have provided novel insights into the pathogenesis of chronic liver allograft rejection, helping to distinguish it from other causes of chronic graft dysfunction and opening new avenues for early detection and targeted therapy (Bellamy et al. 2024). With advances in transplant immunology and diagnostic technologies, the field is poised to shift toward more precise, noninvasive, and dynamic diagnostic tools that will complement or potentially supplant traditional histopathological evaluation in the future.

12.6.1 Tissue-Based Advanced Technologies

Tissue-based gene expression profiling from the INTERLIVER study revealed distinct T cell-mediated rejection (TCMR) signatures in LT biopsies. Notably, some liver allografts also exhibit a natural killer (NK) cell–enriched rejection-like (NKRL) state that is well tolerated, associated with minimal histologic injury, and accompanied by relatively normal liver biochemistry. Genome-wide transcriptomic assessment can detect not only molecular signatures of rejection but also expression patterns associated with parenchymal injury, steatohepatitis, and fibrosis (Madill-Thomsen et al. 2020, 2022, 2025).

In parallel, the recently developed Banff Human Organ Transplant (B-HOT) panel represents a significant advancement in transplant molecular diagnostics. This commercially available panel of 770 genes includes key genes related to rejection, tolerance, viral infections, and both innate and adaptive immune responses. Built on whole-transcriptome technology using the NanoString platform, the B-HOT panel enables quantification of gene expression from formalin-fixed, paraffin-embedded (FFPE) tissue, making it feasible for routine clinical use. The Banff Molecular Diagnostics Working Group has endorsed this molecular, pathway-based approach as a valuable tool for both researchers and clinicians, with the potential to guide therapeutic decision-making and facilitate clinical trials (Mengel et al. 2020).

Engel et al. (2025) identified a distinct molecular phenotype of chronic antibody-mediated rejection (cAMR) in liver transplantation, resembling patterns seen in other solid organ transplants. This phenotype is characterized by gene expression changes associated with inflammation and fibrosis and is linked to a high rate of liver allograft failure. These findings underscore the potential of molecular signatures to differentiate cAMR from other causes of graft injury.

When combined with surveillance liver biopsies, transcriptional profiling offers a powerful strategy for identifying recipients with ongoing alloimmune-mediated liver injury despite normal liver biochemistry, thereby enabling earlier intervention and more personalized immunosuppressive management.

Simultaneously, digital pathology and artificial intelligence (AI) are being increasingly integrated into transplant diagnostics. Machine learning algorithms can analyze whole-slide images to quantify bile duct injury, arteriopathy, and fibrosis with enhanced reproducibility and objectivity. Deep learning-based AI models also have demonstrated promise in detecting subtle histologic features, improving diagnostic accuracy, and reducing interobserver variability (Rahman et al. 2023).

12.6.2 Noninvasive Biomarkers

Hepatologists are often reluctant to perform liver biopsies due to the invasive nature of the procedure, the associated risk of bleeding, and the reluctance of patients to consent to a biopsy. These factors are strong barriers to protocol biopsies for clinically stable allograft recipients with normal liver biochemistry. However, protocol biopsies have demonstrated the presence of significant inflammatory and fibrotic changes even in the setting of unremarkable clinical and biochemical parameters. This discrepancy underscores the need for reliable noninvasive biomarkers suitable for monitoring long-term liver allograft health. The development and validation of such tools could allow protocol biopsies to be reserved for select patient groups, while others could be monitored using noninvasive biomarkers.

A multicenter study showed the utility of elevated ALT, class II donor-specific antibodies (DSAs), and liver stiffness measurements (LSM) as meaningful, noninvasive indicators of subclinical liver allograft injury. The development and validation of two composite models (ALT + LSM and ALT + class II DSAs) offer proof of principle that such practical tools can identify patients with alloimmune damage, despite normal liver tests (Vionnet et al. 2021). By reducing reliance on protocol biopsies, these models support a shift toward a more individualized, less invasive approach to immunosuppression management.

Donor-derived cell-free DNA (dd-cfDNA) has shown considerable promise as a noninvasive biomarker for rejection in LT. Serial elevations in dd-cfDNA have been associated with allograft injury and T cell-mediated rejection, while declining levels may indicate response to treatment (Levitsky et al. 2022). However, further validation is needed before widespread clinical adoption. At present, dd-cfDNA has a limited role in detecting subclinical allograft injury, particularly in the setting of normal liver function tests, and in the diagnosis of antibody-mediated rejection (AMR) (Baumann et al. 2022).

12.6.3 Novel Therapies

Regenerative medicine is an expanding field. It explores the potential for bile duct regeneration using cholangiocyte organoids and stem cell-derived grafts (Choudhary et al. 2017). Though still in experimental stages, these therapies may 1 day offer alternatives to re-transplantation for select patients with ductopenia and stable liver function. Anti-fibrotic therapies, including TGF-beta inhibitors and matrix metalloproteinase modulators, are also under investigation for their potential to limit and reverse fibrotic progression (Demirci et al. 1996).

12.7 Conclusion

Chronic liver allograft rejection is a complex and often irreversible process that remains a major barrier to long-term liver transplant success. It arises from sustained immune-mediated injury, often due to multifactorial humoral and T cell-mediated mechanisms. Once established, treatment options are limited, and re-transplantation remains the only definitive therapy for many patients.

A comprehensive understanding of the immunopathologic mechanisms driving chronic liver allograft rejection is essential in improving long-term outcomes, preventing late graft loss, and tailoring individualized treatment approaches. Looking ahead, the integration of tissue-based molecular diagnostics, genetic and epigenetic profiling using spatial proteomics and transcriptomics, and novel noninvasive biomarkers may soon support a precision medicine approach. Together with emerging therapeutic strategies, these tools are likely to improve early detection of rejection, individualize immunosuppression regimens, and increase long-term allograft and patient survivals after LT.

References

Abraham SC, Freese DK, Ishitani MB, Krasinskas AM, Wu T-T (2008) Significance of central perivenulitis in pediatric liver transplantation. Am J Surg Pathol 32(10):1479–1488. https://doi.org/10.1097/PAS.0b013e31817a8e96

Afford SC, Randhawa S, Eliopoulos AG, Hubscher SG, Young LS, Adams DH (1999) CD40 activation induces apoptosis in cultured human hepatocytes via induction of cell surface fas ligand expression and amplifies fas-mediated hepatocyte death during allograft rejection. J Exp Med 189(2):441–446. https://doi.org/10.1084/jem.189.2.441

Aguilera I, Sousa JM, Gavilan F, Gomez L, Alvarez-Márquez A, Núñez-Roldán A (2011) Complement component 4d immunostaining in liver allografts of patients with de novo immune hepatitis. Liver Transpl 17(7):779–788. https://doi.org/10.1002/lt.22302

Aguilera I, Aguado-Dominguez E, Sousa JM, Nuñez-Roldan A (2018) Rethinking de novo immune hepatitis, an old concept for liver allograft rejection: relevance of glutathione S-transferase T1 mismatch. World J Gastroenterol 24(29):3239–3249. https://doi.org/10.3748/wjg.v24.i29.3239

Angelico R, Sensi B, Manzia TM, Tisone G, Grassi G, Signorello A, Milana M, Lenci I, Baiocchi L (2021) Chronic rejection after liver transplantation: opening the Pandora's box. World J Gastroenterol 27(45):7771–7783. https://doi.org/10.3748/wjg.v27.i45.7771

Bakshi N, Gulati N, Rastogi A, Chougule A, Bihari C, Jindal A (2020) Nodular regenerative hyperplasia—an under-recognized vascular disorder of liver. Pathol Res Pract 216(4):152833. https://doi.org/10.1016/j.prp.2020.152833

Banff Working Group on Liver Allograft Pathology (2012) Importance of liver biopsy findings in immunosuppression management: biopsy monitoring and working criteria for patients with operational tolerance. Liver Transpl 18(10):1154–1170. https://doi.org/10.1002/lt.23481

Banff Working Group, Demetris AJ, Adeyi O, Bellamy COC, Clouston A, Charlotte F, Czaja A, Daskal I, El-Monayeri MS, Fontes P, Fung J, Gridelli B, Guido M, Haga H, Hart J, Honsova E, Hubscher S, Itoh T, Jhala N et al (2006) Liver biopsy interpretation for causes of late liver allograft dysfunction. Hepatology 44(2):489–501. https://doi.org/10.1002/hep.21280

Baumann AK, Beck J, Kirchner T, Hartleben B, Schütz E, Oellerich M, Wedemeyer H, Jaeckel E, Taubert R (2022) Elevated fractional donor-derived cell-free DNA during subclinical graft injury after liver transplantation. Liver Transpl 28(12):1911–1919. https://doi.org/10.1002/lt.26479

Bellamy COC, O'Leary JG, Adeyi O, Baddour N, Batal I, Bucuvalas J, Del Bello A, El Hag M, El-Monayeri M, Farris AB, Feng S, Fiel MI, Fischer SE, Fung J, Grzyb K, Guimei M, Haga H, Hart J, Jackson AM et al (2024) Banff 2022 Liver Group Meeting report: monitoring long-term allograft health. Am J Transplant 24(6):905–917. https://doi.org/10.1016/j.ajt.2024.03.008

Biesterveld BE, Schroder PM, Hitchcock ME, Bolognese A, Kim SC, Al-Adra DP (2023) Nodular regenerative hyperplasia and liver transplantation: a systematic review. Front Transplant 2:1221765. https://doi.org/10.3389/frtra.2023.1221765

Choudhary NS, Saigal S, Bansal RK, Saraf N, Gautam D, Soin AS (2017) Acute and chronic rejection after liver transplantation: what a clinician needs to know. J Clin Exp Hepatol 7(4):358–366. https://doi.org/10.1016/j.jceh.2017.10.003

Dasari BVM, Line P-D, Sapisochin G, Hibi T, Bhangui P, Halazun KJ, Shetty S, Shah T, Magyar CTJ, Donnelly C, Chatterjee D (2025) Liver transplantation as a treatment for cancer: comprehensive review. BJS Open 9(3):zraf034. https://doi.org/10.1093/bjsopen/zraf034

Demetris AJ (2000) Spectrum of chronic hepatic allograft rejection and arteriopathy and the controversy of centrilobular necrosis. Liver Transpl 6(1):102–103. https://doi.org/10.1002/lt.500060106

Demetris AJ, Murase N, Lee RG, Randhawa P, Zeevi A, Pham S, Duquesnoy R, Fung JJ, Starzl TE (1997) Chronic rejection. A general overview of histopathology and pathophysiology with emphasis on liver, heart and intestinal allografts. Ann Transplant 2(2):27–44

Demetris AJ, Murase N, Starzl TE, Fung JJ (1998) Pathology of chronic rejection: an overview of common findings and observations about pathogenic mechanisms and possible prevention. Graft (Georgetown, Tex) 1(2):52–59

Demetris A, Adams D, Bellamy C, Blakolmer K, Clouston A, Dhillon AP, Fung J, Gouw A, Gustafsson B, Haga H, Harrison D, Hart J, Hubscher S, Jaffe R, Khettry U, Lassman C, Lewin K, Martinez O, Nakazawa Y et al (2000) Update of the International Banff Schema for Liver Allograft Rejection: working recommendations for the histopathologic staging and reporting of chronic rejection. An International Panel. Hepatology 31(3):792–799. https://doi.org/10.1002/hep.510310337

Demetris AJ, Bellamy C, Hübscher SG, O'Leary J, Randhawa PS, Feng S, Neil D, Colvin RB, McCaughan G, Fung JJ, Del Bello A, Reinholt FP, Haga H, Adeyi O, Czaja AJ, Schiano T, Fiel MI, Smith ML, Sebagh M et al (2016) 2016 Comprehensive Update of the Banff Working Group on liver allograft pathology: introduction of antibody-mediated rejection. Am J Transplant 16(10):2816–2835. https://doi.org/10.1111/ajt.13909

Demirci G, Nashan B, Pichlmayr R (1996) Fibrosis in chronic rejection of human liver allografts: expression patterns of transforming growth factor-TGFbeta1 and TGF-beta3. Transplantation 62(12):1776–1783. https://doi.org/10.1097/00007890-199612270-00016

Engel B, Alaswad A, Campos-Murguia A, Zoodsma M, Klingbeil A, Ruiz P, Crespo G, Chihab K, Bosselmann EA, Heinrich S, Hartleben B, Jonigk DD, Diaz A, Verboom M, Hallensleben M, Geffers R, Falk CS, Ruff S, Lauber C et al (2025) Molecular signatures discriminating different types of rejection in human liver transplants. J Hepatol. S0168-8278(25)02338-4. https://doi.org/10.1016/j.jhep.2025.06.036

Evans HM, Kelly DA, McKiernan PJ, Hübscher S (2006) Progressive histological damage in liver allografts following pediatric liver transplantation. Hepatology 43(5):1109–1117. https://doi.org/10.1002/hep.21152
Garcia RF, Garcia CE, McMaster P (2000) Chronic rejection of the liver: the role of immunosuppression. BioDrugs 14(5):283–297. https://doi.org/10.2165/00063030-200014050-00002
González I, Lu H-C, Ritter JH, Maluf HM, Dehner LP, He M (2019) Clinicopathologic characteristics of de novo nodular regenerative hyperplasia in pediatric liver transplant. Pediatr Transplant 23(5):e13471. https://doi.org/10.1111/petr.13471
Haas M, Sis B, Racusen LC, Solez K, Glotz D, Colvin RB, Castro MC, David DS, David-Neto E, Bagnasco SM, Cendales LC, Cornell LD, Demetris AJ, Drachenberg CB, Farver CF, Farris AB 3rd, Gibson IW, Kraus E, Liapis H et al (2014) Banff 2013 meeting report: inclusion of c4d-negative antibody-mediated rejection and antibody-associated arterial lesions. Am J Transplant 14(2):272–283. https://doi.org/10.1111/ajt.12590
Haas M, Loupy A, Lefaucheur C, Roufosse C, Glotz D, Seron D, Nankivell BJ, Halloran PF, Colvin RB, Akalin E, Alachkar N, Bagnasco S, Bouatou Y, Becker JU, Cornell LD, Duong Van Huyen JP, Gibson IW, Kraus ES, Mannon RB et al (2018) The Banff 2017 Kidney Meeting Report: revised diagnostic criteria for chronic active T cell–mediated rejection, antibody-mediated rejection, and prospects for integrative endpoints for next-generation clinical trials. Am J Transplant 18(2):293–307. https://doi.org/10.1111/ajt.14625
Hayashi M, Martinez OM, Garcia-Kennedy R, So S, Esquivel CO, Krams SM (1995) Expression of cytokines and immune mediators during chronic liver allograft rejection. Transplantation 60(12):1533–1538. https://doi.org/10.1097/00007890-199560120-00027
Herzog D, Soglio DB-D, Fournet J-C, Martin S, Marleau D, Alvarez F (2008) Interface hepatitis is associated with a high incidence of late graft fibrosis in a group of tightly monitored pediatric orthotopic liver transplantation patients. Liver Transpl 14(7):946–955. https://doi.org/10.1002/lt.21444
Hübscher S (2009) What does the long-term liver allograft look like for the pediatric recipient? Liver Transpl 15(Suppl 2):S19–S24. https://doi.org/10.1002/lt.21902
Hubscher SG, Buckels JA, Elias E, McMaster P, Neuberger J (1991) Vanishing bile-duct syndrome following liver transplantation—is it reversible? Transplantation 51(5):1004–1010. https://doi.org/10.1097/00007890-199105000-00014
Kaltenmeier C, Jorgensen D, Dharmayan S, Ayloo S, Rachakonda V, Geller DA, Tohme S, Molinari M (2021) The liver transplant risk score prognosticates the outcomes of liver transplant recipients at listing. HPB (Oxford) 23(6):927–936. https://doi.org/10.1016/j.hpb.2020.10.002
Kelly D, Verkade HJ, Rajanayagam J, McKiernan P, Mazariegos G, Hübscher S (2016) Late graft hepatitis and fibrosis in pediatric liver allograft recipients: current concepts and future developments. Liver Transpl 22(11):1593–1602. https://doi.org/10.1002/lt.24616
Kerkar N, Hadzić N, Davies ET, Portmann B, Donaldson PT, Rela M, Heaton ND, Vergani D, Mieli-Vergani G (1998) De-novo autoimmune hepatitis after liver transplantation. Lancet (London, England) 351(9100):409–413. https://doi.org/10.1016/S0140-6736(97)06478-7
Krasinskas AM, Demetris AJ, Poterucha JJ, Abraham SC (2008) The prevalence and natural history of untreated isolated central perivenulitis in adult allograft livers. Liver Transpl 14(5):625–632. https://doi.org/10.1002/lt.21404
Kwong AJ, Ebel NH, Kim WR, Lake JR, Smith JM, Schladt DP, Skeans MA, Foutz J, Gauntt K, Cafarella M, Snyder JJ, Israni AK, Kasiske BL (2022) OPTN/SRTR 2020 Annual data report: liver. Am J Transplant 22(Suppl 2):204–309. https://doi.org/10.1111/ajt.16978
Leonard NB, Hale GL, Boylan KE, Ou Z, Zhang C, Kim R, Chandna S, Dong ZM, Evason KJ (2023) Histologic features of allograft livers in patients treated for rejection before biopsy. Hum Pathol 135:11–21. https://doi.org/10.1016/j.humpath.2023.02.005
Levitsky J, Kandpal M, Guo K, Kleiboeker S, Sinha R, Abecassis M (2022) Donor-derived cell-free DNA levels predict graft injury in liver transplant recipients. Am J Transplant 22(2):532–540. https://doi.org/10.1111/ajt.16835

Londoño M-C, Souza LN, Lozano J-J, Miquel R, Abraldes JG, Llovet L-P, Quaglia A, Rimola A, Navasa M, Sánchez-Fueyo A (2018) Molecular profiling of subclinical inflammatory lesions in long-term surviving adult liver transplant recipients. J Hepatol 69(3):626–634. https://doi.org/10.1016/j.jhep.2018.04.012

Lovell MO, Speeg KV, Halff GA, Molina DK, Sharkey FE (2004) Acute hepatic allograft rejection: a comparison of patients with and without centrilobular alterations during first rejection episode. Liver Transpl 10(3):369–373. https://doi.org/10.1002/lt.20094

Lucey MR, Terrault N, Ojo L, Hay JE, Neuberger J, Blumberg E, Teperman LW (2013) Long-term management of the successful adult liver transplant: 2012 practice guideline by the American Association for the Study of Liver Diseases and the American Society of Transplantation. Liver Transpl 19(1):3–26. https://doi.org/10.1002/lt.23566

Ludwig J (1989) Classification and terminology of hepatic allograft rejection: whither bound? Mayo Clin Proc 64(6):676–679. https://doi.org/10.1016/s0025-6196(12)65347-6

Lunz JG 3rd, Contrucci S, Ruppert K, Murase N, Fung JJ, Starzl TE, Demetris AJ (2001) Replicative senescence of biliary epithelial cells precedes bile duct loss in chronic liver allograft rejection: increased expression of p21(WAF1/Cip1) as a disease marker and the influence of immunosuppressive drugs. Am J Pathol 158(4):1379–1390. https://doi.org/10.1016/S0002-9440(10)64089-8

Madill-Thomsen K, Abouljoud M, Bhati C, Ciszek M, Durlik M, Feng S, Foroncewicz B, Francis I, Grąt M, Jurczyk K, Klintmalm G, Krasnodębski M, McCaughan G, Miquel R, Montano-Loza A, Moonka D, Mucha K, Myślak M, Pączek L et al (2020) The molecular diagnosis of rejection in liver transplant biopsies: first results of the INTERLIVER study. Am J Transplant 20(8):2156–2172. https://doi.org/10.1111/ajt.15828

Madill-Thomsen KS, Abouljoud M, Bhati C, Ciszek M, Durlik M, Feng S, Foroncewicz B, Francis I, Grąt M, Jurczyk K, Klintmalm G, Krasnodębski M, McCaughan G, Miquel R, Montano-Loza A, Moonka D, Mucha K, Myślak M, Pączek L et al (2022) The molecular phenotypes of injury, steatohepatitis, and fibrosis in liver transplant biopsies in the INTERLIVER study. Am J Transplant 22(3):909–926. https://doi.org/10.1111/ajt.16890

Madill-Thomsen KS, Gauthier PT, Abouljoud M, Bhati C, Bruno D, Ciszek M, Durlik M, Feng S, Foroncewicz B, Grąt M, Jurczyk K, Levitsky J, McCaughan G, Maluf D, Montano-Loza A, Moonka D, Mucha K, Myślak M, Perkowska-Ptasińska A et al (2025) Defining an NK cell-enriched rejection-like phenotype in liver transplant biopsies from the INTERLIVER study. Transplantation 109(8):1367–1382. https://doi.org/10.1097/TP.0000000000005269

Maor-Kendler Y, Batts KP, Burgart LJ, Wiesner RH, Krom RA, Rosen CB, Charlton MR (2000) Comparative allograft histology after liver transplantation for cryptogenic cirrhosis, alcohol, hepatitis C, and cholestatic liver diseases. Transplantation 70(2):292–297. https://doi.org/10.1097/00007890-200007270-00009

McAlister VC, Haddad E, Renouf E, Malthaner RA, Kjaer MS, Gluud LL (2006) Cyclosporin versus tacrolimus as primary immunosuppressant after liver transplantation: a meta-analysis. Am J Transplant 6(7):1578–1585. https://doi.org/10.1111/j.1600-6143.2006.01360.x

McCaughan GW, Davies SJ, Waugh JA, Bishop AG, Hall BM, Gallagher ND, Thompson JF, Sheil RAG, Painter DM (1990) A quantitative analysis of T lymphocyte populations in human liver allografts undergoing rejection: the use of monoclonal antibodies and double immunolabeling. Hepatology 12(6):1305–1313. https://doi.org/10.1002/hep.1840120610

Mengel M, Loupy A, Haas M, Roufosse C, Naesens M, Akalin E, Clahsen-van Groningen MC, Dagobert J, Demetris AJ, Duong van Huyen J-P, Gueguen J, Issa F, Robin B, Rosales I, Von der Thüsen JH, Sanchez-Fueyo A, Smith RN, Wood K, Adam B, Colvin RB (2020) Banff 2019 Meeting Report: molecular diagnostics in solid organ transplantation-Consensus for the Banff Human Organ Transplant (B-HOT) gene panel and open source multicenter validation. Am J Transplant 20(9):2305–2317. https://doi.org/10.1111/ajt.16059

Miyagawa-Hayashino A, Tsuruyama T, Haga H, Oike F, Il-Deok K, Egawa H, Hiai H, Tanaka K, Manabe T (2004) Arteriopathy in chronic allograft rejection in liver transplantation. Liver Transpl 10(4):513–519. https://doi.org/10.1002/lt.20081

Miyagawa-Hayashino A, Haga H, Egawa H, Hayashino Y, Uemoto S, Manabe T (2009) Idiopathic post-transplantation hepatitis following living donor liver transplantation, and significance of autoantibody titre for outcome. Transpl Int 22(3):303–312. https://doi.org/10.1111/j.1432-2277.2008.00803.x

Nakazawa Y, Jonsson JR, Walker NI, Kerlin P, Steadman C, Lynch SV, Strong RW, Clouston AD (2000a) Fibrous obliterative lesions of veins contribute to progressive fibrosis in chronic liver allograft rejection. Hepatology 32(6):1240–1247. https://doi.org/10.1053/jhep.2000.20350

Nakazawa Y, Walker NI, Kerlin P, Steadman C, Lynch SV, Strong RW, Clouston AD (2000b) Clinicopathological analysis of liver allograft biopsies with late centrilobular necrosis: a comparative study in 54 patients. Transplantation 69(8):1599–1608. https://doi.org/10.1097/00007890-200004270-00014

Nawaz S, Fennell RH (1994) Apoptosis of bile duct epithelial cells in hepatic allograft rejection. Histopathology 25(2):137–142. https://doi.org/10.1111/j.1365-2559.1994.tb01569.x

Neil DAH, Hubscher SG (2002) Histologic and biochemical changes during the evolution of chronic rejection of liver allografts. Hepatology 35(3):639–651. https://doi.org/10.1053/jhep.2002.31726

O'Grady JG, Alexander GJ, Sutherland S, Donaldson PT, Harvey F, Portmann B, Calne RY, Williams R (1988) Cytomegalovirus infection and donor/recipient HLA antigens: interdependent co-factors in pathogenesis of vanishing bile-duct syndrome after liver transplantation. Lancet (London, England) 2(8606):302–305. https://doi.org/10.1016/s0140-6736(88)92356-2

O'Leary JG, Kaneku H, Demetris AJ, Marr JD, Shiller SM, Susskind BM, Tillery GW, Terasaki PI, Klintmalm GB (2014) Antibody-mediated rejection as a contributor to previously unexplained early liver allograft loss. Liver Transpl 20(2):218–227. https://doi.org/10.1002/lt.23788

O'Leary JG, Cai J, Freeman R, Banuelos N, Hart B, Johnson M, Jennings LW, Kaneku H, Terasaki PI, Klintmalm GB, Demetris AJ (2016) Proposed diagnostic criteria for chronic antibody-mediated rejection in liver allografts. Am J Transplant 16(2):603–614. https://doi.org/10.1111/ajt.13476

Oguma S, Belle S, Starzl TE, Demetris AJ (1989a) A histometric analysis of chronically rejected human liver allografts: insights into the mechanisms of bile duct loss: direct immunologic and ischemic factors. Hepatology 9(2):204–209. https://doi.org/10.1002/hep.1840090207

Oguma S, Zerbe T, Banner B, Belle S, Starzl TE, Demetris AJ (1989b) Chronic liver allograft rejection and obliterative arteriopathy: possible pathogenic mechanisms. Transplant Proc 21(1 Pt 2):2203–2207

Rahman MA, Yilmaz I, Albadri ST, Salem FE, Dangott BJ, Taner CB, Nassar A, Akkus Z (2023) Artificial intelligence advances in transplant pathology. Bioengineering (Basel, Switzerland) 10(9):1041. https://doi.org/10.3390/bioengineering10091041

Ros J, Salva F, Dopazo C, López D, Saoudi N, Baraibar I, Charco R, Tabernero J, Elez E (2023) Liver transplantation in metastatic colorectal cancer: are we ready for it? Br J Cancer 128(10):1797–1806. https://doi.org/10.1038/s41416-023-02213-1

Sapisochin G, Javle M, Lerut J, Ohtsuka M, Ghobrial M, Hibi T, Kwan NM, Heimbach J (2020) Liver transplantation for cholangiocarcinoma and mixed hepatocellular cholangiocarcinoma: working group report from the ILTS Transplant Oncology Consensus Conference. Transplantation 104(6):1125–1130. https://doi.org/10.1097/TP.0000000000003212

Saravanan N, Demetris A, Fiel MI, Harrington C, Khurram NA, Schiano TD, Levitsky J (2025) Serum and tissue biomarkers of plasma cell-rich rejection in liver transplant recipients. Am J Transplant 25(2):259–268. https://doi.org/10.1016/j.ajt.2024.10.006

Schotters FL, Beime J, Briem-Richter A, Binder T, Herden U, Grabhorn EF (2021) Impact of donor-specific antibodies on long-term graft survival with pediatric liver transplantation. World J Hepatol 13(6):673–685. https://doi.org/10.4254/wjh.v13.i6.673

Sebagh M, Blakolmer K, Falissard B, Roche B, Emile JF, Bismuth H, Samuel D, Reynès M (2002) Accuracy of bile duct changes for the diagnosis of chronic liver allograft rejection: reliability of the 1999 Banff schema. Hepatology 35(1):117–125. https://doi.org/10.1053/jhep.2002.30085

Sebagh M, Rifai K, Féray C, Yilmaz F, Falissard B, Roche B, Bismuth H, Samuel D, Reynès M (2003) All liver recipients benefit from the protocol 10-year liver biopsies. Hepatology 37(6):1293–1301. https://doi.org/10.1053/jhep.2003.50231

Sebagh M, Yilmaz F, Kounis I, Saliba F, Feray C, Taupin J-L, Cherqui D, Azoulay D, Samuel D, Coilly A, Demetris A-J, Neil D (2023) Evidence for alloimmune sinusoidal injury in de novo nodular regenerative hyperplasia after liver transplantation. Transpl Int 36:11306. https://doi.org/10.3389/ti.2023.11306

Seyam M, Neuberger JM, Gunson BK, Hübscher SG (2007) Cirrhosis after orthotopic liver transplantation in the absence of primary disease recurrence. Liver Transpl 13(7):966–974. https://doi.org/10.1002/lt.21060

Spada M, Bertani A, Sonzogni A, Petz W, Riva S, Torre G, Melzi ML, Alberti D, Colledan M, Segalin A, Lucianetti A, Gridelli B (2001) A cause of late graft dysfunction after liver transplantation in children: de-novo autoimmune hepatitis. Transplant Proc 33(1–2):1747–1748. https://doi.org/10.1016/s0041-1345(00)02826-8

Tariq H, Nasir H (2015) Spectrum of histopathological findings in live donor liver graft biopsies. J Coll Physicians Surg Pak 25(6):412–416

Venturi C, Sempoux C, Bueno J, Ferreres Pinas JC, Bourdeaux C, Abarca-Quinones J, Rahier J, Reding R (2012) Novel histologic scoring system for long-term allograft fibrosis after liver transplantation in children. Am J Transplant 12(11):2986–2996. https://doi.org/10.1111/j.1600-6143.2012.04210.x

Vierling JM (1999) Immunology of acute and chronic hepatic allograft rejection. Liver Transpl Surg 5(4):S1–S20. https://doi.org/10.1053/JTLS005s00001

Vierling JM, Howell CD (1990) Disappearing bile ducts: immunologic mechanisms. Hosp Pract (Off Ed) 25(7):141–150. https://doi.org/10.1080/21548331.1990.11703976

Vionnet J, Miquel R, Abraldes JG, Wall J, Kodela E, Lozano J-J, Ruiz P, Navasa M, Marshall A, Nevens F, Gelson W, Leithead J, Masson S, Jaeckel E, Taubert R, Tachtatzis P, Eurich D, Simpson KJ, Bonaccorsi-Riani E et al (2021) Non-invasive alloimmune risk stratification of long-term liver transplant recipients. J Hepatol 75(6):1409–1419. https://doi.org/10.1016/j.jhep.2021.08.007

Ward SC, Schiano TD, Thung SN, Fiel MI (2009) Plasma cell hepatitis in hepatitis C virus patients post-liver transplantation: case-control study showing poor outcome and predictive features in the liver explant. Liver Transpl 15(12):1826–1833. https://doi.org/10.1002/lt.21949

Wiesner RH, Ludwig J, van Hoek B, Krom RA (1991) Current concepts in cell-mediated hepatic allograft rejection leading to ductopenia and liver failure. Hepatology 14(4 Pt 1):721–729. https://doi.org/10.1016/0270-9139(91)90064-3

Wight DA (1996) Chronic liver transplant rejection: definition and diagnosis. Transplant Proc 28(1):465–467

Wozniak LJ, Hickey MJ, Venick RS, Vargas JH, Farmer DG, Busuttil RW, McDiarmid SV, Reed EF (2015) Donor-specific HLA antibodies are associated with late allograft dysfunction after pediatric liver transplantation. Transplantation 99(7):1416–1422. https://doi.org/10.1097/TP.0000000000000796

Yi Z, Mayorga ME, Orman ES, Wheeler SB, Hayashi PH, Barritt AS (2017) Trends in characteristics of patients listed for liver transplantation will lead to higher rates of waitlist removal due to clinical deterioration. Transplantation 101(10):2368–2374. https://doi.org/10.1097/TP.0000000000001851

Yilmaz F, Aydin U, Nart D, Zeytunlu M, Karasu Z, Kaya T, Ozer I, Yuce G, Aydogdu S, Kilic M (2006) The incidence and management of acute and chronic rejection after living donor liver transplantation. Transplant Proc 38(5):1435–1437. https://doi.org/10.1016/j.transproceed.2006.02.108

Younossi ZM, Stepanova M, Al Shabeeb R, Eberly KE, Shah D, Nguyen V, Ong J, Henry L, Alqahtani SA (2024) The changing epidemiology of adult liver transplantation in the United States in 2013–2022: the dominance of metabolic dysfunction-associated steatotic liver disease and alcohol-associated liver disease. Hepatol Commun 8(1):e0352. https://doi.org/10.1097/HC9.0000000000000352

Zarrinpar A, Busuttil RW (2013) Liver transplantation: past, present and future. Nat Rev Gastroenterol Hepatol 10(7):434–440. https://doi.org/10.1038/nrgastro.2013.88

MIX
Papier aus verantwortungsvollen Quellen
Paper from responsible sources
FSC® C105338

If you have any concerns about our products,
you can contact us on
ProductSafety@springernature.com

In case Publisher is established outside the EU,
the EU authorized representative is:
Springer Nature Customer Service Center GmbH
Europaplatz 3, 69115 Heidelberg, Germany

Printed by Libri Plureos GmbH
in Hamburg, Germany